Energy Sources

Energy Sources

Fundamentals of Chemical Conversion Processes and Applications

Balasubramanian Viswanathan
National Centre for Catalysis Research
Indian Institute of Technology Madras
Chennai, India

ELSEVIER
AMSTERDAM • BOSTON • HEIDELBERG • LONDON • NEW YORK • OXFORD
PARIS • SAN DIEGO • SAN FRANCISCO • SINGAPORE • SYDNEY • TOKYO

Elsevier
Radarweg 29, PO Box 211, 1000 AE Amsterdam, Netherlands
The Boulevard, Langford Lane, Kidlington, Oxford OX5 1GB, United Kingdom
50 Hampshire Street, 5th Floor, Cambridge, MA 02139, United States

Library of Congress Cataloging-in-Publication Data
A catalog record for this book is available from the Library of Congress

British Library Cataloguing-in-Publication Data
A catalogue record for this book is available from the British Library

ISBN: 978-0-444-56353-8

For information on all Elsevier publications
visit our website at https://www.elsevier.com/

Publisher: John Fedor
Acquisition Editor: Kostas KI Marinakis
Editorial Project Manager: Sarah Jane Watson
Production Project Manager: Maria Bernard
Cover Designer: Victoria Pearson

Typeset by TNQ Books and Journals

Contents

Preface

Many exercises in the documentation of energy sources focus on availability, sustainability, and economics. Documentation of the scientific aspects of energy conversion has been buried in primary publications. In addition, a comparative evaluation is rarely seen of this aspect, although economic comparisons are available. This situation has hampered in introducing this subject in the curriculum of education, although some individual energy conversions such as fossil fuel–based energy conversions find an appropriate place in the educational curriculum. Keeping in mind this situation and also the directions that future attempts of energy conversion processes will take, a single source dealing with these aspects is desirable.

Also, electrochemical energy conversion is sustainable and environmentally acceptable. These energy sources have been poised to become acceptable energy conversion devices for over 4 decades. Reasons for the delay in adopting these energy sources have to be known to students and learners in the community so that appropriate remedial measures in the knowledge domain may be taken.

A hydrogen-based economy has been talked about. However, the two components of this economy, namely production and storage, have not seen any remarkable progress in reaching acceptable levels despite intense research over the past 4–5 decades. This situation has to be carefully analyzed and the current situation explicitly stated and documented so that the younger generation will take up suitable remedial action in this research area.

On the whole, this compilation is only for education purposes and to build a base for further research and developments in this emerging field. If this compilation helps even to a small measure to introduce this subject into our educational curriculum and promotes the acquisition of knowledge in this domain, then the purpose of this exercise will have been accomplished to a limited extent.

Many shortcomings may have crept inadvertently or otherwise into this compilation. They will be gratefully acknowledged. Any suggestions for improvement of this compilation will improve and aid education in this domain.

Special thanks for the patience of the enthusiastic editorial team of Elsevier and to my colleagues, especially Mr. Hariprasad Narayanan, at the National

Center for Catalysis Research (NCCR), Indian Institute of Technology, Madras, for their support and understanding. Special thanks are due to the Department of Science and Technology, Ministry of New and Renewable Energy, Government of India, for their generous support of the activities of NCCR.

Chennai 600036
May 3, 2016

B. Viswanathan

Chapter 1

Introduction

WHY ANOTHER BOOK ON ENERGY?

A number of printed documents are available on energy sources. Only a few typical examples are given in references [1−13]. In this case the listing is not comprehensive. In this context one can argue about why another book on energy sources might be needed. This question has been raised not to advertise the current volume but to introduce what one can expect at most from this volume. In the available volumes on energy, one can find a comprehensive treatment on certain energy sources and possibly more, at best some comparison with other energy sources. Today energy conversion processes are really at a crossroads, because where process technology is mature and well practiced, they are being threatened by the extent of availability of the raw materials (for example, fossil fuels) and hence require the introduction of newer processing methodologies to extend the life span of these energy sources or alterations introduced into these energy sources, such as shale gas or the bottom of the crudes. In the case of some renewable sources such as hydrogen, all three technical aspects, namely generation, storage, and distribution, appear not to be satisfactory. In such cases one has to assess available technical progress and postulate some practical solutions to make these technologies viable for the energy availability for society. A critical evaluation of available technologies and the development required to exploit each energy source is necessary to benefit society. In most situations such as for hydrogen and solar energy, intense research efforts have expanded in the past five decades, but the solution still appears to be far from realized. This presentation considers the following aspects of energy sources:

1. Critically examine various available energy sources from the points of view of sustainability and environmental acceptability.
2. How do these energy sources rely on resources available on earth?
3. Which energy sources among available sources are considered renewable and nonrenewable on a human time scale, and why?
4. Because fossil fuels (oil, natural gas, and coal) are considered the main sources of energy, their origin, exploration, and exploitation need to be addressed critically from several points of view [14].

Energy Sources. http://dx.doi.org/10.1016/B978-0-444-56353-8.00001-0

5. In the case of promising energy sources such as hydrogen and solar energy, the reason for the delay in exploiting them universally, such with solar cells, is to find an appropriate material that will give the desired level of efficiency and be cost-effective [15].
6. Scientific society is obsessed with materials and is totally engaged in designing and fabricating materials with desired functionalities and efficiencies. With all of these skills, in the case of fuel production (such as hydrogen from water), solar energy conversion to electricity, or conversion of so-called waste products (such as CO_2) into value-added products, performing materials have yet to be identified. Reasons for this failure or blind search need to be examined [16a–c].

Among the different scientific activities of the human race, energy conversion occupies a preeminent position. There are various reasons for this selective concentration in research and development among the communities; the main reason is that it provides an edge and superiority over other communities in addition to conventional economic benefits. In fact, it is tacitly assumed that most conflicts on earth have arisen as a result of competition and anxiety regarding establishing priority in exploiting available and exploitable natural energy sources.

Over the decades since the oil crisis in 1973, there has been intense research in establishing alternate energy conversion processes, refining existing fossil fuel conversion processes, or even finding substitutes for conventional fossil fuel sources in the form of so-called biodiesel. One attempt has been to harness energy sources from so-called shale oil and other sources. It appears that it may be necessary to examine and establish the viability of these alternate routes based on fossil fuel sources. In addition, a variety of other energy conversion options have been proposed and experimented upon, and developments are at various stages of adaptation. It is appropriate to consider an example to assess the hectic activity in this sphere.

In general, one needs to clearly understand the term "energy." As a definition, energy is considered to be the ability to carry out work. Hence, energy can be found in various forms such as chemical energy (in all states of matter), electrical energy, heat (thermal energy), light (radiant energy), mechanical energy, and nuclear energy. Essentially energy is divided into two categories: If it is in stored form, it is designated as **potential energy**; energy in motion is called **kinetic energy**. Essentially there is a variety of sources of energy:

1. nuclear fission in the sun
2. gravity generated by the earth and moon, possibly the sources for wind and ocean wave energy
3. nuclear fission and fusion reactions
4. energy stored in the interior of the earth in all three states of matter, so-called fossil fuels
5. energy in chemical bonds

It is believed that over 75% of the energy needs of the earth are provided by fossil fuels and that among the three major fossil fuel sources (oil, natural gas, and coal, all three states of matter), oil occupies a preeminent position. Let us consider the status of oil first, because it appears to be the preferred choice of energy source despite of alarm regarding its long-term availability. Basic data (by country) on available reserves of oil are collected in Table 1.1A. Corresponding data for other fossil fuel sources such as gas and coal are compiled in Tables 1.1B and 1.1C, respectively. The preferences for fossil fuel sources could be due to various reasons, possibility transportability.

Energy sources have been classified into two categories: renewable and nonrenewable, or conventional and unconventional. Both of these classifications have limitations because even fossil fuels are renewable for extended

TABLE 1.1A Oil Reserve Amounts in Millions of Barrels

S. No	Country	Reserves (MMbbl)
1	Saudi Arabia	268,350
2.	Canada	173,000
3	Iraq	144,211
4	Kuwait	104,000
5	United Arab Emirates	97,800
6	Russia	80,000
7	Libya	48,014
8	Nigeria	37,200
9	United States	30,529
10	Kazakhstan	30,002
12	China	25,585
13	Qatar	25,385
14	Brazil	13,986
15	Bolivia	13,200
16	Angola	10,470
17	Mexico	10,364
18	Nearly another 80 countries	<10,000–0.4

MMbbl, millions of barrels.
Data extracted from http://en.wikipedia.org./wiki/list_of_countries-by-proven-oil-reserves.

TABLE 1.1B Main Gas Reserves, in Trillion Cubic Meters

Country	Proved Reserves, in Trillion Cubic Meters	Percent World Reserves
World	186	100
Iran	34	18
Russian Federation	31	17
Qatar	25	13
Turkmenistan	17	9
United States	9	5
Saudi Arabia	8	4
United Arab Emirates	6	3
Venezuela	6	3
Nigeria	5	2.5
Algeria	5	2.5

Data extracted from http://knoema.com/smsfgud/world-reserves-of-fossil-fuels.

TABLE 1.1C Major Coal Reserves in Various Countries (in Million Tons)

Country	Total Proved Reserves	Percent World Reserves
World	891,531	100
United States	237,295	27
Russian Federation	157,010	18
China	114,500	13
Australia	76,400	9
India	60,600	7
Germany	40,548	5
Ukraine	33,873	4
Kazakhstan	33,600	4
South Africa	30,156	3
Indonesia	28,017	3

Data extracted from http://knoema.com/smsfgud/world-reserves-of-fossil-fuels.

periods of time. Similarly, wind or ocean energy cannot be considered unconventional because these have been realized for a long time. Although we do not want to discard these classifications, energy sources can be classified on the basis of their sources. Available energy sources produce heat, power, living beings move objects, or produce electricity. Energy consumption has grown steadily; its increase is estimated to be nearly 110 times that of early human beings.

Most energy needs today are met by fossil fuels (stored solar energy). However, fossils fuels are nonrenewable on a human time scale and cause environmental damage such as the greenhouse effect and other environmental degradation. In any case, the harnessing of all energy sources (except direct solar heating) depends on the availability of suitable materials [16a–c]. Energy stored in chemical bonds has been harnessed for a long time. When chemical reactions take place, energy is either released or absorbed. If it is absorbed, it is stored in the chemical bond for later use. If it is released, it can produce useful heat energy, electricity, and light. Biomass energy is one such example. It involves burning (a chemical reaction) wood or other organic byproducts. These organic materials are produced by photosynthesis, a natural chemical process that derives energy from the sun.

In the case of fossil fuels, it may be appropriate to consider resources rather than reserves. This subtle distinction is necessary because it reflects possible accessibility to the market. Table 1.2 lists data on total possible fossil fuel reserves and probable resources.

Fossil fuel energy resources are considered to last for a limited period (although there is uncertainty regarding exactly how long they last) but the technology has been undergoing periodic changes. For example, gasoline and diesel fuel specifications have been undergoing rapid changes throughout the world especially with respect to sulfur content owing to regulations imposed to

TABLE 1.2 Estimated Fossil Fuel Reserves and Resources

	Reserves	Resources
Conventional oil	1.3 trillion barrels	2.7 trillion barrels
Unconventional oil	400 billion barrels	3.2 trillion barrels
Conventional gas	220 tcm	460 tcm
Unconventional gas	–	~330 tcm
Coal	730 Gt (3.6 tboe)	18 trillion tons
Lignite	280 Gt (0.7 tboe)	4 trillion tons

GT, gigaton; *tboe*, trillion barrel oil equivalents; *tcm*, trillion cubic meters.

safeguard from environmental degradation. Accordingly refining operations have also undergone changes in the extent of removal of sulfur or metal-containing molecules or in the extent of handling or mixing of heavier crudes. Refinery operations will continue to be practiced because these products well satisfy wide-ranging energy requirements, although there is a strong desire to replace them with renewable resources. It is therefore necessary that developments in this well-practiced technology (oil refining) be visited periodically especially when the feedstock quality and grades undergo remarkable changes.

Similarly, technology for other fossil fuel energy sources such as natural gas and coal will undergo considerable alterations especially when they are to be used as substitutes for oil.

Natural gas can be associated, nonassociated or coal bed gas. Natural gas may contain a significant amount of ethane, propane, butane, and pentane and some nonhydrocarbon gases such as CO_2, N_2, He, and H_2S. Some typical compositions of natural gas are given in Table 1.3.

Shale gas is natural gas produced from shale. This gas has to be harnessed through fractures that allow the gas to flow through. All shale gas reserves have to be provided via artificial fractures made by hydraulic fracturing. Shale gas exploration has been pursued by a number of countries since 2000.

There are other forms of gas used as fuel. One is the town gas made by the destructive distillation of coal; it contains a variety of calorific gases such as hydrogen, CO, methane, and other volatile hydrocarbons with some non-calorific gases such as CO_2 and nitrogen, and is used in the same way as natural gas.

Other sources of natural gas are landfill gas, biogas, and methane hydrate. When methane is formed by the anaerobic decay of nonfossil organic matter (biomass), it is termed "natural biogas."

Methane hydrates exist under sediment on offshore continental shelves and on land in arctic regions. The cost of extracting natural gas from this crystallized form is high, although the Japan Oil Gas and Metals Nationals Corporation announced the possibility of a cheaper process.

Fig. 1.1 shows a simplified diagram of activities involved in the production and consumption of natural gas. Natural gas processing has been undergoing changes depending on the end use of the material required, such as value-added chemicals. One main thrust is in the coupling reaction (possibly a modified Fischer–Tropsch process) or oxidative coupling to produce liquid fuel, which is a convenient form for today's technological needs.

Coal is the oldest fuel used to smelt copper. Nearly 40% of the world's electricity needs are generated by burning coal. There are various ranks of coal. The lowest rank is called peat, which in the dehydrated form is an effective absorbent for fuel and oil spills on land and water. Lignite or brown coal is the lowest rank of coal and is mostly used for electric power generation. Subbituminous coal is the primary fuel for steam-electricity generation and is also a

TABLE 1.3 Compositions of Some Available Forms of Natural Gas in Volume Percent From Different Sources

Location	CO_2	N_2	H_2S	CH_4	C_2H_6	C_3H_8	C_4H_{10}	C_5^+	He
Oldenburg, Germany	6.9	7.0	7.1	78.9	0.1	–	–	–	–
Groningen, Holland	0.9	14.3	–	81.3	2.8	0.4	0.2	0.1	0.04
Lacq, France	9.3	0.2	15.8	69.3	3.0	0.9	0.6	0.8	<0.001
Siberia	–	1.2	–	97.8	0.7	0.3	-	–	–
Kirkouk, Iraq	2.7	–	–	59.6	19.8	5.5	1.9	1.1	–
Texas, United States	–	25.6	–	65.8	3.8	1.7	0.8	0.9	1.8
New Mexico, United States	0.8	1.4	–	77.3	11.2	5.8	2.3	1.2	–
Alberta, Canada	8.2	2.5	35.8	52.3	0.4	0.1	0.2	0.4	–
Assam, India (associated)	0.2–1.4	0.2–1.0	–	77.3–88.3	5.1–7.0	2.9–6.0	0.7–5.5	0.3–2.8	–
Assam, India (nonassociated)	0.1–0.4	0.6–1.7	–	91.3–93.2	3.7–5.4	0.4–2.1	0.5–1.1	0.1–0.4	–

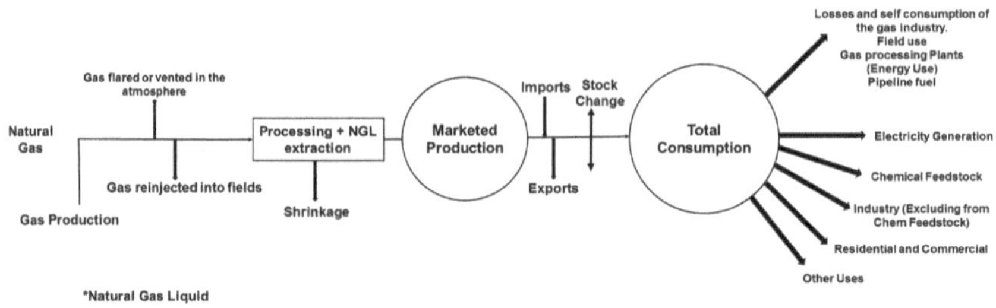

FIGURE 1.1 Scheme of activities in the production and consumption of natural gas.

source of light aromatic hydrocarbons. The next ranked material, bituminous coal, is the main source fuel for heat and power generation applications. The highest ranked coal, anthracite, is employed for residential and commercial space heating. The chemical composition of coal varies widely: volatiles, 7–65%; carbon, 60% to >91.5%; hydrogen <3.75–6%; oxygen, <2.5–17%; and sulfur, 0.5–1%, with an average heat content of about 30,000 kJ/kg.

There are three main coal employment technologies in addition to coke, which is used for metallurgical processes:

1. direct combustion to generate electric power. In this process, coal is primarily used as a solid fuel to generate electricity and heat through combustion. Coal is pulverized and then combusted to generate heat, which is used to generate steam. This is then used in a turbine to turn generators and produce electricity;
2. coal gasification to produce gaseous fuel for direct use: for example, synthetic natural gas or synthesis gas (which is essentially a mixture of CO and hydrogen). Synthetic gas is used for liquefaction or as chemical feedstock; and
3. coal liquefaction, either direct or indirect, in which the molecular structure of coal is converted via a catalyzed process into the desired hydrocarbons, eg, gasoline or the molecular structure is completely broken down and rearranged. A simplified flow diagram for the indirect liquefaction of coal is shown in Fig. 1.2.

The extraction of chemicals from coal is an interesting area of development. There is a variety of ways in which this objective can be achieved. One important path of development is chemicals from synthetic gas. A simplified flowchart for this path is given in Fig. 1.3.

It is anticipated that coal-derived products will amount to tens of millions of tons per year. To meet this challenge, chemists and engineers have to harness all of their skills and knowledge to propose new and designed processes in the future.

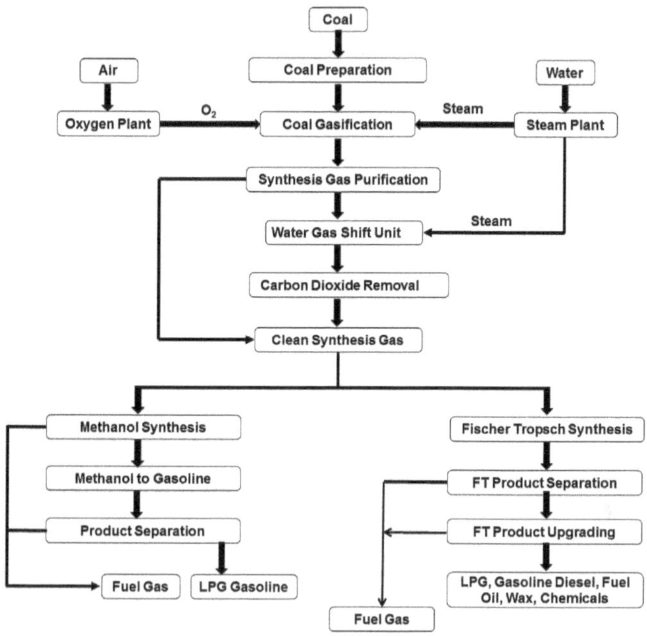

FIGURE 1.2 Simplified flowchart for the two main indirect coal liquefaction processes. *FT*, Fischer–Tropsch; *LPG*, liquefied petroleum gas.

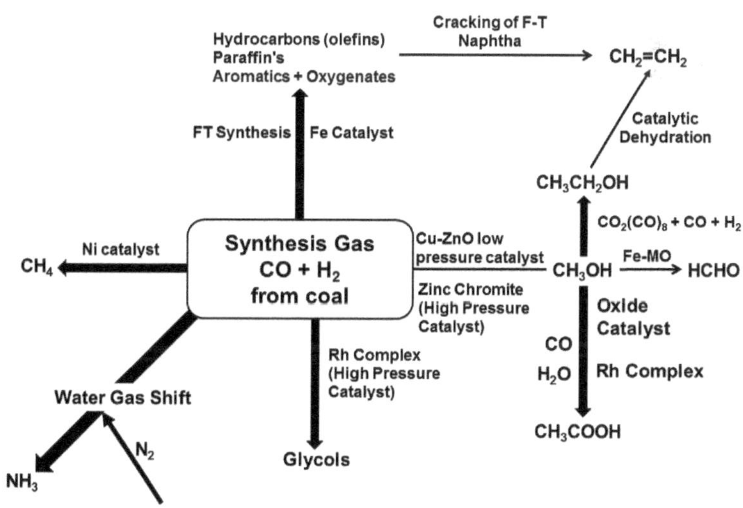

FIGURE 1.3 Possible routes for the production of chemicals from synthetic gas.

NUCLEAR ENERGY

This is one possible route for energy conversion that has caused considerable debate on both sides. Without entering into that debate, and also without entering into the question of the desirability or not of this technology, let us consider only the scope of this energy conversion as an option for energy needs of humanity. There are essentially two principal methods of nuclear energy harvesting: fission and fusion. Between them, fusion technology may require a larger time scale to achieve a viable process. Fission has been pursued on different levels and hence it has proven to be a viable process without the considerations of safety and other environmental aspects.

In simple terms, a nuclear fission reaction involves using a fissionable nuclear fuel and neutrons, thus splitting the heavier nucleus into roughly (a variety of a combination of nuclei are formed as a result of fission) half and releasing energy that is harnessed as electricity. In some sense nuclear energy is similar to thermal **power** stations that generate electricity by harnessing thermal **energy** released from the burning of fossil fuels. **Nuclear power** plants convert **energy** released from the nucleus of an atom via **nuclear** fission that takes place in a **nuclear** reactor, although the magnitude of the heat liberated and energy generated can be of different scales. An example of a typical nuclear reaction can be represented using ^{235}U as a possible fissionable nucleus.

$$^{235}_{92}\mathrm{U} + ^{1}_{0}n \longrightarrow ^{236}_{92}\mathrm{U} \longrightarrow \underbrace{^{89}_{36}\mathrm{Kr} + ^{144}_{56}\mathrm{Ba}}_{\text{Fussion products}} + 3^{1}_{0}n + 200\ \mathrm{MeV}$$

Fission reactions are exothermic. Although we have shown a possible nuclear fission reaction by a single nuclear reaction scheme, one can write nearly 30 different fission schemes for this process. The average number of neutrons produced per fission of ^{235}U by thermal neutrons is approximately 2.47. These neutrons are termed "prompt neutrons," but there can also be some small fraction of delayed neutrons emitted in the fission reaction. These neutrons carry an average energy of 2 MeV.

A simple hypothetical equation for the multiplication factor and conversion factor in a typical reactor is shown in Table 1.4. The data given are only hypothetical; actual values can vary but they will be adjusted in the thermal reactor operation.

Some 70 neutrons are absorbed by the breeding material whereby 70 new fissile nuclei are formed. Therefore the conversion rate is 70/100 = 0.7. There is a variety of reactors such as light water, heavy water, gas-cooled, and fast breeder. Details about these reactors are essential to judging which conversion technology is appropriate. The choice of these reactors has to be made based on the site(s) chosen. Some aspects of fuel processing are also essential for a proper understanding of this energy conversion technology.

TABLE 1.4 Simple Hypothetical Example of a Chain Reaction in a Thermal Reactor For a Stationary State Operation Controlling in a Nuclear Reactor (Multiplication Factor Is Assumed to Be Equal to 1 and Conversion Factor Is 0.7)

100 → Neutrons causing fission in the fuel

↓

230 Neutrons originating from the fission reacton as prompt neutrons

→ 20 Neutrons escape as fast neutrons

→ 40 Neutrons are absorbed in the breeding material

↓

170 Neutrons are obtained as thermal neutrons by moderation

→ 20 neutrons escape as thermal neutrons

→ 30 Neutrons are absorbed in the breeding material

→ 20 neutrons are absorbed in structural material in moderator and in fission products

↓

100 neutrons promote fission reaction in the fuel

Therefore the multiplication factor is 100/100 = 1

Nuclear fission is a kind of chain reaction. The neutrons released in the first fission reaction can be used for a further fission reaction. A simplified pictorial representation is shown in Fig. 1.4.

Like nuclear fission, nuclear fusion is controlled by factors such as the nuclear reaction cross-section and energy distribution in the nuclei involved in the reaction. Because nuclear fusion depends on the Coulomb barrier between the two nuclei fusing, mostly light nuclei alone are considered. These aspects have to be considered to understand the possibility of using this energy conversion process. There is also a concept called cold fusion in which D–D (deuterium) and D–T (tritium) reactions are supposed to take place in storage media. Their feasibility and prospects may have to be carefully evaluated; and this aspect will be considered in a separate section (Fig. 1.5).

SOLAR ENERGY

The energy needs of the earth ultimately have to be met through solar energy. There are a number of ways by which solar energy can be harnessed; among the various possibilities, some have already reached a stage of adaptation. Harnessing solar energy in the form of thermal energy has been a well-known and well-practiced conversion method for many centuries. However the route involving a thermal path as one of the energy conversion steps is limited by the so-called **Carnot** limitation.

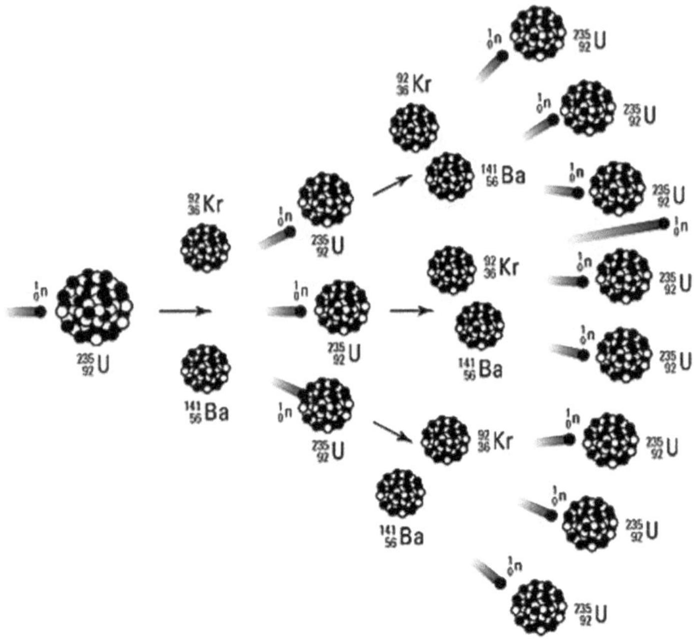

FIGURE 1.4 Schematic representation of a typical nuclear fission reaction as a chain reaction. *Adopted from https://docs.google.com/presentation/d/1d8gsQCyDkwt8 TzfUZ_72 NmJq2C1kOZzh CopfDR_cjPg/edit#slide=id.p13.*

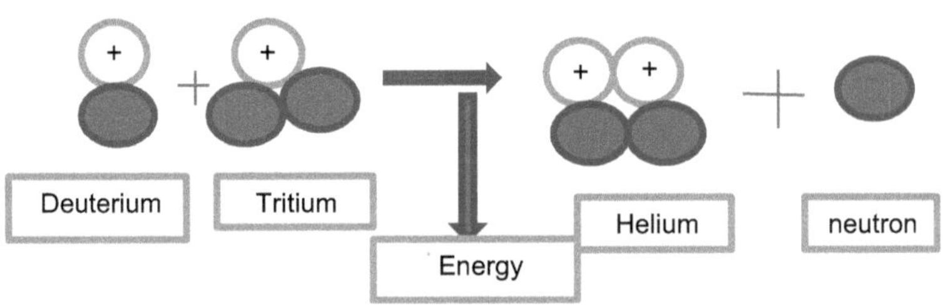

FIGURE 1.5 One of the typical nuclear fusion reaction involving Deuterium and Tritium.

Therefore, there can be some efficiency issues in selecting this route in which solar or chemical energy is converted into useful energy involving thermal energy as one of the steps, as shown in Fig. 1.6.

In addition to the thermal route, there are a number of ways solar (photon) energy can be harnessed:

1. photovoltaic (direct solar energy to electricity)
2. photoelectrochemical (solar energy and chemicals to electricity)
3. photobiochemical (using photons in biochemical reactions)
4. hybrid of photon, electrochemical, and biochemical routes

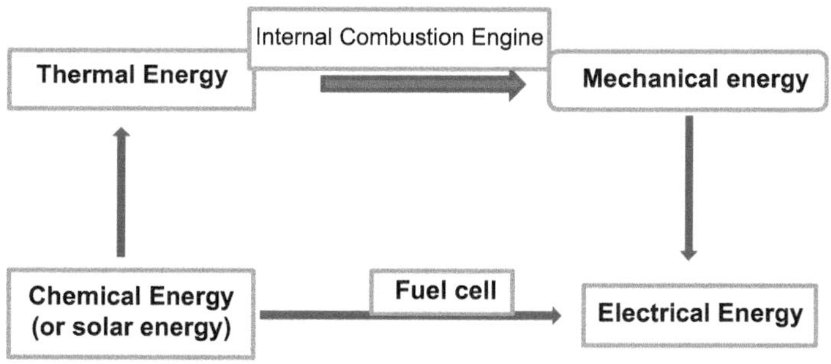

FIGURE 1.6 Possible energy conversion routes for chemical (and solar) energy conversion.

For example, photovoltaic devices (converting light energy directly into electricity) have undergone much development in terms of materials (solar cells). One is the replacement of dye-sensitized solar cells (DSSCs) [17] by an inorganic organic hybrid system using methyl ammonium triiodide plumbate in place of organic dyes or inorganic molecular systems for light harvesting. This development can be considered another variation of so-called DSSCs. These developments have been poised to become possible commercial substitutes for over a decade because the efficiency of DSSCs has been consistently improving. Details of these developments will be dealt within a subsequent chapter. Despite these remarkable research efforts, the developments and discoveries of economically viable energy conversions using solar energy appear to remain a distant dream. With existing fossil fuel price fluctuations and social and economic imbalances between countries causing social inequalities, the need for a universally acceptable conversion device has become imperative, especially when the source of energy is universally available.

The transport industry and small-scale energy conversions have been predominantly dominated by internal combustion engines for so long that it seems inconceivable that they could ever be replaced. Society's reliance on gasoline has serious implications for energy security and economic consequences. With the transport sector completely reliant on oil, future availability and possible price shocks are major policy concerns.

Regarding the concerns of oil, Roger Anderson, director of the Energy Research Center at Columbia's Lammont–Doherty Earth Observatory, says, "If you pay people enough money, they will figure out all sort of ways to get oil you need." Chevron worked with a consortium of other oil companies and drilled in exploration in Mexican waters 17,000 ft deep, more than 3 km below the surface. The problem they faced was after spending $20 million in drilling, the well was said to have come up dry. This was not unusual; about half of prospective wells come to naught. Nonetheless, oil companies consider the risky investment in deep-sea drilling to be money well spent. There is lot of oil to be found at that depth, predicts Anderson.

Efforts to diminish the environmental damage of rapidly growing automobile use for the past 50 years initially focused on adding control devices to the internal combustion engines and then on producing cleaner gasoline. But as more countries become industrialized, it seems likely that a billion cars will soon be on the road. Given these figures, it is doubtful whether further improvements in conventional engines will be enough to ward off global warming and health risks resulting from traffic-related pollution. The dangers of air pollution were first experienced in the densely populated cites of California, where the Zero Emission Vehicle mandate was introduced in 1994. According to the mandate's original target 2% of all new vehicles had to produce no emissions by 1998, which increased to 5% by 2000 and to 10% by 2003. These targets were subsequently delayed, largely because battery-powered vehicles, which were originally seen as the solution, did not have sufficient driving range. The development of storage or rechargeable batteries has seen considerable progress in past decades, although no direct substitute for existing fossil fuel resources has yet emerged. The developments and limitations that were experienced need to be carefully analyzed. However, the partial story of this development will reveal the level of scientific activity in this field and, in general, in the energy conversion process itself. Driven by the motivation that the efficiency of any energy conversion device should be better than the **Carnot efficiency**, between 2006 and 2008 Miyasaka and coworkers reported that perovskite sensitized solar cells employing methyl ammonium lead trihalide and measured a full sun power conversion efficiency between 0.4% and 2% for solid state and electrolyte-based solar cells [18,19]. The first journal-recorded perovskite-sensitized solar cell was reported with an efficiency of 3.5% in 2009 [20]. Then Park and coworkers increased the efficiency of these solar cells to 6.5% by improving the titania surface and the processing procedure of the perovskite [21]. These initial efforts were followed by using [2,2(7,7) tetrakis-(*N*,*N*-dipmethoxyphenylamine)9,9(spirobifluorene)] (Spiro-OMeTAD) as the whole transporter and an efficiency of 10% was achieved [22,23]. The trend of research in this area shows that soon the power conversion efficiency will supersede that of copper indium gallium selenide (20%) and that of crystalline silicon (25%). All of these developments have been achieved in a span of less than 10 years, which shows the intense research activity in this field [17].

In 2013, the planar and sensitized architectures of these types of solar cells experienced a number of developments such as a new deposition technique for the sensitized architecture exceeding 15% efficiency [24]. In the same period, Liu et al. fabricated planar solar cells by thermal evaporation, also achieving more than 15% efficiency [25]. Perovskite solar cells in the typical "organic solar cell" architecture, an "inverted" configuration with the hole transporter below and the electron collector above the perovskite planar film, was also demonstrated at the time [26].

A variety of deposition techniques and higher efficiencies were reported in these perovskite-sensitized solar cells in 2014. A reverse-scan efficiency of 19.3% was claimed by Yang using the planar thin-film architecture [27]. In November 2014, at the sixth World Conference on Photovoltaic Energy Conversion in Japan, the achievement of a single-junction perovskite solar cell with a power-conversion efficiency of 24% was reported. This short mention of the hectic activity in this sphere is meant to describe the feverish activity to achieve efficiencies overriding the Carnot efficiency and to bring home that it appears to be feasible and may be achieved within the time scale of the exhaustion of fossil fuels, even though many have expressed alarm regarding the situation for life on earth owing to the depletion of fossil fuel sources in the near future.

HYDROGEN ENERGY

Hydrogen energy depends on the availability of hydrogen. Hydrogen is currently produced by the steam reforming of naphtha, which is dependent on fossil fuels. Hydrogen can also be obtained from water through a number of ways: thermochemical, photochemical, photoelectrochemical, electrochemical, biochemical, and biophotochemical routes. It may be necessary to considers some of these routes to change over to a hydrogen economy. There is a parallel suggestion based on a methanol economy proposed by Olah [28a–d].

Among various methods for hydrogen generation, photoelectrochemical cells appear to be a promising technology from several points of view. In this mode, instead of generating electricity directly, chemical energy is converted into fuels, mostly hydrogen, although other fuel molecules can also be obtained in this mode for such as photoelectrochemical cells. Among available fuel molecules, hydrogen is a better fuel with good combustion energy. It has many desirable attributes, such as that it is abundantly available and efficient, and generates no emissions when it is employed as a fuel; it is nontoxic and can be generated from renewable sources. Hydrogen may be an alternative fuel that can reduce dependence on imports for some countries and also can significantly reduce greenhouse gases.

These are various reasons for considering hydrogen as a primary energy carrier. It has the simultaneous advantages of being used as a transportation fuel in an internal combustion engine and as a fuel for power generation in a fuel cell. Fuel cells have the advantage of unbelievable efficiency compared with internal combustion engines. It is appropriate at this stage to compare the electrochemical data of various fuels to realize the superiority of hydrogen as a fuel (Table 1.5).

There is a variety of ways in which hydrogen can be generated. Fig. 1.7 shows some economically viable options. It can be seen that hydrogen can be generated using a variety of fossil fuel sources, nuclear energy (by either electrolysis or thermochemical cycles), or solar energy (so-called

TABLE 1.5 Electrochemical Data on Various Fuels

Fuel	ΔG^0 (k cal/mol)	E^0_{theor} (V)	$E^0_{maximum}$ (V)	Energy (kWh/kg)
Hydrogen	−56.7	1.23	1.15	32.67
Methanol	−166.8	1.21	0.98	6.13
Ammonia	−80.8	1.17	0.62	5.52
Hydrazine	−143.9	1.56	1.28	5.22
Formaldehyde	−124.7	1.35	1.15	4.82
Carbon monoxide	−61.60	1.33	1.22	2.04
Formic acid	−68.2	1.48	1.14	1.72
Methane	−195.5	1.06	0.58	–
Propane	−503.2	1.08	0.65	–

photoelectrochemical cells) from biomass, and there can still be some controllable chemical routes which have not yet been tested. Development of some of these technologies may have reached advanced stages and it may be necessary to deal with them in some detail in a book on energy.

Among the various available technologies, electrolysis is an established procedure and needs only improvement in the design and fabrication of the

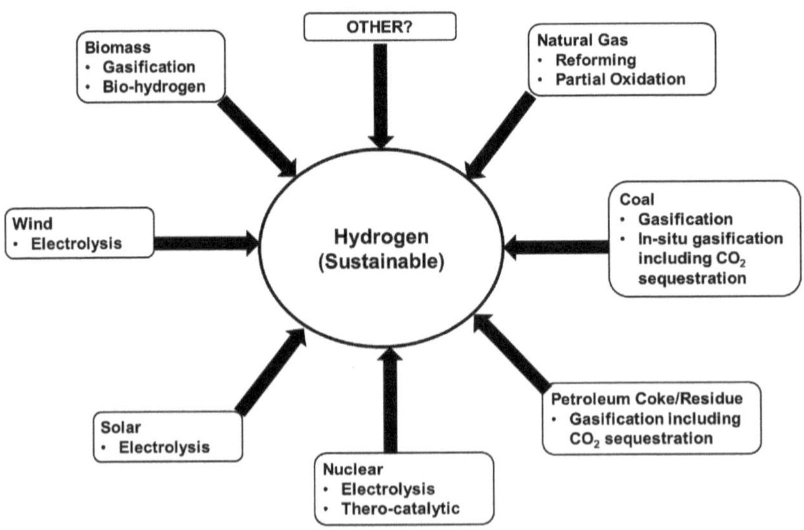

FIGURE 1.7 Possible routes for the production of hydrogen from various sources.

electrodes, the configuration of the electrolysis cell, and the electrolyte medium to be employed for commercial-scale production. Another type of technology that has been extensively researched with success is the thermochemical cycle. A variety of thermochemical cycles have been theoretically proposed, among which two cycles (sulfur-iodine and iron halogen cycles) at least are possible for practical use in some of the countries. Hence it is desirable for some fundamental knowledge about these two cycles and the basis for the formulation of other possible cycles to be considered.

Both photoelectrochemical devices (photovoltaic and photoelectrochemical cells) have undergone considerable developments, especially in materials in the past four to five decades, although they have not yet reached the level of commercial exploitation, especially from the point of energy conversion efficiency. However, these two energy conversion devices may tilt the balance in favor of so-called **developing countries** in relation to so-called developed countries even though technically all countries are only developing. The science and technology of photoelectrochemical cells have been undergoing considerable progress [28a–d], possibly in parallel with the development of solar cells; this aspect will be considered in detail.

In the context of hydrogen, another possible device that will directly convert chemical fuel into electricity is the so-called **fuel cell**, which has been poised to become a possible viable energy conversion device ever since its introduction in 1839 and has also been used in space missions since the 1960s. Efficiencies exceeding Carnot limits have been claimed for these devices, but the economics of the device are not favorable owing to the use of noble metal electrodes and the use of proprietary membranes. The development of this device is an interesting story spanning 170 years and deserves a separate treatment.

Fuel cells have long been considered the most likely alternative to conventional internal combustion engines because they provide ultra-low emission and high efficiencies. Unlike batteries, fuel cells do not require charging and do not lose much energy when converting between electrical and chemical energy. Indeed, energy storage in some form of fuel is more effective than in any type of rechargeable battery, because it improves the power density of the system and ultimately the driving range of the vehicle. Hydrogen storage in a solid state has been a challenge been faced by the research community ever since hydrogen energy has been considered as an alternative energy source. Hydrogen can be stored in a gaseous or liquid form but this is not desirable because the process itself requires energy expansion and it will considerably reduce the availability for energy conversion.

Hydrogen in some form is all that is required for a fuel cell. In vehicles, hydrogen can be stored as a cryogenic liquid or a pressurized gas. However, liquefying hydrogen is expensive and storing this extremely cold fuel in a vehicle is a difficult engineering task. Storing hydrogen requires significant energy expenditure for compression, stringent safety precautions, and bulky, heavy storage tanks. Natural gas is used as a hydrogen source in some fuel cell designs for large stationary electricity generating stations. However, this also prevents many of the drawbacks of cryogenic liquefaction or compression when considering mobile applications with weight and space limitations.

Metal hydride storage of hydrogen is possible. However, weight constraints currently limit the potential range of a vehicle with this particular hydrogen storage technology. Carbon nanofibers or tubes as hydrogen storage also have been proposed; however, this technology is far from commercial development.

Intermetallics, carbon materials, porous solids, and metal organic frameworks are some options being considered for hydrogen storage purposes. This is also another area of research which is at the crossroads without a proper direction toward the desired goal. The desired goal is the weight percent storage capacity (taken to be around 6.5 wt% on the basis of various considerations). This expected percentage of hydrogen storage capacity has not been achieved in any of the solid-state storage materials under the desired temperature and pressure conditions (namely near room temperature and atmospheric pressure conditions). Fig. 1.8 shows the hydrogen storage capacities of various options available and hydrogen storage in conventional fuels in terms of both volumetric and gravimetric density. The desired storage capacity is around 6.5 wt%.

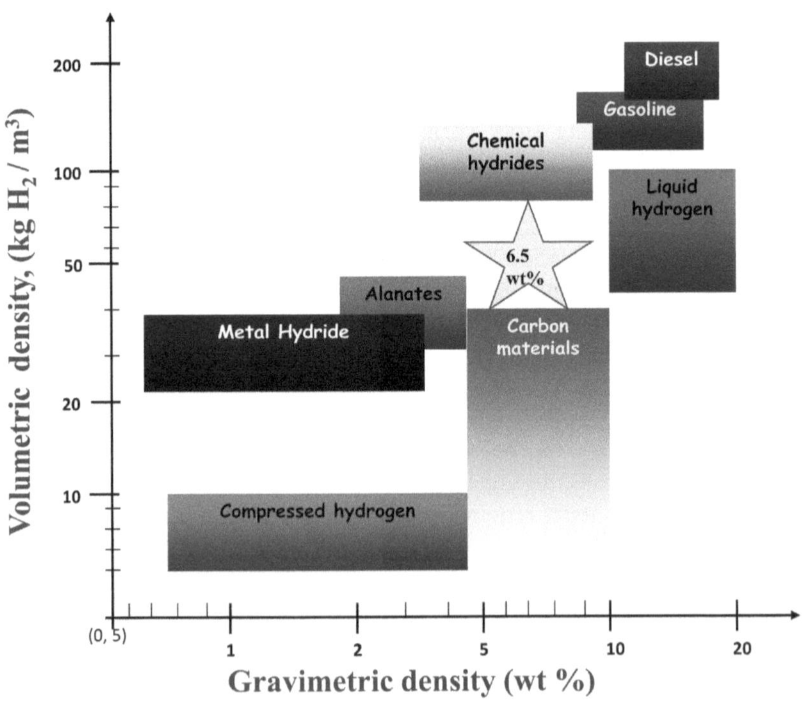

FIGURE 1.8 Hydrogen storage capacity in various available storage materials and conventional fuels.

Typical metal hydrides with enough hydrogen storage are MgH_2, BeH_2, and TiH_2 but their decomposition temperatures are not favorable. The main intermetallics in which hydrogen storage has been attempted are AB (FeTi), A_2B (Mg_2Ni, ZrV_2), and AB_5 ($LaNi_5$) but the reported storage capacity is only less than 3 wt%. Complex metal hydrides such as alanates and borohydrides can store the desired amount but the decomposition to release hydrogen requires an appropriate catalyst, the kinetics is not favorable, and the decomposition is not a single-step process. Some porous materials such as metal-organic framework, porous oxides such as zeolites, and glass micro-spheres have also been proposed as possible hydrogen storage media but the experimental parameters for this storage are not favorable for economic exploitation. A pictorial representation of these types of materials and how hydrogen is stored in these typical materials is shown in Fig. 1.9.

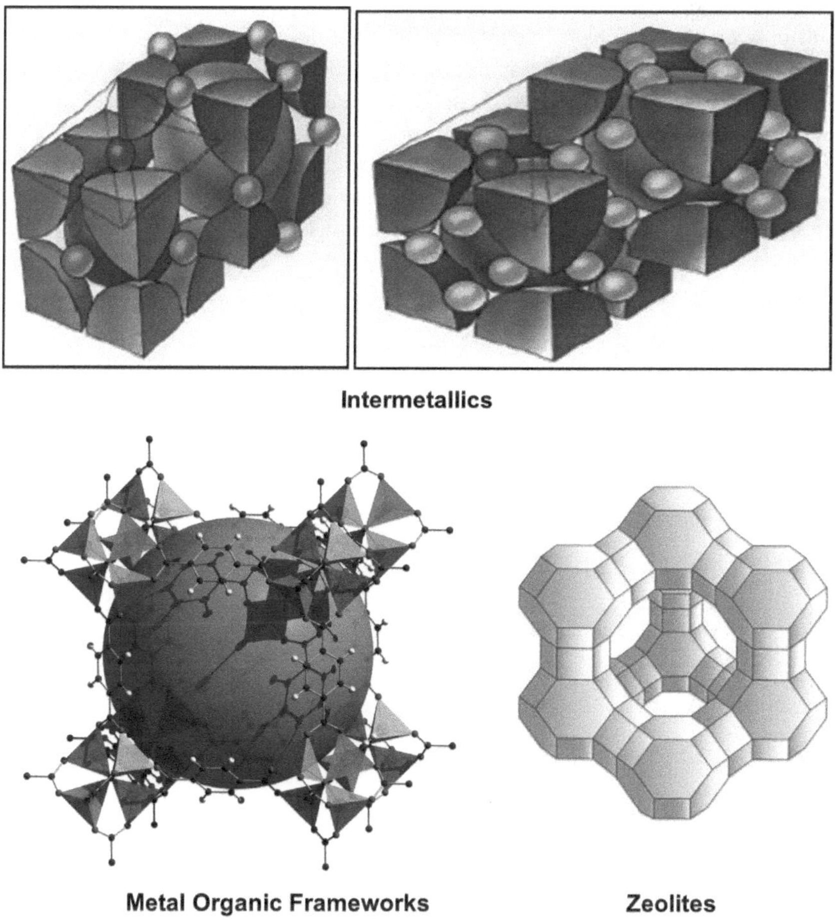

FIGURE 1.9 Possible solid-state hydrogen storage materials.

Despite intense research activities in this field (this will be dealt with in detail in a separate chapter), the desired levels of storage (at least approximately 6.25 wt%) of hydrogen have not yet been achieved. This is a question that will be taken up for analysis in a separate chapter.

FUEL CELLS

Among the energy conversion devices using hydrogen, fuel cells occupy a unique place. Even employing hydrogen as fuel, there can be a variety of possible configurations of fuel cells, classified on the basis of the electrolyte used or the temperature of the operation. One such simple representation of the configurations of fuel cells is given in Fig. 1.10.

Methanol emerges as an ideal hydrogen carrier for transport application because it is liquid at room temperature and ambient pressure. Releasing hydrogen from its bound state in a methanol molecule is easier to accomplish than for other available liquid fuels. Moreover, methanol fuel contains no sulfur, which is a fuel cell contaminant, in addition, because no carbon to carbon atomic bonds exists, which are hard to break. In addition, methanol has a high hydrogen to carbon ratio. In fact, a gallon of methanol fuel contains even more hydrogen than the amount of cryogenic liquid hydrogen.

The proton exchange membrane (PEM) fuel cell is a leading contender for vehicle applications [6,7]. There are two principal PEM fuel cell types under commercial development. The first, the gaseous PEM, operates in hydrogen

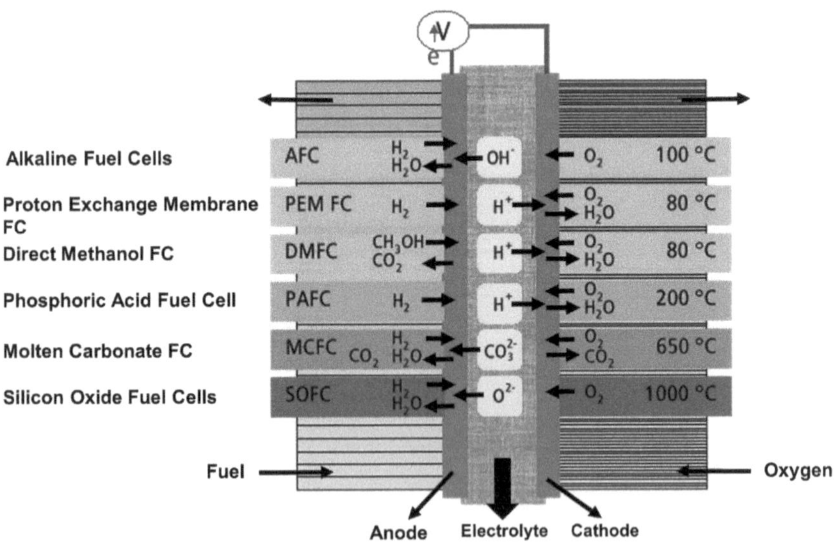

FIGURE 1.10 Various possible configurations of fuel cells.

gas. The second type, or liquid PEM fuel cell, feeds methanol directly into the fuel cell.

Gaseous Proton Exchange Membrane Fuel Cell

In this type of fuel cell, hydrogen gas is required for the fuel cell to operate. Hydrogen can be stored directly onboard the vehicle. However, the gaseous fuel imposes several penalties in weight and cost. Therefore, developers focus on hydrogen-containing liquid fuels. These fuels require reforming to deliver hydrogen to the fuel cell, and methanol is one possible choice.

Prime advantages of this technology are that these fuel cells operate at relatively low temperatures and the reformer easily splits the methanol to produce hydrogen needed for the stack. It is easy to distribute and store on the vehicles themselves.

Main disadvantages are that the reformer may produce trace emissions as it burns some of the methanol and hydrogen supply to create the heat necessary for the reaction. The reformer adds weight, complexity, and cost to the overall system.

Liquid Proton Exchange Membrane Fuel Cell

The second type of PEM fuel cell is the direct methanol fuel cell (DMFC) developed by NASAs Jet Propulsion Laboratory in California, the University of Southern California, the California Institute of Technology, and others. No reformer is needed in this fuel cell; methanol is injected directly into the fuel cell. Owing to its simplicity, compact size, and near-zero emissions, the potential for the DMFC is great. It is distinctly possible that the DMFC will become a universally available power source in coming years.

Challenges

The challenges facing PEM developers are threefold:

1. One has to reduce the cost to build the electrode plates.
2. There is a need to reduce the amount of expensive platinum used as the catalyst.
3. It is necessary to design a cheap and effective membrane.

There has been considerable progress in each of these areas. A few decades ago, the fuel stack, which was only one part of the whole fuel cell power system, cost a prohibitive $5000 per kilowatt!

The whole fuel cell system cost (fuel cell stack, methanol reformer, and associative components) has come down to less than $500 per kilowatt and developers are currently targeting full power system costs in the range of $50 per kilowatt.

Ballard Power Systems led the way, reducing plate costs from $100 per plate to about $1. Working with Johnson Matthey, Ballard Power Systems has dropped its platinum cost to about $140 per car, which is not much more than the cost per platinum used in the catalytic converter in an internal combustion engine's exhaust system.

The manufacturer of membranes, DuPont, has announced that future membranes will cost $10 or less when large production volumes are achieved.

Many of the challenges facing reformer developers are being met with equal zeal. For a cold start, a reformer needs to produce hydrogen quickly. Much progress has been made in this area. For example, Johnson Matthey's hot spot methanol reformer has achieved a start up time of 20 s for 50% hydrogen production and full production in only 50 s. Its fuel processor system is highly efficient, releasing 89% of the hydrogen contained in the methanol fuel. Daimler Benz has reduced the weight of the evaporator that supplies the reformer from 300 kg to just 3 kg while increasing process efficiency. There are many leading players such as Daimler Benz, Toyota, General Motors, and Volkswagen. Research experts estimate that by 2020 the total fleet of DMFC vehicles on the road may reach or even surpass a few million.

Developments will be required:

1. The loading of the noble metals needs to be minimized to improve the long-term performance of the anode and better catalysts than Pt–Ru need to be searched for based on combinatorial analysis.
2. Membrane properties need to be improved in terms of water balance sensitivity and by inhibiting the cross-diffusion of methanol (so-called methanol crossover). Although perfluorinated and partially fluorinated ionomer membranes exhibited the best performance and the highest longevity for fuel cell applications, the high cost and methanol permeability through the membrane may limit their use in future.
3. A new or modified manufacturing technology for nafion has to be created.
4. The ultimate goal of research in membrane technology is to avoid high-cost membranes for transport and low-power fuel cell systems with ease of production on a large scale at low cost.
5. To reduce the use of precious metals and thus cost, the best possible performance has to be extracted from a given amount of the catalyst. This involves mainly catalyst preparation, pretreatment, electrode fabrication, and, in the case of the supported catalyst, the choice of the suitable and durable support.
6. Electronic devices (especially fuel cells) could exploit the role of bifunctional catalysts in the future. It was claimed that the poisoning of platinum anodes by CO in hydrogen–oxygen fuel cells could be circumvented by using very short current pulses capable of removing poisoning residues and restoring the initial catalytic activity. Similar approaches are likely to be effective for DMFCs.

7. The question of whether there an alternate to platinum for electrode applications has been raised repeatedly in the literature owing to its cost and its poisonous nature.
8. It is also necessary for a better (in terms of the kinetics) catalyst for oxygen reduction cathode reaction to be identified and adopted in a reasonable time frame.
9. Some engineering issues associated with the design and operation of DMFC stack systems should be addressed.

In liquid fuel cells, there are problems associated with methanol mass transfer from anode and carbon dioxide release, which disturbs the maintenance of required stack operating concentration. The use of vapor fuel cells results in a superior power density performance and avoids the problem associated with methanol stripping from the fuel as a result of carbon dioxide generation.

Although there are various types and options available, the possibility of DMFC being the correct one is considered from various considerations, although this fuel cell may not be the best choice. In this context, two factors may supersede in consideration:

1. the possibility of creating the necessary infrastructure and distribution system within the available means and time frame; and
2. the technology may provide the opportunity to make this adaptation viable within a short time frame for both transport and stationary applications.

A simple design of a DMFC is shown in Fig. 1.11. This design is the simplest in terms of the components: the electrodes, electrolyte, and membrane. The possible options for these three components are considered.

The cell reactions can be represented as:

$$\text{Anode reaction: } CH_3OH + H_2O \rightarrow CO_2 + 6H^+ + 6e^-$$

$$\text{Cathode reaction: } 3/2O_2 + 6H^+ + 6e^- \rightarrow 3H_2O$$

$$\text{Cathode corrosion reaction: } CH_3OH + H_2O \rightarrow CO_2 + 6H^+ + 6e^-$$

$$\text{Overall reaction: } 3/2O_2 + CH_3OH \rightarrow CO_2 + 2H_2O$$

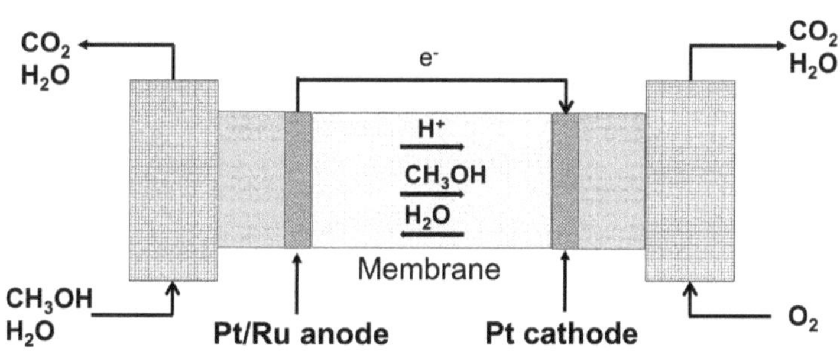

FIGURE 1.11 Schematic diagram of typical direct methanol fuel cell.

Various aspects of fuel cells demand intense research and development. However, one issue that has to be handled is an understanding of the cathode reaction (the so-called oxygen reduction reaction) and creating an appropriate and active electrode that will promote this reaction at the same or similar rate as the anode reaction in a fuel cell. This is an aspect that deserves an increased level of attention in fuel cell development.

PHOTOSYNTHESIS AS A POSSIBLE ENERGY SOURCE

Photosynthesis is nature's way of harnessing energy. This is an energy-saving process although it may not be truly energy efficient. One can learn certain key issues by considering the basic principles of photosynthesis, and this will aid in our attempts to harness at least solar energy to some extent. Nature conserves energy in a certain pathway by suitably adjusting the redox potentials of the redox species through which electron transfer is facilitated. One such simple electron relay system in the electrochemical sense is shown in Fig. 1.12. In our attempts to mimic the natural process, we need to formulate methods to self-adjust the redox potentials of the species in situ instead of

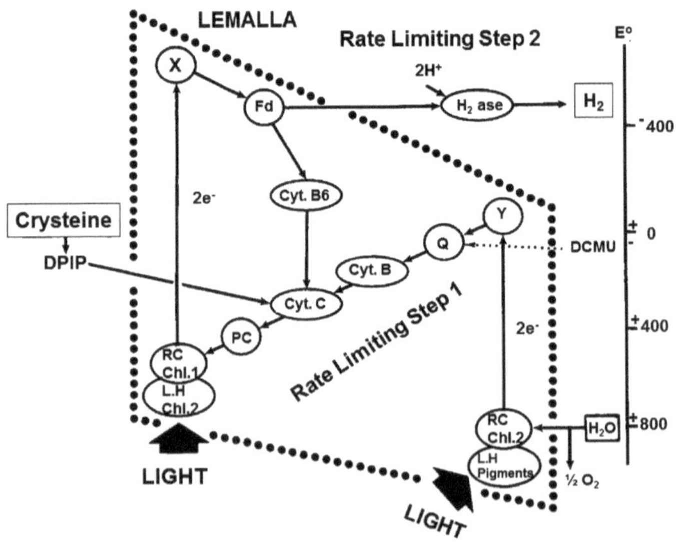

FIGURE 1.12 Electron transport system and hydrogen photo production. *LH Pigments*, light-harvesting pigment proteins; *R.C. ChL1* and *R.C. ChL2*, reaction center chlorophyll protein of photosystems 1 and 2, respectively; *Q*, plastoquinone; *Cyt. B*, cytochrome B; *Cyt. C*, high–redox potential algal cytochrome C; *PC*, plastocyanine; *Fd*, ferredoxin; *H_2ase*, hydrogenase; E_0', redox potential in millivolts at pH 7. Components within dotted lines represent the core of the photosynthetic electron transport system. *Reproduced from T. Ohta, Solar-Hydrogen Energy Systems, Pergamon Press, 1979.*

using only chemical species with fixed redox potentials. Although the methodology is well-known, realizing them in real chemical systems appears to be a long way off. However insufficient the presentation will be, it is our intention to include a section on this topic.

It is well-known that other forms of renewable energy sources go under the name: wind energy, ocean thermal energy conversion, and other geothermal sources. This presentation will not go deeply into these sources for various reasons. The main reason is the author's absolute lack of knowledge on these topics, which does not mean that the author is not competent. The situation is only at the origin for these topics and at the start in the case of other topics.

Biochemical Conversions

Biomass can be converted into gaseous (for example, biogas) or liquid fuels (bioethanol) by the action of bacteria, microorganisms, and enzymes. There are essentially two biochemical technologies for energy conversion: anaerobic digestion and fermentation. Anaerobic digestion involves the decomposition of organic material in multiple metabolic steps in a depleted oxygen environment. Anaerobic digestion is the process by which wet organic wastes in the absence of air are transformed into biofertilizer or biogas that can be used for the production of electricity and heat. A simple flowchart is given in Fig. 1.13.

Liquid and solid nutrients can be processed for use as fertilizers. A variety of fuels can be obtained from waste forestry and agricultural resources and other man-made wastes such as ethanol, methanol, biodiesel, Fischer–Tropsch diesel, and gaseous fuels such as hydrogen and methane. Ethanol can be produced from a variety of lignocellulosic biomass wastes. At the same time, this can be considered one ways to dispose of these wastes. A simple process flow scheme is shown in Fig. 1.14.

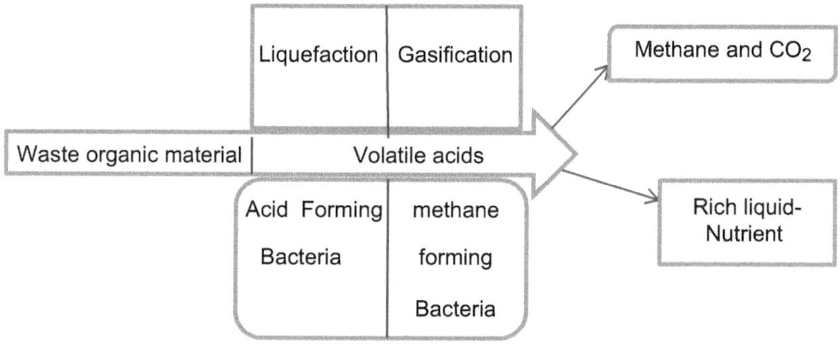

FIGURE 1.13 Flow diagram showing how anaerobic digestion works in 5–20 days (depending on the temperature).

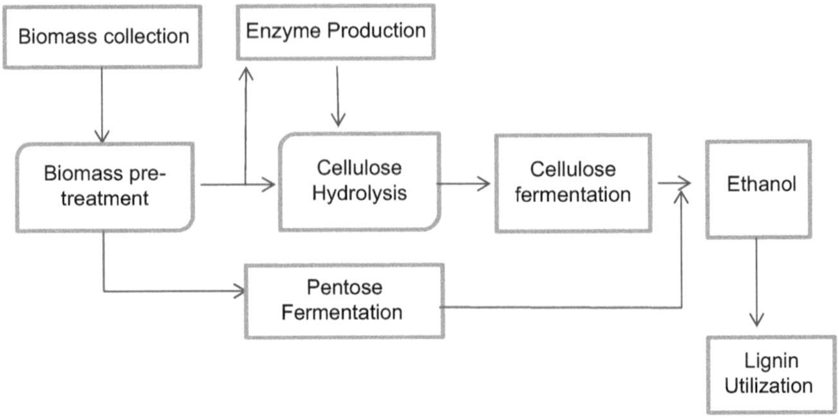

FIGURE 1.14 Process flow diagram for waste treatment for fuels.

PROPOSED SCOPE OF THIS PRESENTATION

It is inevitable that in an exercise like this, one has to impose certain self-imposed limitations; otherwise various sections will be of different standards. Second, it is well-known that a variety of documents are available on this topic. Thus, in what sense is this compilation different from other sources or available documents? Considering the knowledge base, the economics aspects of energy as it is currently understood has been completely kept out of bounds of this compilation. Statistical data have also been kept at a minimum because this is an exercise in comprehending technologies adopted in the energy sector and how and why certain energy technologies (such as fuel cells) have taken an unduly long time period for development.

Instead of listing what one cannot find in this volume, a list of some of what one can hope to find covered (at least in parts) in this compilation is provided:

1. Instead of the adopted boundaries (such as conventional and unconventional, and renewable or nonrenewable, because these classifications have shortcomings), the energy vector is treated in terms of its scope of availability and the adaptation of technological advancements.
2. The available knowledge base on the practiced technologies is given its due place in this compilation.
3. While dealing with the available knowledge base, emphasis is given to reasons why certain technologies have taken longer periods for development.
4. As we all know, materials have assumed a central place in the development of technology. Identifying appropriate materials and their fine-tuning has been a major endeavor in the energy conversion sector; this aspect has been given due attention [29].
5. In some energy conversion technologies such as the splitting of water for the production of hydrogen, fuel production from carbon dioxide, and

enlisting fuel cells as power sources, progress appears to be frustrating, probably because the paths chosen have not been critically examined in view of the anxiety for a quick solution.

6. It is obvious that the extent of coverage on each topic is not uniform for various reasons; it is hoped that this will be tolerated.
7. There are certain limitations to measurement techniques and the evaluation of data in energy conversion research. These were previously pointed out in the literature [30].
8. Energy conversion processes using fossil fuels have undergone changes owing to some alternatives being exploited, such as the feedstock including shale gas, and also increasing refining of the bottom of the barrel.
9. The educational curriculum dealing with energy needs to be drawn with care and attention from the point of view of knowledge rather than economy although this is also important.
10. The new generation of knowledge in the energy sector has been explosive. For example, consider photovoltaics: we have moved away from solid state to other configurations [31].

REFERENCES

[1] V. Dusastre (Ed.), Materials for Sustainable Energy, World Sciencitific Publishing Co., 2011.

[2] A. Hernández-Ramírez, I. Medina-Ramírez (Eds.), Photo-Catalytic Semiconductors, Springer, 2015.

[3] X. Feng (Ed.), Nano-Carbons for Advanced Energy Storage, vol. 1, Wiley VCH, 2015.

[4] L. Guczi, A. Erdôhelyi (Eds.), Catalysis for Alternative Energy Generation, Springer, 2012.

[5] C. Zamfirescu, I. Dincer, G.F. Naterer, Hydrogen Production From Nuclear Energy, Springer, 2013.

[6] B. Viswanathan, R. Subramanian, J.S. Lee (Eds.), Materials and Processes for Solar Fuel Production, Springer, 2014.

[7] D. Stolten, V. Scherer (Eds.), Transition to Renewable Energy Systems, Wiley, 2013.

[8] J.G. Speight, The Chemistry and Technology of Petroleum, CRC Press, 2007.

[9] B. Viswanathan, M. Aulice Scibioh, Fuel Cells: Principles and Applications, Taylor & Francis Group, 2007.

[10] R. Narayan, B. Viswanathan, Chemical and Electrochemical Energy Systems, University Press, 1998.

[11] C.A. Grimes, O.K. Varghese, S. Ranjan, Light, Water, Hydrogen: The Solar Generation of Hydrogen by Water Photo-Electrolysis, Springer, 2008.

[12] R. van de Krol, M. Gratzel (Eds.), Photo-Electrochemical Hydrogen Production, Springer, 2012.

[13] S.J.C. Irvine (Ed.), Materials Challenges Inorganic Photovoltaic Solar Energy, The Royal Society of Chemistry, 2015.

[14] http://www.tulane.edu/~sanelson/eens1110/energy.htm.

[15] H. Savin, P. Repo, G. von Gastrow, P. Ortega, E. Calle, M. Garín, R. Alcubilla, Black silicon solar cells with interdigitated back-contacts achieve 22.1% efficiency, Nat. Nanotechnol. (May 18, 2015). http://dx.doi.org/10.1038/NNANO 2015.89.

[16] [a] https://nccr.iitm.ac.in/Course%20Material1.pdf.
[b] F.E. Osterloh, Inorganic nanostructures for photoelectrochemical and photocatalytic water splitting, Chem. Soc. Rev. 42 (2013) 2294–2320.
[c] F.E. Osterloh, Inorganic materials as catalysts for photochemical splitting of water, Chem. Mater. 20 (2008) 35–54.

[17] H. Snaith, Perovskites: the emergence of a new erafor low cost high efficiency solar cells, J. Phys. Chem. Lett. 4 (2013) 3623–3630 and subsequent papers in this particular journal like, D.A. Egger, E. Edri, D. Cahen, G. Hodes, J. Phys. Chem. Lett. 6 (2015) 279–282.

[18] A. Kojima, K. Teshima, Y. Shirai, T. Miyasaka, Novel photoelectrochemical cell with mesoscopic electrodes sensitized by lead halide compounds, in: 210th ECS Meeting, Cancun, Mexico, October 29–November 3, 2006.

[19] A. Kojima, K. Teshima, Y. Shirai, T. Miyasaka, Novel photoelectrochemical cell with mesoscopic electrodes sensitized by lead halide compounds, in: 214th ECS Meeting, Honolulu, Hawaii, October 12–17, 2008.

[20] A. Kojima, K. Teshima, Y. Shirai, T. Miyasaka, Organometal halide perovskites as visible light sensitizers for photovoltaic cells, J. Am. Chem. Soc. 131 (2009) 6050.

[21] J.H. Im, C.R. Lee, J.W. Lee, N.G. Park, 6.5% efficient perovskite quantum dot sensitized solar cell, Nanoscale 3 (2011) 4088–4093.

[22] H. Kim, C.R. Lee, J.H. Im, K.B. Lee, T. Moehl, A. Marchioro, S.J. Moon, R. Humphry-Baker, J.H. Yum, J.E. Moser, M. Gratzel, N.G. Park, Lead iodide perovskite sensitized all solid state submicron think film mesoscopic solar cell with efficiency exceeding 9%, Sci. Rep. 2 (2012) 591.

[23] M.M. Lee, J. Teuscher, T. Miyasaka, T.N. Murakami, H.J. Snaith, Efficient hybrid solar cells based on meso-superstructured organometal halide perovskites, Science 338 (2012) 643–647.

[24] J. Burschka, N. Pellet, S.-J. Moon, R. Humphry-Baker, P. Gao, M.K. Nazeeruddin, M. Grätzel, Sequential deposition as a route to high-performance perovskite-sensitized solar cells, Nature 499 (7458) (July 10, 2013) 316–319.

[25] M. Liu, M.B. Johnston, H.J. Snaith, Efficient planar heterojunction perovskite solar cells by vapour deposition, Nature 501 (7467) (September 11, 2013) 395–398.

[26] P. Docampo, J.M. Ball, M. Darwich, G.E. Eperon, H.J. Snaith, Efficient organometal trihalide perovskite planar-heterojunction solar cells on flexible polymer substrates, Nat. Commun. 4 (November 12, 2013).

[27] H. Zhou, Q. Chen, G. Li, S. Luo, T.-B. Song, H.-S. Duan, Z. Hong, J. You, Y. Liu, Y. Yang, Interface engineering of highly efficient perovskite solar cells, Science 345 (6196) (July 31, 2014) 542–546.

[28] [a] G.A. Olah, Beyond oil and gas: the methanol economy, Angew. Chem. Int. Ed. 44 (18) (2005) 2636–2639.
[b] G.A. Olah, The methanol economy, Chem. Eng. News 81 (38) (2003) 5.
[c] G.A. Olah, G.K. Suray Prakash, A. Goeppert, Chemical recycling of carbon dioxide to methanol and dimethyl ether: from greenhouse gas to renewable, environmentally carbon neutral fuels and synthetic hydrocarbons, J. Org. Chem. 74 (2) (2009) 487–498.
[d] G.A. Olah, A. Goeppert, G.K. Surya Prakash, Beyond Oil and Gas: The Methanol Economy, Wiley-VCH, 2006.

[29] J. Li, N. Wu, Semiconductor-based photocatalysts and photoelectrochemical cells for solar fuel generation: a review, Catal. Sci. Technol. 5 (2015) 1360–1384.

[30] H. Kisch, D. Bahnemann, Best practice in photo-catalysis: comparing rates or apparent quantum yields? J. Phys. Chem. Lett. 6 (2015) 1907–1910.

[31] F. Bella, C. Gerbaldi, C. Barolo, M. Gratzel, Aqueous dye-sensitized solar cells, Chem. Soc. Rev. 44 (2015) 3431.

Chapter 2

Petroleum

INTRODUCTION

Petroleum is an oily, flammable, thick dark brown or greenish liquid that occurs naturally in deposits, usually beneath the surface of the earth; it is also called crude oil. Petroleum means "rock oil" (*petra* meaning "rock" and *elaion* meaning "oil" in Greek, and *oleum* meaning "oil" in Latin). The name came from its discovery in sedimentary rocks. It is mostly used to produce fuel oil, which is the primary energy source today. Petroleum is also the raw material for many chemical products, including solvents, fertilizers, pesticides, and plastics. Because of its high demand in our daily life, it is also called "black gold."

Oil has generally been used since early human history to keep fires ablaze, and also for warfare. Its importance in the world economy evolved slowly. Its detailed historical background was outlined in the work of Speight. Wood and coal were used to heat and cook whereas whale oil was used for lighting. However, whale oil produced a black, smelly, thick liquid known as tar or rock oil and was seen as a substance to avoid.

When the whaling industry hunted the sperm whale almost to extinction and the industrial revolution needed a fuel to run generators and engines, a new source of energy was needed. In the search for new products, it was discovered that kerosene could be extracted from crude oil or petroleum and used as a light and heating fuel. Petroleum was in great demand by the end of the 17th century, which led to the creation of the vibrant petroleum industry.

Petroleum is often considered the lifeblood of nearly almost all other industries. Because of its high energy content (Table 2.1) and ease of use, petroleum remains the primary energy source.

Oil accounts for nearly over 40% of the US energy demand and a comparable percentage of the world's energy supply. The United States currently consumes about 7.5 billion barrels (1.2 km^3; 1 barrel = 159 L or 35 gallons) of oil per year, whereas the world at large consumes approximately 33 billion barrels (4.8 km^3). It is possible actual figures per year may differ because these are only estimates. Petroleum is distributed unequally throughout the world. The United States and most other countries are net importers of this source.

Energy Sources. http://dx.doi.org/10.1016/B978-0-444-56353-8.00002-2

TABLE 2.1 Energy Density of Different Fossil Fuels

Fuel	Energy Density
Petroleum or crude oil	45 MJ/kg
Coal	24 MJ/kg
Natural gas	34–38 MJ/m^3

ORIGIN OF PETROLEUM

Biogenic Theory

Most geologists view crude oil, like coal and natural gas, as the product of compression and heating of ancient vegetation over geological time scales. According to this theory, it is formed from the decayed remains of prehistoric marine animals and terrestrial plants. Over many centuries, this organic matter, mixed with mud, is buried under thick sedimentary layers of material. The resulting high levels of heat and pressure cause the remains to metamorphose, first into a waxy material known as kerogen and then into liquid and gaseous hydrocarbons in a process known as catagenesis. These then migrate through adjacent rock layers until they become trapped underground in porous rocks called reservoirs, forming an oil field from which the liquid can be extracted by drilling and pumping. One hundred fifty meters is generally considered the "oil window." Although this corresponds to different depths for different locations around the world, a "typical" depth for an oil window might be 4–5 km. Three situations must be present for oil reservoirs to form: a rich source rock, a migration conduit, and a trap (seal) that forms the reservoir.

Reactions that produce oil and natural gas are often modeled as a first-order breakdown, in which kerogen breaks down to oil and natural gas by another set of reactions.

Abiogenic Theory

In 1866, Berthelot proposed that carbides are formed by the action of alkali metal on carbonates. These carbides react with water to give rise to large quantities of acetylene, which in turn is converted to petroleum at elevated temperatures and pressures. For example, one can write the sequence as follows:

$$CaCO_3 \rightarrow CaC_2 \rightarrow HC{\equiv}CH \rightarrow \text{Petroleum}$$

Mendelejeff proposed another reaction sequence involving acetylene in the formation of petroleum. He proposed that dilute acids or hot water react with the carbides of iron and manganese to produce a mixture of hydrocarbons from which petroleum could have evolved. The reaction sequence according to the proposal of Mendelejeff is:

$$\underset{\text{Iron carbide}}{Fe_3C} + \underset{\text{Manganese carbide}}{Mn_3C} \xrightarrow{H^+/H_2O} \text{Hydrocarbons} \rightarrow \text{Petroleum}$$

These postulates based on inorganic chemicals, although interesting, may not be completely accepted because:

1. One often finds optical activity in petroleum constituents that could not have been present if the source of petroleum were only these inorganic chemicals.
2. The presence of thermo-labile organic constituents (biomarkers) in petroleum cannot be accounted for in terms of their origin from these inorganic chemicals.
3. It is known that oil is exclusively found in sedimentary rocks, which would not have been the case if the origin of oil could be attributed to processes involving only these inorganic chemicals.

The theory is a minority opinion among geologists. This theory often pops up when scientists are not able to explain apparent oil inflows into certain oil reservoirs. These instances are rare.

In 1911, Engler proposed that an organic substance other than coal was the source material of petroleum. He proposed the following three stages of development:

1. In the first stage, animal and vegetable deposits accumulate at the bottom of island seas and are then decomposed by bacteria. The water-soluble components are removed and fats, waxes, and other fat-soluble and stable materials remain.
2. In the second stage, high-temperature and pressure cause carbon dioxide to be produced from carboxyl-containing compounds, and water is produced from the hydroxyl acids and alcohols to yield a bituminous residue. There can also be a little cracking, producing a liquid product with high olefin content (petro-petroleum).
3. In the third stage, the unsaturated compounds are polymerized to naphthenic and/or paraffinic hydrocarbons. Aromatics are presumed to be formed by either cracking and cyclization or decomposition of petroleum.

The elements of this theory have survived; the only objection to it is that the end products obtained from the same sequence of experiments, namely, paraffins and unsaturated hydrocarbons, differ from those of petroleum.

COMPOSITION OF PETROLEUM

Petroleum is a combination of gaseous, liquid, and solid mixtures of many alkanes. It principally consists of a mixture of hydrocarbons, with traces of

TABLE 2.2 Overall Elemental Composition of Petroleum[a]

Element	Percentage Composition
Carbon	83.0–87.0
Hydrogen	10.0–14.0
Nitrogen	0.1–2.0
Sulfur	0.05–6.0
Oxygen	0.05–1.5

[a]*Petroleum also contains trace levels of nickel and vanadium (≈ 1000 parts per million).*

various nitrogenous and sulfurous compounds. Gaseous petroleum consists of lighter hydrocarbons with an abundant methane content and is termed "natural gas." Liquid petroleum not only consists of liquid hydrocarbons but includes dissolved gases, waxes (solid hydrocarbons), and bituminous material. Solid petroleum consists of heavier hydrocarbons, and this bituminous material is usually referred to as bitumen or asphalt. Along with these, petroleum also contains smaller amounts of nickel, vanadium, and other elements.

Large deposits of petroleum have been found in widely different parts of the world and their chemical composition varies greatly. Consequently the elemental composition of petroleum varies greatly from one source of crude oil to another. It is unsurprising that the composition varies, because the local distribution of plant, animal, and marine life is varied and presumably was similarly varied when the petroleum precursors formed. Furthermore, the geological history of each deposit is different and allows for varying chemistry to have occurred as the organic matter originally deposited matured into petroleum. Typically, the specific gravity of the crude petroleum, depending on the location, is in the range 0.800–0.950. Data on the typical elemental composition of petroleum are given in Table 2.2.

Petroleum can also be classified according to the chemical composition. One simple classification is given in Table 2.2A.

The composition of petroleum can also be represented in a simple triangular diagram. One such representation is shown in Fig. 2.1.

It is also possible to classify the composition of petroleum in terms of the C/H or H/C ratio. Accordingly, the classification can be represented as given in Table 2.2B.

Because petroleum is a mixture of widely varying constituents, it would be better to classify its products as natural, derived, and manufactured materials. One such classification is given in Table 2.3.

TABLE 2.2A Classification Based on Chemical Composition

	Composition of 250–350°C Fraction				
Type of Crude	% Paraffinic	% Naphthenic	% Aromatic	% Wax	% Asphalt
Paraffinic	46–61	22–32	12–25	1.5–10	0–6
Paraffinic-naphthenic	42–45	38–39	16–20	1–6	0–6
Naphthenic	15–26	61–76	8–13	Trace	0–6
Paraffinic-naphthanic-aromatic	27–35	36–47	26–33	0.5–1	0–10
Aromatic	0–8	57–78	20–25	0–0.5	0–20

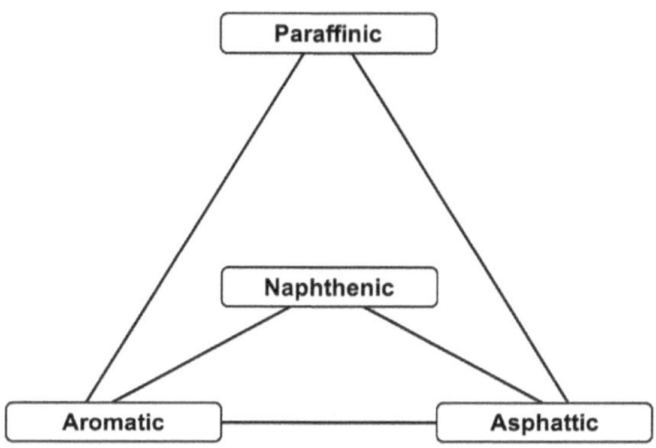

FIGURE 2.1 Composition diagram for petroleum.

TABLE 2.2B Classification Based on C/H and H/C Ratio in Petroleum

Type of Species	C/H Ratio	H/C Ratio
Simple aromatics (eg, benzene)	1.00	1.00
Condensed aromatics (eg, naphthalene)	1.25	0.80
Polycondensed aromatics	>1.40	<0.75
General aliphatic hydrocarbons	Ca 0.50	Ca 2.00
Paraffins	Ca 0.50	Ca 2.00

TABLE 2.3 Classification of Petroleum-Based Products

Natural Materials	Derived Materials	Manufactured Materials
Natural gas	Saturates	Synthetic crude oil
Petroleum	Aromatics	Distillates
Heavy oil	Resins	Lubricating oil
Bitumen	Asphaltenes	Wax
Asphatite		Residuum
Asphaltoid		Asphalt
Coal		Coke
		Tar
		Pitch

PRODUCTION OR EXTRACTION OF PETROLEUM

Locating an oil field is the first obstacle to be overcome. Today, petroleum engineers use instruments such as gravimeters and magnetometers in the search for petroleum. Generally, the first stage in the extraction of crude oil is to drill a well into the underground reservoir. Often many wells (called multilateral wells) are drilled into the same reservoir, to ensure that the extraction rate will be economically viable. Also, some wells (secondary wells) may be used to pump water, steam, acids, or various gas mixtures into the reservoir to raise or maintain the reservoir pressure, and so maintain an economical extraction rate.

Primary Oil Recovery

If the underground pressure in the oil reservoir is sufficient, the oil will be forced to the surface under this pressure. Gaseous fuels or natural gas are usually present, which also supply needed underground pressure. In this situation, it is sufficient to place a complex arrangement of valves on the well head to connect the well to a pipeline network for storage and processing. This is called primary oil recovery. Usually, only about 20% of the oil in a reservoir can be extracted in this way.

Secondary Oil Recovery

Over the lifetime of the well, the pressure will fall, and at some point there will be insufficient underground pressure to force the oil to the surface. If it is economical, and it often is, the remaining oil in the well is extracted using secondary oil recovery methods. Secondary oil recovery uses various techniques to aid in recovering oil from depleted or low-pressure reservoirs. Sometimes pumps, such as beam pumps and electrical submersible ones, are used to bring the oil to the surface. Other secondary recovery techniques increase the reservoir's pressure by water injection, natural gas reinjection, and gas lift, which inject air, carbon dioxide, or some other gas into the reservoir. Together, primary and secondary recovery allows 25–35% of the reservoir's oil to be recovered.

Tertiary Oil Recovery

Tertiary oil recovery reduces oil's viscosity to increase oil production. Tertiary recovery is started when secondary oil recovery techniques are no longer enough to sustain production, but only when the oil can still be extracted profitably. This depends on the cost of the extraction method and the current price of crude oil. When prices are high, previously unprofitable wells are brought back into production, and when they are low, production is curtailed. Thermally enhanced oil recovery methods (TEOR) are tertiary recovery techniques that heat the oil and make it easier to extract.

- Steam injection is the most common form of TEOR and is often done with a cogeneration plant. In this type of cogeneration plant, a gas turbine is used to generate electricity and the waste heat is used to produce steam, which is then injected into the reservoir.
- In situ burning is another form of TEOR, but instead of steam some of the oil is burned to heat the surrounding oil.
- Occasionally, detergents are also used to decrease oil viscosity.

Tertiary recovery allows another 5–15% of the reservoir's oil to be recovered.

PETROLEUM REFINING

The petroleum industry can be divided into two broad groups: upstream producers (exploration, development, and production of crude oil or natural gas) and downstream transporters (tanker and pipeline transport), refiners, retailers, and consumers.

Raw oil or unprocessed crude oil is not useful in the form in which it comes in out of the ground. It needs to be broken down into parts and refined before use in a solid material such as plastics and foams, or as petroleum fossil fuels, as in the case of automobile and airplane engines. An oil refinery is an industrial process plant in which crude oil is processed in three ways into useful petroleum products.

1. Separation: separates crude oil into various fractions
   Oil can be used in so many ways because it contains hydrocarbons of varying molecular masses and lengths such as paraffins, aromatics, naphthenes (or cycloalkanes), alkenes, dienes, and alkynes. Hydrocarbons are molecules of varying length and complexity made of hydrogen and carbon. The trick in separating different streams in the oil refinement process is the

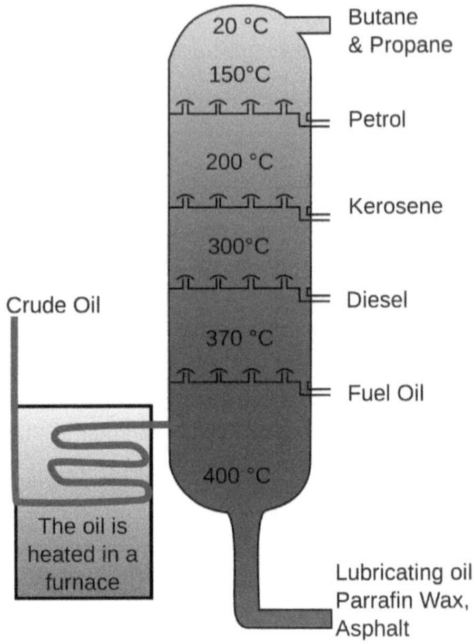

FIGURE 2.2 Schematic of the distillation of crude oil.

difference in boiling points between the hydrocarbons, which means that they can be separated by distillation. Fig. 2.2 shows the typical distillation scheme of an oil refinery.

2. Conversion: conversion to saleable products by skeletal alteration
Once it is separated and any contaminants and impurities have been removed, the oil can be sold with no further processing or smaller molecules such as isobutene and propylene or butylenes can be recombined to meet specified octane number requirements by processes such as alkylation or, less commonly, dimerization. The octane number requirement can also be improved by catalytic reforming, which strips hydrogen out of hydrocarbons to produce aromatics, which have higher octane ratings. Intermediate products such as gas oils can even be reprocessed to break a heavy, long-chained oil into a lighter short-chained one, by various forms of cracking such as fluid catalytic cracking (FCC), thermal cracking, and hydrocracking. The relevance of these processes and in what sequence these processes were developed can be followed from data given in Table 2.4. In fact, oil processing technology has been undergoing steady development almost at regular intervals (even every year), and only the most important milestones are assembled in Table 2.4.

As the final step in gasoline production, petroleum is the blending of fuels with different octane ratings, vapor pressures, and other properties to meet product specifications. The processing technology involves a variety

TABLE 2.4 History of Development of Processes in Refining of Petroleum

Year	Name of the Process	Main Product	By-products
1862	Atmospheric distillation	Kerosene	Naphtha and cracked residue
1870	Vacuum distillation	Lubricants	Asphalt, residua
1913	Thermal cracking	Gasoline	Fuel oil and residua
1916	Sweetening	Sulfur removal	Sulfur
1930	Thermal reforming	Octane number increase	Residua
1932	Hydrogenation	Sulfur removal	Sulfur
	Coking	Gasoline	Coke
1933	Solvent extraction	Viscosity index alteration	Aromatics
1935	Solvent dewaxing	Pour point improvement	Wax
	Catalytic polymerization	Octane number increase	Petrochemical feedstock
1937	Catalytic cracking	Gasoline	Petrochemical feedstock
1939	Visbreaking	Viscosity reduction	Distillate yield increase
1940	Alkylation	Octane number increase	Aviation fuel
	Isomerization	Alkylation feedstock	Naphtha
1942	Fluid catalytic cracking	Gasoline	Petrochemical feedstock
1950	Deasphalting	Cracker feedstock	Asphalt
1952	Catalytic reforming	Low-quality naphtha conversion	Aromatics
1954	Hydrodesulphurization	Sulfur removal	Sulfur
1956	Inhibitor sweetening	Mercaptans removal	Sulfur and disulfides
1957	Catalytic isomerization	High-octane products	Alkylation feedstocks
1960	Hydrocracking	Quality improvement and reduction of sulfur	Alkylation feedstocks
1974	Catalytic dewaxing	Pour point improvement	Wax
1975	Residua hydrocracking	Gasoline yield increase	Cracked residua

Adopted from J.G. Speight, The Chemistry and Technology of Petroleum, fourth ed., CRC Press, 2007.

TABLE 2.5 Common Process Units in Oil Refinery

Unit Process	Function
Atmospheric distillation unit	Distillation of crude oil into fractions
Vacuum distillation unit	Further distils residual bottoms after atmospheric distillation
Hydrotreater unit	Desulfurizes naptha from atmospheric distillation before sending to catalytic reformer unit
Catalytic reformer unit	Reforms paraffins to aromatics, olefins, and cyclic hydrocarbons, which have high octane number
Fluid catalytic cracking	Breaks down heavier fractions into lighter, more valuable products by catalytic system
Hydrocracker unit	Breaks down heavier fractions into lighter, more valuable products by steam
Alkylation unit	Produces high-octane component by increasing branching or alkylation
Dimerization unit	Smaller olefinic molecules of lower octane number are converted to molecules of higher octane number by dimerization of smaller olefins
Isomerization unit	Straight chain normal alkanes of lower octane number are isomerized to branched-chain alkane of higher octane number

of important steps, and each and every step converts the basic molecule into value-added molecules by addition, cracking, or the removal of unwanted elements, and so on. These steps are summarized in Table 2.5.

3. Finishing: purification of product streams

Details of Unit Processes

The unit processes briefly given in Table 2.5 are considered next in some detail.

Hydrotreater

A hydrotreater uses hydrogen to saturate aromatics and olefins as well as to remove undesirable compounds of elements such as sulfur and nitrogen.

Common major elements of a hydrotreater unit are a heater, a fixed-bed catalytic reactor, and a hydrogen compressor. The catalyst promotes the reaction of the hydrogen with sulfur compounds such as mercaptans to produce hydrogen sulfide, which is then usually bled off and treated with amine in an amine treater. Hydrogen also saturates hydrocarbon double bonds, which helps raise the stability of the fuel.

Catalytic Reforming

A catalytic reforming process converts a feed stream containing paraffins, olefins, and naphthenes into aromatics to be used either as a motor fuel blending stock or as a source for specific aromatic compounds such as benzene, toluene and xylene for use in petrochemical production. The product stream of the reformer is generally referred to as a reformate. Reformate produced by this process has a high octane rating. Significant quantities of hydrogen are also produced as a by-product. Catalytic reforming is normally facilitated by a bifunctional catalyst that is capable of rearranging and breaking long-chain hydrocarbons as well as removing hydrogen from naphthenes to produce aromatics. This process is different from steam reforming, which is also a catalytic process that produces hydrogen as the main product.

Cracking

In an oil refinery, cracking processes allow the production of light products [such as liquefied petroleum gas (LPG) and gasoline] from heavier crude oil distillation fractions (such as gas oils) and residues. FCC produces a high yield of gasoline and LPG whereas hydrocracking is a major source of jet fuel, gasoline components, and LPG. Thermal cracking is currently used to upgrade very heavy fractions or to produce light fractions or distillates, burner fuel, and/or petroleum coke. Two extremes of thermal cracking in terms of product range are represented by the high-temperature process called steam cracking or pyrolysis (≥750–900°C), which produces valuable ethylene and other feedstocks for the petrochemical industry, and the milder-temperature delayed coking (500°C), which under the right conditions can produce valuable needle coke, a highly crystalline petroleum coke used in the production of electrodes for the steel and aluminum industries.

Fluid Catalytic Cracking

Initial process implementations were based on a low-activity alumina catalyst and a reactor in which the catalyst particles were suspended in a rising flow of feed hydrocarbons in a fluidized-bed. In newer designs, cracking takes place using an active zeolite-based catalyst in a short-contact time, vertical or upward sloped pipe called the "riser." Preheated feed is sprayed into the base of the riser via feed nozzles where it contacts the extremely hot fluidized catalyst at 665–760°C. The hot catalyst vaporizes the feed and catalyzes the cracking reactions that break down the high–molecular weight oil into lighter components including LPG, gasoline, and diesel. The catalyst–hydrocarbon mixture flows upward through the riser for just a few seconds and then the mixture is separated via cyclones. The catalyst-free hydrocarbons are routed to a main fractionator for separation into fuel gas, LPG, gasoline, and light-cycle oils used in diesel and jet fuel, and heavy fuel oil.

The catalytic cracking process involves the presence of acid catalysts (usually solid acids such as silica-alumina and zeolites) that promote the heterolytic (asymmetric) breakage of bonds yielding pairs of ions of opposite charges, usually a carbocation and the unstable hydride anion.

During the trip up the riser, the cracking catalyst is "spent" by reactions that deposit coke on the catalyst and greatly reduce activity and selectivity. The "spent" catalyst is disengaged from the cracked hydrocarbon vapors and sent to a stripper, where it is contacted with steam to remove hydrocarbons remaining in the catalyst pores. The "spent" catalyst then flows into a fluidized-bed regenerator, where air (or in some cases air and oxygen) is used to burn off the coke to restore the catalyst and provide heat necessary for the next reaction cycle, because cracking is an endothermic reaction. The "regenerated" catalyst then flows to the base of the riser, repeating the cycle.

Hydrocracking

Hydrocracking is a catalytic cracking process assisted by the presence of an elevated partial pressure of hydrogen. The products of this process are saturated hydrocarbons; depending on the reaction conditions (temperature, pressure, and catalyst activity), these products range from ethane and LPG to heavier hydrocarbons mostly composed of isoparaffins. Hydrocracking is normally facilitated by a bifunctional catalyst capable of rearranging and breaking hydrocarbon chains as well as adding hydrogen to aromatics and olefins to produce naphthenes and alkanes. Major products from hydrocracking are jet fuel, diesel, relatively high-octane rating gasoline fractions, and LPG. All of these products have a low content of sulfur and contaminants.

Steam Cracking

Steam cracking is a petrochemical process in which saturated hydrocarbons are broken down into smaller, often unsaturated hydrocarbons. It is the principal industrial method for producing the lighter alkenes (commonly olefins), including ethane (ethylene) and propene (propylene).

In steam cracking, a gaseous or liquid hydrocarbon feed such as naphtha, LPG, or ethane is diluted with steam and then briefly heated in a furnace (obviously without the presence of oxygen). Typically, the reaction temperature is very hot, around 850°C, but the reaction is allowed to take place only briefly. In modern cracking furnaces, the residence time is even reduced to milliseconds (resulting in gas velocities reaching speeds beyond the speed of sound) to improve the yield of desired products. After the cracking temperature has been reached, the gas is quickly quenched to stop the reaction in a transfer line exchanger.

The products produced in the reaction depend on the composition of the feed, the hydrocarbon to steam ratio, and on the cracking temperature and furnace residence time. Light hydrocarbon feeds (such as ethane, LPGs, or light naphthas) give product streams rich in the lighter alkenes, including ethylene, propylene, and butadiene. Heavier hydrocarbon (full-range and heavy naphthas as well as other refinery products) feeds give some of these, but also give products rich in aromatic hydrocarbons and hydrocarbons suitable for inclusion in gasoline or fuel oil. The higher cracking temperature (also referred to as severity) favors the production of ethane and benzene, whereas lower severity produces relatively higher amounts of propene, C4-hydrocarbons, and liquid products.

The thermal cracking process follows a hemolytic mechanism; that is, bonds break symmetrically and thus pairs of free radicals are formed. The main reactions that take place include petroleum *initiation reactions*, in which a single molecule breaks apart into two free radicals. Only a small fraction of the feed molecules actually undergoes initiation, but these reactions are necessary to produce the free radicals that drive the rest of the reactions. In steam cracking, initiation usually involves breaking a chemical bond between two carbon atoms, rather than the bond between a carbon and a hydrogen atom.

$$CH_3CH_3 \rightarrow 2CH_3^{\bullet}$$

In *hydrogen abstraction*, a free radical removes a hydrogen atom from another molecule, turning the second molecule into a free radical.

$$CH_3^{\bullet} + CH_3CH_3 \rightarrow CH_4 + CH_3CH_2^{\bullet}$$

In *radical decomposition*, a free radical breaks apart into two molecules, one an alkene, the other a free radical. This process results in the alkene products of steam cracking.

$$CH_3CH_2^{\bullet} \rightarrow CH_2{=}CH_2 + H^{\bullet}$$

Radical addition is the reverse of radical decomposition, in which a radical reacts with an alkene to form a single, larger free radical. These processes are involved in forming the aromatic products that result when heavier feedstocks are used.

$$CH_3CH_2^{\bullet} + CH_2{=}CH_2 \rightarrow CH_3CH_2CH_2CH_2^{\bullet}$$

Termination reactions happen when two free radicals react with each other to produce products that are not free radicals. Two common forms of termination are recombination, in which the two radicals combine to form one larger molecule, and *disproportionation*, in which one radical transfers a hydrogen atom to the other, giving an alkene and an alkane.

$$CH_3^{\bullet} + CH_3CH_2^{\bullet} \rightarrow CH_3CH_2CH_3$$

$$CH_3CH_2^{\bullet} + CH_3CH_2^{\bullet} \rightarrow CH_2{=}CH_2 + CH_3CH_3$$

The process also results in the slow deposition of coke, a form of carbon, on the reactor walls. This degrades the effectiveness of the reactor, so reaction conditions are designed to minimize this. Nonetheless, a steam cracking furnace can usually run for only a few months at a time between decokings.

Alkylation

Alkylation is the transfer of an alkyl group from one molecule to another. The alkyl group may be transferred as an alkyl carbocation, a free radical, or a carbanion.

In a standard oil refinery process, alkylation involves low–molecular weight olefins (primarily a mixture of propylene and butylenes) with isobutene in the presence of a catalyst, either sulfuric or hydrofluoric acid. The product is called alkylate and is composed of a mixture of high octane, branched-chain paraffin hydrocarbons. Alkylate is a premium gasoline blending stock because it has exceptional antiknock properties and is clean burning.

Most crude oils contain only 10–40% of their hydrocarbon constituents in the gasoline range, so refineries use cracking processes that convert high–molecular weight hydrocarbons into smaller and more volatile compounds. Polymerization converts small gaseous olefins into liquid gasoline-size hydrocarbons. Alkylation processes transform small olefin and isoparaffin molecules into larger isoparaffins with a high octane number. Combining cracking, polymerization, and alkylation can result in a gasoline yield representing 70% of the starting crude oil.

Isomerization

Isomerization is a process by which straight-chain alkanes are converted to branched-chain alkanes that can be blended in petrol to improve its octane rating (in the presence of finely dispersed platinum on an aluminum oxide catalyst).

A diagrammatic representation of these processes is shown in Fig. 2.3.

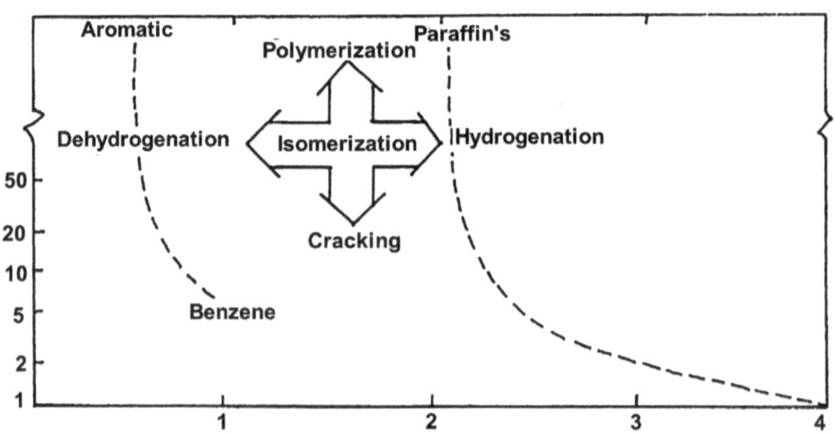

FIGURE 2.3 Diagrammatic representation of the changes occurring in the conversion process.

PRODUCTS OF OIL REFINERY

Asphalt

The term "asphalt" is often used as an abbreviation for asphalt concrete. Asphalt is a sticky, black, highly viscous liquid or semisolid that is present in most crude petroleum and in some natural deposits. Asphalt is composed almost entirely of bitumen. Asphalt is sometimes confused with tar, which is an artificial material produced by the destructive distillation of organic matter. Tar is also predominantly composed of bitumen; however, the bitumen content of tar is typically lower than that of asphalt. Tar and asphalt have different engineering properties.

Asphalt can be separated from the other components in crude oil (such as naphtha, gasoline, and diesel) by the process of fractional distillation, usually under vacuum conditions. A better separation can be achieved by further processing of the heavier fraction of the crude oil in a deasphalting unit that uses either propane or butane in a processing made possible by "blowing" the product, that is, by reacting it with oxygen. This process makes the product harder and more viscous. Asphalt is hard to transport in bulk (it hardens unless kept very hot). Thus it is sometimes mixed with diesel oil or kerosene before shipping. Upon delivery, these lighter materials are separated out of the mixture. This mixture is often called bitumen feedstock.

The largest use of asphalt is for making asphalt concrete for pavements. Roofing shingles account for most of the remaining asphalt consumption. Other uses include cattle sprays, fence post treatments, and waterproofing for fabrics. Ancient Middle East natural asphalt deposits were used for mortar between bricks and stones, ship caulk, and waterproofing.

Diesel Fuel

Petroleum-derived diesel is composed of about 75% saturated hydrocarbons (primarily paraffins including *n*, iso-, and cycloparaffins), and 25% aromatic hydrocarbons (including naphthalenes and alkylbenzenes). The average chemical formula for common diesel fuel is $C_{12}H_{26}$, ranging from approximately $C_{10}H_{22}$ to $C_{15}H_{32}$.

Diesel is produced from petroleum and is sometimes called petrodiesel when there is a need to distinguish it from diesel obtained from other sources. As a hydrocarbon mixture, it is obtained in the fractional distillation of crude oil between 250 and 350°C at atmospheric pressure.

Petrodiesel is considered a fuel oil and is about 18% denser than gasoline. The density of diesel is about 850 g/L whereas gasoline has a density of about 720 g/L, or about 18% less. Diesel is generally simpler to refine than gasoline and often costs less.

However, diesel fuel often contains higher quantities of sulfur. High levels of sulfur in diesel are harmful to the environment. Sulfur prevents the use of catalytic diesel particulate filters to control diesel particulate emissions, as well as more advanced technologies such as nitrogen oxide (NO_x) absorbers, to reduce emission. However, lowering sulfur also reduces the lubricity of the fuel, meaning that additives must be put into the fuel to help lubricate engines. Biodiesel is an effective lubricant. Diesel contains approximately 18% more energy per unit of volume than gasoline, which, along with the greater efficiency of diesel engines, contributes to fuel economy.

Synthetic Diesel

Wood, straw, corn, garbage, and sewage-sludge may be dried and gasified. After purification, the Fischer–Tropsch process is used to produce synthetic diesel. Other attempts use enzymatic processes and are also economic in case of high oil prices.

Biodiesel

Biodiesel can be obtained from vegetable oil and animal fats (biolipids using transesterification). Biodiesel is a non–fossil fuel alternative to petrodiesel. There have been reports that a diesel–biodiesel mix results in lower emissions that either can achieve alone. A small percentage of biodiesel can be used as an additive in low-sulfur formulations of diesel to increase the lubricity lost when the sulfur is removed.

Chemically, most biodiesel consists of alkyl (usually methyl) esters instead of the alkanes and aromatic hydrocarbons of petroleum-derived diesel. However, biodiesel has combustion properties similar to petrodiesel, including combustion energy and cetane ratings. Paraffin biodiesel also exists. Because of the purity of the source, it has a higher quality than petrodiesel.

Fuel Oil

Fuel oil is a fraction obtained from petroleum distillation as either a distillate or a residue. Broadly speaking, fuel oil is any liquid petroleum product that is burned in a furnace for the generation of heat or used in an engine for the generation of power. Fuel oil is made of long hydrocarbon chains, particularly alkanes, cycloalkanes, and aromatics. Factually and in a stricter sense, the term "fuel oil" is used to indicate the heaviest commercial fuel that can be obtained from crude oil, heavier than gasoline and naphtha. Fuel oil is classified into six classes, according to its boiling temperature, composition, and purpose. The boiling point ranges from 175 to 600°C, and the carbon chain length from 20 to 70 carbon atoms. These are mainly used in ships, with varying blending proportions.

Gasoline

Gasoline (or petrol) is a petroleum-derived liquid mixture consisting primarily of hydrocarbons, used as fuel in internal combustion engines. Gasoline is separated from crude oil via distillation, called natural gasoline; it will not meet the required specifications for modern engines (in particular, the octane rating), but these streams will form the blend. The bulk of a typical gasoline consists of hydrocarbons with between 5 and 12 carbon atoms per molecule.

The various refinery streams produce gasoline of different characteristics. Some important streams are:

- reformate, produced in a catalytic reformer with a high octane and high aromatics content, and very low olefins (alkenes);
- catalytically cracked gasoline or catalytically cracked naphtha, produced from a catalytic cracker, with a moderate-octane, high-olefins (alkene) content, and moderate aromatics level;
- product from a hydrocracker, contains medium to low octane and moderate aromatic levels;
- natural gasoline, directly from crude oil contains low octane, low aromatics (depending on the crude oil), some naphthenes (cycloalkanes), and zero olefins (alkenes);
- alkylate, produced in an alkylation unit, with high octane and which is pure paraffin (alkane), mainly branched chains; and
- isomerate, made by isomerizing natural gasoline to increase its octane rating, and which is low in aromatics and benzene content.

Overall a typical gasoline is predominantly a mixture of paraffins (alkanes), naphthenes (cycloalkanes), aromatics, and olefins (alkenes). The exact ratios can depend on:

- the oil refinery that is making the gasoline, because not all refineries have the same set of processing units;
- the crude oil used by the refinery on a particular day; and
- the grade of gasoline, in particular the octane.

Octane Rating

The octane number is a figure representing the resistance of gasoline to premature detonation when exposed to heat and pressure in the combustion chamber of an internal combustion engine. Such detonation is wasteful of the energy in the fuel and potentially damaging to the engine; premature detonation is indicated by knocking or ringing noises that occur as the engine operates. If an engine running on a particular gasoline makes such noises, they can be lessened or eliminated by using a gasoline with a higher octane number. The octane number of a sample of fuel is determined by burning the gasoline in an engine under controlled conditions, eg, of spark timing, compression, engine speed, and load, until a standard level of knock occurs. The engine is next operated on a fuel blended from a form of isooctane (octane number 100) that is very resistant to knocking and a form of heptane (octane number 0) that knocks very easily. When a blend is found that duplicates the knocking intensity of the sample under testing conditions, the percentage of isooctane by volume in the blended sample is taken as the octane number of the fuel. Octane numbers greater than 100 are determined by measuring the amount of tetraethyl lead that must be added to pure isooctane so as to duplicate the knocking of a sample fuel. Factors that can increase the octane number are more branching: 2-methylbutane is less likely to autoignite than is pentane. Pentane is less likely to autoignite than is heptane.

Additives to Gasoline for Value Addition

Additives have been added to increase the value of either gasoline's octane number or combustion capacity.

To Increase the Octane Number

The discovery that lead additives reduced the knocking property of gasoline in internal combustion engine resulted in widespread adoption of the practice in the 1920s and therefore more powerful higher-compression engines. The most popular additive was **tetraethyl lead**. However, with recognition of the environmental damage caused by lead and the incompatibility of lead with catalytic converters found on virtually all automobiles since 1975, this practice began to wane in the 1980s. Most countries are phasing out leaded fuel; different additives have replaced the lead compounds. The most popular additives include **aromatic hydrocarbons**, **ethers**, and **alcohol** (usually **ethanol** or **methanol**).

To Increase Combustion Capacity

Oxygenate blending increases oxygen to the fuel in oxygen-bearing compounds such as methyl tert-butyl ether (MTBE), ethanol, and ethyl tert-butyl ether (ETBE) and so reduces the amount of carbon monoxide and unburned fuel in the exhaust gas, thus reducing smog. Use of MTBE is being phased out in some countries owing to issues with the contamination of groundwater. Ethanol and to a lesser extent the ethanol-derived ETBE are common replacements. Especially ethanol derived from biomatter such as corn, sugar cane or grain is frequently used; this is often referred to as bioethanol. An ethanol–gasoline mix of 10% ethanol mixed with gasoline is called gasohol.

Energy Content

The heat content of gasoline is about 45 mega joules per kilogram (MJ/kg) or 135 MJ/US gallon. A high octane fuel such as LPG has a lower energy content than lower-octane gasoline, resulting in an overall lower power output at the regular compression ratio of an engine that runs on gasoline. However, with an engine tuned to the use of LPG (ie, via higher compression ratios such as 12:1 instead of 8:1), this lower power output can be overcome. This is because higher octane fuels allow for higher compression ratio.

TABLE 2.6 Energy Content of Different Fuels Obtained From Petroleum

Fuel Type	MJ/L	MJ/kg
Gasoline	29.0	45
Liquefied petroleum gas	22.16	34.39
Ethanol	19.59	30.40
Methanol	14.57	22.61
Gasohol (10% ethanol + 90% gasoline)	28.06	43.54
Diesel	40.9	63.47

The volumetric energy density of some fuels compared with gasoline is given in Table 2.6.

Kerosene

Kerosene is a colorless, flammable hydrocarbon liquid. Kerosene is obtained from the fractional distillation of petroleum between 150 and 275°C (carbon chains from C12 to C15 range). Typically, kerosene directly distilled from crude oil requires some treatment in a hydrotreater to reduce its sulfur content.

At one time it was widely used in kerosene lamps, but now it is mainly used in aviation fuel for jet engines. A form of kerosene known as RP-1 is burned with liquid oxygen as rocket fuel. Its use as a cooking fuel is mostly restricted to some portable stoves in less developed countries, where it is usually less refined and contains impurities and even debris. It can also be used to remove lice from hair and can be dangerous on skin. Most of these uses of kerosene create thick black smoke because of the low temperature of combustion. It is also used as an organic solvent.

Liquefied Petroleum Gas

LPG is manufactured during the refining of crude oil or extracted from oil or gas streams as they emerge from the ground. LPG (also called liquid petroleum gas, LP gas, or auto gas) is a mixture of hydrocarbon gases used as a fuel in cooking, heating appliances, and vehicles, and is increasingly replacing fluorocarbons as an aerosol propellant and a refrigerant to reduce damage to the ozone layer. Varieties of LPG bought and sold include mixes that are primarily propane, mixes that are primarily butane, and mixes including both propane and butane, depending on the season. Propylene and butylenes are usually also present in small concentrations. A powerful odorant, ethane thiol, is added so that leaks can be detected easily.

At normal temperatures and pressures, LPG will evaporate. Because of this, LPG is supplied in pressurized steel bottles. To allow for thermal expansion of the contained liquid, these bottles should not be filled completely; typically, they are filled to between 80% and 85% of their capacity.

Lubricant

A lubricant is introduced between two moving surfaces to reduce the friction and wear between them. A lubricant provides a protective film which allows for two touching surfaces to be separated, thus lessening the friction between them.

Typically lubricants contain 90% base oil (most often petroleum fractions, called mineral oils) and less than 10% additives. Vegetable oils or synthetic liquids such as hydrogenated polyolefins, esters, silicone, fluorocarbons, and many others are sometimes used as base oils. Additives deliver reduced friction and wear, increased viscosity, and resistance to corrosion and oxidation, aging, or contamination.

In developed nations, lubricants contribute to nearly 25% of total pollution released into environment. Spent lubricants are referred to as used oil or waste oil. As a liquid waste, 1 L of used oil can contaminate 1 million liters of water.

Paraffin

Paraffin is a common name for a group of high−molecular weight alkane hydrocarbons with the general formula C_nH_{2n+2}, where n is greater than 20. It is also called paraffin wax. Paraffin is also a technical name for an alkane in general, but in most cases it refers specifically to a linear or normal alkane, whereas branched or isoalkanes are also called isoparaffins.

It is mostly a white, odorless, tasteless, waxy solid, with a typical melting point between 47 and 65°C. It is insoluble in water but soluble in ether, benzene, and certain esters. Paraffin is unaffected by most common chemical reagents but burns readily.

Liquid paraffin has a number of names including nujol, mineral spirits, adepsine oil, alboline, glymol, liquid paraffin oil, saxol, and USP mineral oil. It is often used in infrared (IR) spectroscopy because it has a relatively uncomplicated IR spectrum.

Paraffin is used:

- in candle making
- in coatings for waxed paper or cloth
- in coatings for many kinds of hard cheese
- as anticaking, moisture repellents, and dust binding coatings for fertilizers
- for preparing specimens for histology
- as solid propellant for hybrid rockets
- to seal jars, cans, and bottles

- in dermatology, as an emollient (moisturizer)
- in surfing, for grip on surfboards as a component of surf wax
- as the primary component of glide wax, used on skis and snowboards
- as a food additive
- in forensics to detect granules of gunpowder in the hand of a shooting suspect
- in food grade in some candies to make them look shiny
- in impure mixtures of mostly paraffin wax in wax baths for beauty and therapy purposes

Mineral Oil

Mineral oil is a by-product in the distillation of petroleum to produce gasoline. It is a chemically inert, transparent, colorless oil composed mainly of alkanes and cyclic paraffins related to white petroleum. Mineral oil is a substance of relatively low value and is produced in large quantities. Mineral oil is available in light and heavy grades and can often be found in drugstores. It is used in various forms:

- Refined mineral oil is used as transformer oil.
- Mineral oil is used to store and transport alkali metals. The oil prevents the metals from reacting with atmospheric moisture.
- It is used in personal care.
- Mineral oil is sometimes taken orally as a laxative. It works by lubricating feces and the intestinal mucus membranes.
- Mineral oil with added fragrance is marketed as "baby oil" in the United States and United Kingdom.
- It is used as an ingredient in baby lotions, cold creams, ointments, and other pharmaceuticals and cosmetics.
- It can also be used for eyelashes and can generally be used to prevent brittleness and/or breaking of eyelashes.
- It is used for lubrication.
- It is also used as a coolant.
- Low-viscosity mineral oil is a traditional preservative for wooden cutting boards and utensils.
- A coating of mineral oil is excellent for protecting metal surfaces from moisture and oxidation.
- Food-preparation butcher block surfaces are often conditioned periodically with mineral oil.
- Light mineral oil is used in the textile industries and used as a jute batching oil.
- Mineral oil is used as a sealer for soapstone countertops.
- It is sometimes used in the food industry (particularly for candies).
- It is used as a cleaner and solvent for inks in fine art printmaking.

Tar

Tar is viscous black liquid derived from the destructive distillation of organic matter. Most tar is produced from coal as a by-product of coke production, but it can also be produced from petroleum, peat, or wood. The use of the word "tar" is frequently a misnomer. Naturally occurring "tar pits" actually contain asphalt, not tar, and are more accurately called asphalt pits. Tar sand deposits contain bitumen rather than tar.

Tar, of which petroleum tar is surprisingly the most effective, is used to treat psoriasis. Tar is a disinfectant substance and is used is as such. Petroleum tar was also used in ancient Egyptian mummification around 1000 BC.

Tar was a vital component of the first sealed or "tarmac" roads. It was also used as seal for roofing shingles and to seal the hulls of ships and boats. It was also used to waterproof sails, but today sails made from naturally waterproof synthetic substances have negated the need for sail sealing.

Wood tar is still used to seal traditional wooden boats and the roofs of historical shingle roofed churches. Wood tar is also available diluted as tar water, which has numerous uses:

- flavoring for candies and alcohol
- scent for saunas
- antidandruff agent in shampoo
- as a component of cosmetics

Bitumen

Bitumen is a category of organic liquids that are highly viscous, black, sticky, and wholly soluble in carbon disulfide. Asphalt and tar are the most common forms of bitumen.

Bitumen in the form of asphalt is obtained by the fractional distillation of crude oil. Because bitumen is the heaviest and the fraction with the highest boiling point, it appears as the bottom-most fraction. Bitumen in the form of tar is obtained by the destructive distillation of organic matter, usually bituminous coal.

Bitumen is primarily used for paving roads. It is also the prime feedstock for petroleum production from tar sands currently under development in Alberta, Canada. In the past, bitumen was used to waterproof boats, and even as a coating for buildings. For example, the city of Carthage was easily burned down owing to the extensive use of bitumen in construction.

Most geologists believe that naturally occurring deposits of bitumen are formed from the remains of ancient, microscopic algae and other once-living things. These organisms died and their remains were deposited in the mud on the bottom of the ocean or lake where they lived. Under the heat and pressure of burial deep in the earth, the remains were transformed into materials such a bitumen, kerogen, or petroleum. A minority of geologists, proponents of the theory of abiogenic petroleum origin, believe that bitumen and other hydrocarbons heavier than methane originally derive from deep inside the mangle of the earth rather than biological detritus.

Pitch (Resin)

Pitch is the name for any of a number of highly viscous liquids that appear to be solid. Pitch can be made from petroleum products or plants. **Petroleum-derived pitch** is also called bitumen. **Pitch produced from plants** is also known as **resin** or rosin.

Tar pitch appears to be solid and can be shattered with a hard impact, but it is actually a liquid. Pitch flows at room temperature, but extremely slowly. Pitch has a viscosity approximately 100 billion (10^{11}) times that of water.

Pitch was traditionally used to help caulk the seams of wooden sailing vessels. It was heated and put into a container with a very long spout. The word "pitcher" is said to derive from this long-spouted container used to pour hot pitch.

PETROCHEMICALS

According to crude oil's composition and demand, refineries can produce different shares of petroleum products. The largest share of oil products is used as energy carriers: various grades of fuel oil and gasoline. Refineries also produce other chemicals, some of which are used in chemical processes to produce plastics and other useful materials. Because petroleum often contains a few percent of sulfur, large quantities of sulfur are also often produced as a petroleum product. Carbon and hydrogen may also be produced as petroleum products. The hydrogen produced is often used as an intermediate product for other oil refinery processes such as hydrocracking and hydrodesulfurization.

A petrochemical is any chemical derived from fossil fuels. These include purified fossil fuels such as methane, propane, butane, gasoline, kerosene, diesel fuel, aviation fuel, or fuel oil and include many agricultural chemicals such as pesticides, herbicides and fertilizers, and other items such as plastics, asphalt, and synthetic fibers. Also, a wide variety of industrial chemicals are petrochemicals. Because petroleum products are feedstocks for many industries, chemical plants are frequently sited adjacent to a refinery, using intermediate products of the refinery as feedstocks for the production of specialized materials such as plastics or agrochemicals (Table 2.7).

A simplified flow diagram of how typical petrochemicals are obtained from one of the raw materials, natural gas, shown in Fig. 2.4. A similar and more extensive flowchart is available in Narayan and Viswanathan.

Petroleum serves as an extensive source for energy needs and as a feedstock for a spectrum of industries and a host of products. Petroleum is a nonrenewable natural resource and the industry is faced with the inevitable eventual depletion of the world's oil supply. By the very definition of nonrenewable resources, oil exploration alone will not stave off future shortages of the resource. Resource economists argue that oil prices will rise as demand increases relative to supply and that this will spur further exploration and development. However, this process will not increase the amount of oil in

TABLE 2.7 Partial List of Major Commercial Petrochemicals Derived From Petroleum Sources

Ethylene	Polyethylene		
	Ethylene oxide	Ethylene glycols	Polyesters
		Glycol esters	
		Ethoxylates	
	Vinyl acetate		
	1,2-Dichlorethane	Trichloroethylene	
		Tetrachloroethylene	
		Vinyl chloride	Polyvinyl chloride
	Ethyl benzene	Styrene	Polystyrene
			Synthetic rubbers
	Higher olefins	Detergent alcohols	
Propylene	Cumene	Acetone	
		Bisphenol A	Epoxy resins
			Polycarbonate
		Solvents	
	Isopropyl alcohol		
	Acrylonitrile		
	Polypropylene		
	Propylene oxide	Propylene glycol	
		Glycol esters	
	Acrylic acid	Allyl chloride	Epichlorohydrin
			Epoxy resins
Butadiene	Synthetic rubbers		

TABLE 2.7 Partial List of Major Commercial Petrochemicals Derived From Petroleum Sources—cont'd

Benzene	Ethyl benzene	Styrene	Polystyrene
			Synthetic rubber
	Cumene	Phenol	
		Bisphenol A	Epoxy resins
			Polycarbonate
	Cyclohexane	Adipic acid	Nylons
		Caprolactam	Nylons
	Nitrobenzene	Aniline	Methylene diphenyl Diisocyanate (MDI)
			Polyurethanes
	Alkyl benzene	Detergents	
	Chlorobenzene		
Toluene	Benzene		
	Toluene isocyanate	Polyurethanes	
	Benzoic acid	Caprolactam	Nylon
Mixed xylenes	Ortho xylene	Phthalic anhydride	
	Para xylene	Dimethyl terephthalate	Polyesters
		Terephthalic acid	Polyesters

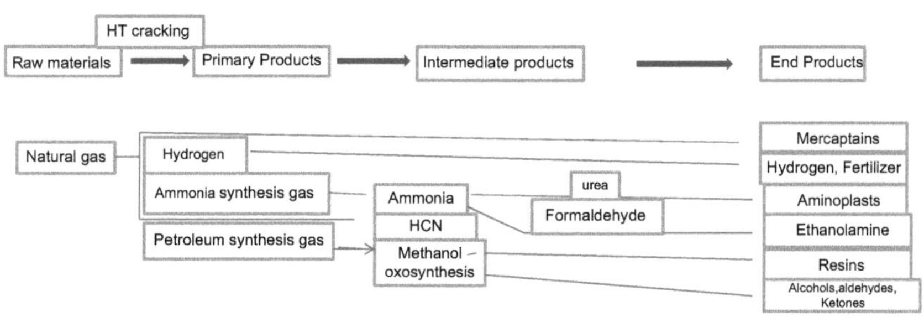

FIGURE 2.4 Typical variety of chemicals that can be produced from one of the raw materials from petroleum (similar flow diagrams can be made for other raw materials as well).

the ground but will rather temporarily prolong production as higher prices make it economical to extract oil that was not previously recoverable economically.

Available crude henceforth will become heavier with increased sulfur and other contaminants. Thermal cracking processes have to be preferred to crack heavy feedstock because it involves a low operating pressure despite high temperatures. Commonly employed residuum cracking processes are visbreaking and delayed coking, according to economic considerations.

The essential features of these processes are compared in Table 2.8.

Visbreaking (viscosity reduction or viscosity breaking) is a mild thermal operation with or without blending with lighter heating oils to reduce the viscosity of the residua to obtain fuel oil. Variations in this process (soaker or coil) depend on the temperature, residence time, and decoking of the heater tubes. In this process, crude oil residuum or tar sand bitumen is processed through a furnace and then passed through a flash distillation chamber and fractioned as overhead product naphtha and the liquid products are cooled and vacuum fractionated into gas oil distillate and a residuum.

Delayed coking is preferred in many refineries for processing bottom of the barrel upgradation. This process is amenable for complete rejection of metals and precursors to coke while providing partial or complete conversion to naphtha or diesel. Delayed coking is a semicontinuous process in which the

TABLE 2.8 Comparison of Visbreaking Process With Delayed Coking and Fluid Cracking

Thermal Cracking
Conversion to volatile and nonvolatile components in various types of process configurations
Visbreaking
Main objective is to reduce viscosity and not conversion levels; essentially conversions were kept low without allowing equilibrium to be reached; boiling fraction is mostly below 220°C.
Delayed Coking
Major aim is to obtain maximum yield of distillates and reactions are allowed to proceed to completion. Process is carried out until coking is complete and coke is removed hydraulically. Coke yield is in the range of 20–40 wt%
Fluid Cracking
This process employs severe conditions of temperature and pressure to produce maximum yield of distillate products under conditions of completion of the process. The process gives high yields of light ends (<C_5)

heated charge is transferred to soaking drums where the residence times are sufficient to allow the cracking reactions go to completion. Fluid coking is a process in which the residuum feedstock is decomposed by spraying the feedstock into a fluidized bed of hot coke particles. **Flexicoking** is a similar process to the fluid coking with the additional gasification section. There are a number of variations for the treatment of heavy feedstocks. One is **aqua-conversion**, which is hydrovisbreaking using catalyst-promoted hydrogen transfer from water to the feedstock. **Asphalt coking** technology is a combination of delayed coking and the deep solvent deasphalting process. There are other variations of these processes for the residuum treatment:

Cherry-P is an upgrading process of heavy petroleum residues at conditions between those of conventional visbreaking and delayed coking.

Decarbonizing is a process similar to delayed coking but at lower temperatures and pressures. ET-II is a thermal cracking process for the production of distillates and residuum for metallurgical coke. It takes up heavy oils, atmospheric residua, and vacuum residua. The Eureka process is also a thermal cracking process to produce a cracked oil and aromatic residuum from heavy residual materials. Fluid thermal cracking is also an upgrading process in which the feedstock is thermally cracked to produce distillate and coke, which is gasified to fuel gas. High conversion soaker cracking is a cracking process designed for conversion levels higher than visbreaking and lower than that of coking, thus providing higher yields of distillate. Mixed phase cracking is a thermal decomposition process employing rapid heating of the feedstock; the products from the flash chamber are fractionated to gasoline components and the bottoms are withdrawn as heavy fuel oil. Naphtha cracking, selective cracking, and shell thermal cracking are other variations for upgrading low-octane fractions of catalytic naphtha or converting high-boiling fractions to low-boiling products.

Residuum fluidized catalytic cracking (RFCC) has also evolved as an established method for treating the heavier fractions into a high-octane gasoline blend. The asphalt residual treating process aims to increase the production of transportation fuel and reduce heavy fuel oil production without hydrocracking. In the residue fluid catalytic cracking process, the residuum is first desulfurized and the nonvolatile fraction is charged to the RFCC unit. The amount of steam is adjusted for the additional heat required for vaporization of volatile component in the heavy feedstock.

Heavy oil treating is a process in which topped crude oil, vacuum residua, and solvent deasphalted bottoms are treated with iron ore particles.

The FCC process itself has been modified to treat heavy feedstocks in the R2R process and also in the residue crude oil conversion (RCC) process. In the R2R process the feedstock is vaporized upon contact with the hot regenerated catalyst at the base of the reactor and the conversion takes place when they (catalyst and feed) travel together up the riser. The product stream flows into the fractionation unit where the catalyst particles (entrained) deactivated with coke are let into regenerator. In the RCC process produces high-quality gasoline blending stock and petrochemical feedstock. The feedstock is injected at an appropriate position in the riser and then the products are steam stripped and the catalyst is returned to the bottom of the riser. There are two other commercial versions, the shell FCC process and S and W FCC process, which can maximize the production of distillates from residua.

It is necessary to consider some aspects of the catalysts for this process. Cracking of crude oil fractions with desired products is obtained with hydrated alumina silicates as catalysts. These catalysts are used as such or with additives of oxides of zirconium, boron, or thorium. These catalysts can be used in various shapes and must withstand the action of CO_2, air, nitrogen compounds, and steam. Normally the proportion of alumina is in the range 10–15%. Because the cracking function is mostly attributed to the property of acidity on the surface, these natural or synthetic catalysts are poisoned by proton-accepting vanadium. Generally the cracking process is operated in a catalyst to oil volume ratio of 10:1, although for a moving bed operation this ratio can be substantially lower than 10:1. Because metal impurities accumulate on the catalyst surface and affect the yield and selectivity of the desired product, suitable treatment processes are adopted. The removal of vanadium and nickel is achieved by the use of a sulfurous compound followed by chlorination (for the removal of vanadium) and the subsequent removal of nickel using an aqueous solution of a suitable chelating agent. An ion exchange process can also be adopted.

A variety of catalysts are employed in the petroleum industry for refining purposes. A simplified listing is given in Table 2.9. Readers are referred to authoritative monographs on catalysis for a detailed account of each of these catalyst systems and the way they function in the oil-refining process.

It is true that many warnings have been sounded in the literature about the exhaustible nature of petroleum and other petroleum-based chemicals. This

TABLE 2.9 Typical Reactions and Catalysts Used in Refining Processes

Reaction Type	Common Catalysts Employed
Cracking **1.** Dehydrogenation **2.** Dehydrocyclization	Amorphous silica-alumina, zeolite, molecular sieve Chromia on alumina
Hydrogenation **1.** Hydrocracking **2.** Hydrotreating	Dual-function catalysts Silica-alumina containing oxides of Pt and W, Ni for hydrogenation Metals such as Ni, Pt, Pd, Co, Fe, copper chromite, and Co-Mo-Al_2O_3
Isomerization	$AlCl_3$ promoted by HCl, Pt-containing catalyst
Alkylation	$AlCl_3$ acid catalyst

should not prevent humanity from considering petroleum's role as an energy source over the next few years. Moreover, one can obtain valuable information and usable technology from the petroleum industry for possible adaptation in the future for altered energy resources. One such possibility already being talked about is the creation of biorefineries. The starting materials for the petrochemical industry stem directly from crude petroleum or are generated during refining operations. The various chemicals that may be produced from the petroleum industry are shown in an abridged flowchart in Fig. 2.4.

SUGGESTED READING

[1] J.G. Speight, The Chemistry and Technology of Petroleum, fourth ed., CRC Press, 2007.
[2] R. Narayan, B. Viswanathan, Chemical and Electrochemical Energy Systems, University Press, Hyderabad, India, 1998.
[3] R. Prasad, Petroleum Refining Technology, Khanna Publishers, New Delhi, 2002.
[4] W.L. Leffler, Petroleum Refining in Non-technical Language, fourth ed., Penn Well, 2008.
[5] J. Sa (Ed.), Fuel Production With Heterogeneous Catalysis, CRC Press, 2015.

Chapter 3

Natural Gas

INTRODUCTION

As seen in Chapter 2, the dependence on petroleum appears to be strong despite the accompanying environmental and political consequences, because the political stability of the sources of petroleum appears to be vulnerable. Alternate energy sources mean that energy security should be ensured, which means that there should be a nearly equitable distribution of energy sources among the various nations. The search for an alternate energy source should also ensure good air quality; good options for storage; possibly more renewable, decreased emissions; and all other desirable and acceptable properties. Natural gas is the next option that is already satisfying nearly about 25% of the energy needs of humanity [1].

Natural gas is a naturally occurring gas mixture consisting mainly of methane. The location-specific composition is provided in Chapter 1. In Table 3.1 typical components of natural gas are given in typical values and ranges that can be expected depending on the site. The system average heating values will depend on the mix of gas supplies, and therefore can vary from the values listed in this table [2].

For consideration as an energy source, natural gas has to be compared with other fossil fuel sources such as petroleum and coal. This comparison has to be based on the emissions level from these sources; typical values are given in Table 3.2.

The heating properties of natural gas are also important from the point of view of the energy source; values are given in Table 3.3.

Natural gas is like petroleum, a fossil fuel source formed when buried plants and animals are subjected to intense heat and pressure over long periods. The energy that the plants originally acquired from the sun is stored in chemical bonds in components of natural gas. Natural gas is a so-called nonrenewable resource.

Natural gas is found in deep underground rock formations or is associated with other hydrocarbon reservoirs in coal bed sand as methane clathrates. It is also found with petroleum, another fuel resource, and natural gas is often found in close proximity to petroleum. Most natural gas was created over time by two mechanisms: biogenic and thermogenic. Biogenic gas is created by

Energy Sources. http://dx.doi.org/10.1016/B978-0-444-56353-8.00003-4

TABLE 3.1 Typical Analysis and Range of Composition of Natural Gas and Typical Heating Values

Component	Analysis (mol%)	Range (mol%)
Methane	95.0	87–97
Ethane	3.2	1.5–7
Propane	0.2	0.1–1.5
Isobutane	0.03	0.01–0.3
Normal butane	0.03	0.01–0.3
Isopentane	0.01	Trace to 0.04
Normal pentane	0.01	Trace to 0.04
Hexanes plus	0.01	Trace to 0.06
Nitrogen	1.0	0.2–5.5
Carbon dioxide	0.5	0.1–1.0
Oxygen	0.02	0.01–0.1
Hydrogen	Traces	Trace to 0.02
Specific gravity	0.58	0.57–0.62
Gross heating value (MJ/m^3)	38.0	36–40.2

TABLE 3.2 Fossil Fuel Emission Levels

Pollutant	Natural Gas	Oil	Coal
Carbon dioxide	117,000	164,000	208,000
Carbon monoxide	40	33	208
Nitrogen oxide	92	448	457
Sulfur dioxide	1	1122	2591
Particulates	7	84	2744
Mercury	0.000	0.007	0.016

Values are per billion BTU of energy input.
Reproduced from Natural Gas Issues and Trends (1998).

TABLE 3.3 Typical Heating Values of Natural Gas

Property Considered	Typical Value
Ignition point (4–16 vol% in air)	593°C
Flame temperature	1960°C
Maximum flame velocity	0.3 m/s
Relative density	0.58
Wobbe number (MJ/m^3)	49.9

Data are from the North American Combustion Handbook, vol. 1, third ed., North American Mfg. Co. (9186), and the Union Gas Material Safety Data Sheet; and https://www.uniongas.com/about-us/about-natural-gas/chemical-composition-of-natural-gas.

methanogenic organisms in marshes, bogs, landfills, and shallow sediments. Deeper in the earth, at greater temperature and pressure, thermogenic gas is created from buried organic material. The ancient people of Greece, Persia, and India discovered natural gas many centuries ago. A brief timeline of this source is given in Table 3.4.

Before natural gas can be used as a fuel, it must be processed to remove impurities, including water, to meet the specifications of marketable natural gas. By-products of this processing include ethane, propane, butanes, pentanes, and higher–molecular weight hydrocarbons, hydrogen sulfide (which may be converted into sulfur), carbon dioxide, water vapor, and sometimes helium and nitrogen.

Natural gas is often informally referred as "gas," especially in the context of fossil fuels. It should not be confused with the term "gas" for "gasoline" in some parts of the world.

TABLE 3.4 Timeline for the Recent History of Natural Gas

Year	Natural Gas Usage
1816	First used in America to illuminate Baltimore
1821	William Hart dug the first successful American natural gas well in Fredonia, New York
1858	Fredonia Gas Light Company opened its doors in 1858 as the nation's first natural gas company
1900	Natural gas discovered in 17 states
Present	Natural gas accounts for about a quarter of the energy we use.

Recorded information shows that the Chinese could have used natural gas at about 500 BC. They discovered a way to transport gas seeping from the ground in crude pipelines of bamboo to where it was used to boil saltwater to extract the salt, in the Ziliujing District of Sichuan. The world's first industrial extraction of natural gas started in Fredonia, New York, in 1825. By 2009, 66 trillion cubic meters (or 8%) had been used of the total 850 trillion cubic meters of estimated remaining recoverable reserves of natural gas. Based on data of the world consumption rate of about 3.4 trillion cubic meters of gas per year, the total estimated remaining economically recoverable reserves of natural gas would last for 250 years. However, if consumption were to increase, this period of longevity would reduce considerably.

Natural gas has emerged as a promising fuel owing to its environmentally friendly nature, efficiency, and cost-effectiveness. Natural gas is considered to be the most eco-friendly fuel based on available information. An economically natural gas is more efficient because only 10% of the produced gas is wasted before consumption and it does not need to be generated from other fuels. Moreover natural gas is used in its normal state. Natural gas has a high heat content of about 1000−11,000 BTU per standard cubic foot for pipeline-quality gas and it has a high flame temperature. Natural gas is easy to handle and convenient to use and an energy-equivalent basis, it has been price controlled below its competitor oil. It is also a suitable chemical feedstock for the petrochemical industry. Hence natural gas can substitute for oil in both sectors, fuels (industry and domestic) and chemicals (fertilizer petrochemicals and organic chemicals) [3].

SOURCES OF NATURAL GAS

Natural gas can be difficult to find because it can be trapped in porous rocks deep underground. However, various methods have been developed to find natural gas deposits:

1. looking at surface rocks to find clues about underground formations,
2. setting off small explosions or dropping heavy weights on the surface and recording the sound waves as they bounce back from the rock layers underground, and
3. measuring the gravitational pull of rock masses deep within the earth.

Scientists are also researching for new ways to obtain natural (methane) gas from biomass as a fuel source derived from plant and animal wastes. Methane gas is naturally produced whenever organic matter decays. Coal beds and landfills are other sources of natural gas.

Natural gas resources are widely distributed around the globe; possibly their use will reduce conflicts regarding energy distribution among different countries. It is estimated that a significant amount of natural gas remains to be discovered. In short, the world's largest reserves are held by the former Soviet Union (about 38% of the total world reserves) and the Middle East holds about 35%.

PHYSICAL PROPERTIES OF NATURAL GAS

Natural gas is a mixture of light hydrocarbons including methane, ethane, propane, butanes, and pentanes (the typical composition is given elsewhere). Other compounds found in natural gas include CO_2, helium, hydrogen sulfide, and nitrogen. The composition of natural gas is never constant; however, the primary component of natural gas is methane (typically at least 90%). Methane is highly flammable; it burns easily and almost completely. It emits little air pollution. Natural gas is neither corrosive nor toxic, its ignition temperature is high, and it has a narrow flammability range, which makes it an inherently safe fossil fuel compared with other fuel sources. In addition, because of its specific gravity (0.60), which is lower than that of air (1.00), natural gas rises if escaping, and thus dissipates from the site of any leak.

CLASSIFICATION OF NATURAL GAS

In terms of occurrence, natural gas is classified as nonassociated gas, associated gas, dissolved gas, and gas cap.

Nonassociated Gas

Nonassociated natural gas is found in reservoirs in which there is no, or at best minimum amounts of, crude oil. Nonassociated gas is usually richer in methane but is markedly leaner in terms of higher paraffinic hydrocarbons and condensate material. Unlike associated gas, nonassociated gas could be kept underground as long as required. This is therefore discretionary gas to be tapped according to economic and technological conditions.

Associated Gas

Natural gas found in crude oil reservoirs and produced during the pumping of crude oil is called associated gas. It exists as a free gas (gas cap) in contact with crude petroleum and also as a "dissolved natural gas" in the crude oil. Associated gas is usually is leaner in methane than nonassociated gas but it will be richer in the higher−molecular weight hydrocarbons. Nonassociated gas can be produced at higher pressures whereas associated gas (free or dissolved gas) must be separated from petroleum at lower separator pressures, which usually involves increased expenditure for compression.

Classification Based on Gas Composition

Natural gas can also be classified on the basis of composition; one such classification is given in Table 3.5.

NATURAL GAS PRODUCTS

Natural gas and/or its constituent hydrocarbons is marketed in the form of different products such as lean natural gas; wet natural gas [liquefied natural gas (LNG)]; compressed natural gas (CNG); natural gas liquids; liquefied petroleum gas (LPG); natural gasoline; natural gas condensate; ethane, propane, and ethane–propane fraction; and butanes.

Natural Gas Liquids

Natural gas liquids are ethane, propane, and ethane–propane fraction, LPG, and natural gasoline. There are also standards for the natural gas liquids that are usually set by mutual agreement between the buyer and the seller, but such specifications vary widely and can be given only approximate limits. For example, ethane may have a maximum methane content of 1.58% by volume and a maximum carbon dioxide content of 0.28% by volume. On the other hand, propane will be specified to have a maximum of 95% propane by volume, a maximum of 1–2% butane, and a maximum vapor pressure that limits ethane content. For butane, the percentage of one of the butane isomers is usually specified along with the maximum amounts of propane and pentane.

Other properties that may be specified are vapor pressure, specific gravity, corrosivity, dryness, and sulfur content. The specifications for the propane–butane mixtures will have limits on the amount of the nonhydrocarbons; in addition, the maximum isopentane content is usually stated.

TABLE 3.5 Classification Based on Natural Gas Composition

Name	Components
Lean gas	Methane
Wet gas	With considerable amounts of higher–molecular weight hydrocarbons
Sour gas	With hydrogen sulfide
Sweet gas	Little hydrogen sulfide
Residue gas	Natural gas from which higher–molecular weigh hydrocarbons have been extracted
Casing head gas	Derived from petroleum but is separated at separation facility at well head

The LPG is usually composed of propane, butanes, and/or mixtures of them; small amounts of ethane and pentane may also be present as impurities. On the other hand, the natural gasoline (such as refinery gasoline) mostly consists of pentane and higher–molecular weight hydrocarbons. The term "natural gasoline" has also been applied to mixtures of LPG, pentanes, and higher–molecular weight hydrocarbons. Natural gasoline may be sold on the basis of vapor pressure or of the actual composition, which is determined from the Reid vapor pressure composition curves prepared for each product source (ASTM D323).

Natural Gas Processing

Natural gas produced at the well contains contaminants and natural gas liquids that have to be removed before the product is sent to the consumers. These contaminants can cause operational problems, pipe rupture, or pipe deterioration. In addition to avoiding this damage, it is important to obtain maximum heating value, so the natural gas has to be processed before it is distributed as a fuel. A typical sequence of a processing unit is shown in Fig. 3.1.

Natural Gas Chain

Exploration: Geologists have a central role in identifying natural gas formations. They evaluate the structure of the soil and compare it with other areas

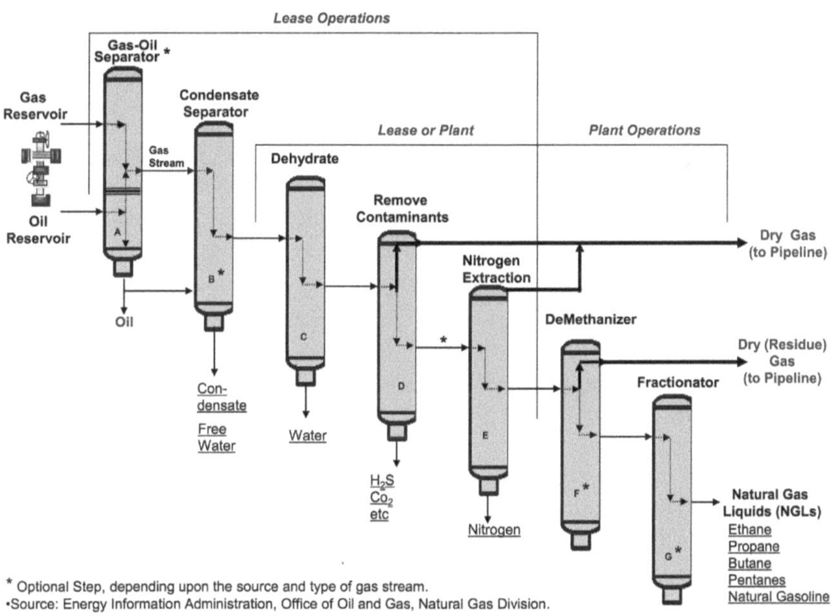

FIGURE 3.1 Natural gas processing. *Reproduced from G. Whitney, C.E. Behrens. Energy: Natural Gas: The Production and Use of Natural Gas. See also Energy Information Administration, Office of Oil and Gas (2006).*

where natural gas has been found. They carry out specific tests such as studying top ground rock formations where natural gas traps may have been formed The more accurate these techniques are, the higher the probability is of finding gas when drilling.

Extraction: Natural gas is captured by drilling a hole into the reservoir rock. Drilling can be onshore or offshore. Equipment used for drilling depends on the location of the natural gas trap and the nature of the rock. Once natural gas has been found, it has to be recovered efficiently. The most efficient recovery rate is characterized by the maximum quantity of gas that can be extracted during a period of time without damaging the formation. Several tests must be taken at this stage. Most often, the natural gas is under pressure and will come out of the hole on its own. In some cases, pumps and other more complicated procedures are required to remove the natural gas from the ground.

Processing: Processing has been carried out to remove contaminate from the natural gas and to convert it into useful energy for its different applications. This processing first involves extracting natural gas liquids from the natural gas stream and then fractioning the natural gas liquids into their separate components.

TRANSPORTATION

Natural gas normally reaches the consumer through a pipeline usually made of steel piping that measures 20–42 inches in diameter. Because gas moves at high pressures, there are compressor stations along the pipeline to maintain the level of pressure needed. Compared with other energy sources, natural gas transportation is efficient because the portion of energy lost from the origin to the destination is low.

Transported as Liquefied Natural Gas

Natural gas can also be transported by sea. In this case, it is transformed into LNG. The liquefaction process removes oxygen, carbon dioxide, sulfur compounds, and water. A full LNG chain consists of a liquefaction plant, low-temperature and pressurized transport ships, and a regasification terminal.

Sector-Wise Exploitation of Natural Gas

Residential Use

Natural gas is used in cooking, washing and drying, warming water, and air conditioning. The operating costs of natural gas equipment are generally lower than those of other energy sources.

Commercial Use

Natural gas can also be used for commercial purposes. A simplified flow diagram for the processing of natural gas is shown in Fig 3.2.

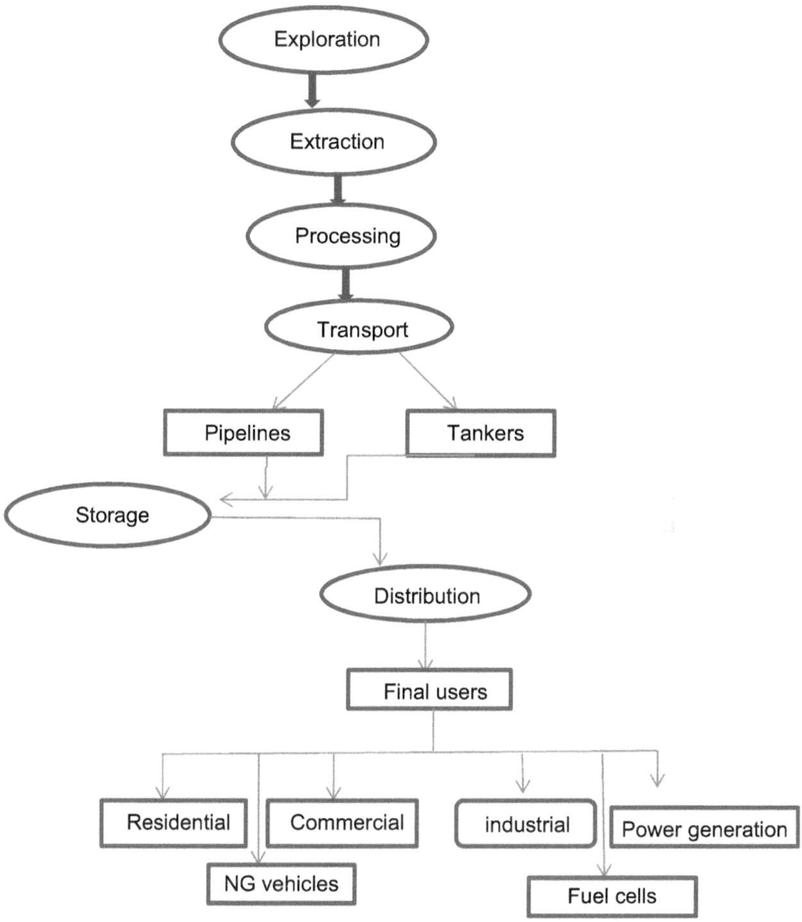

FIGURE 3.2 Simplified flow diagram for processing and uses of natural gas. *NG*, Natural gas.

Industrial Use of Natural Gas

Various fields in which natural gas is effectively used are in the manufacture of pulp and paper, metals, chemicals, stone, clay, and glass, and to process certain foods. Gas is also used to treat waste materials and for incineration, drying, dehumidification, heating and cooling, and CO generation. It is also a suitable chemical feedstock for the petrochemical industry. Natural gas has a multitude of industrial uses, including providing the base ingredients for such varied products as plastic, fertilizer, antifreeze, and fabrics. In fact, industry is the largest consumer of natural gas, accounting for 43% of natural gas use across all sectors. Natural gas is the second most used energy source in industry, behind only electricity. Lighting is the main use of energy in the industrial

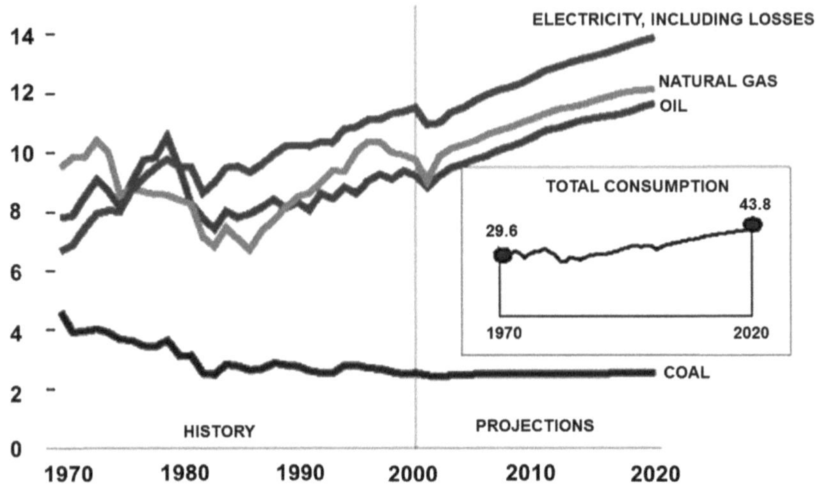

FIGURE 3.3 Industrial primary energy consumption by fuel, 1970–2020. *Energy Information Administration Annual Energy Outlook 2002, with projections to 2020.*

sector, which accounts for the tremendous electricity requirements of this sector. Fig. 3.3 shows current as well as projected energy consumption by fuel type in the industrial sector.

Natural gas as a feedstock is commonly found as a building block for methanol, which in turn has many industrial applications. Natural gas is converted to synthesis gas, which is a mixture of hydrogen and carbon oxides formed through a process known as steam reforming. In this process, natural gas is exposed to a catalyst that causes oxidation of the natural gas when brought into contact with steam. Once formed, this synthesis gas may be used to produce methanol, which in turn is used to produce such substances as formaldehyde, acetic acid, and methyl tertiary butyl ether (MTBE), which is used as an additive for cleaner-burning gasoline. Methanol may also be used as a fuel source in fuel cells.

Power Generation

Natural gas works more efficiently and emits less pollution than other fossil fuel power plants. Because of economic, environmental, and technological changes, natural gas has become the fuel of choice for new power plants. In fact, in 2000, 23,453 MW of new electric capacity was added in the United States. Of this, almost 95%, or 22,238 MW, was natural gas-fired additions. According to the Energy Information Administration, natural gas fired electricity generation is expected to increase dramatically over the next 20 years, as shown in Fig. 3.4. The data include all of the new capacity that is currently being constructed as it comes online. Steam generation units, centralized gas

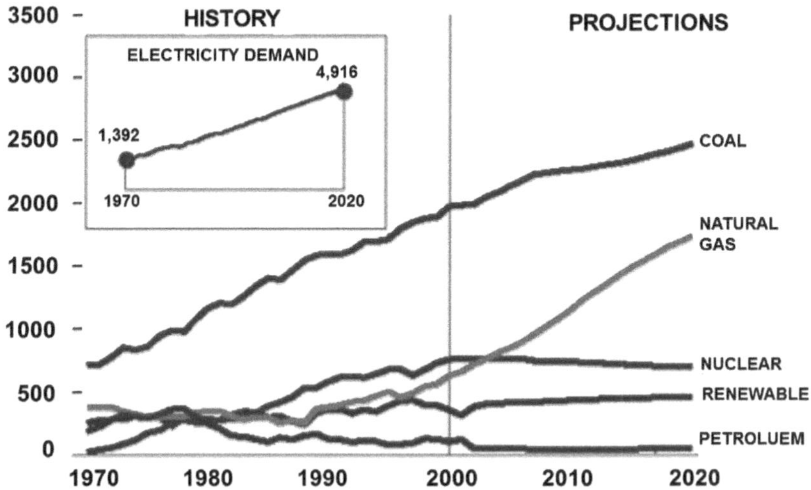

FIGURE 3.4 Electricity generation by fuel, 1970–2020 (per billion kilowatt hours).

turbines, micro turbines, combined cycle units, and distributed generation are the other examples in which natural gas is used.

Transportation

Natural gas can be used as a motor vehicle fuel in two ways: as CNG, which is the most common form, and as LNG. Cars using natural gas are estimated to emit 20% less greenhouse gases than gasoline or diesel cars. In many countries natural gas–powered vehicles are introduced to replace buses, taxis, and other public vehicle fleets. Natural gas in vehicles is inexpensive and convenient.

Most natural gas vehicles operate using CNG. This compressed gas is stored in a fashion similar to a car's gasoline tank, attached to the rear, top, or undercarriage of the vehicle in a tube-shaped storage tank. A CNG tank can be filled in a similar manner and amount of time as a gasoline tank.

(Fuel cells are dealt with in a separate chapter.) Natural gas is one of multiple fuels upon which fuel cells can operate. Fuel cells are becoming an increasingly important technology for the generation of electricity. They are like rechargeable batteries, except instead of using an electric recharger, they use a fuel such as natural gas to generate electric power even when they are in use. Fuel cells for distributed generation systems offer a multitude of benefits and are an exciting area of innovation and research for distributed generation applications. One major technological innovation with regard to electric generation, whether distributed or centralized, is the use of combined heat and power systems. These systems make use of heat that is normally wasted in the electric generation process, thereby increasing the energy efficiency of the total system.

CHEMICALS FROM NATURAL GAS: NATURAL GAS AS A FEEDSTOCK FOR PRODUCTION OF VALUE-ADDED PRODUCTS/CHEMICALS

The production of chemicals and power and transportation fuel from natural gas is shown in Schemes 3.1 and 3.2.

Natural gas has applications as a feedstock in the chemical industry for producing a number of methane-based and syngas-based products. Natural gas

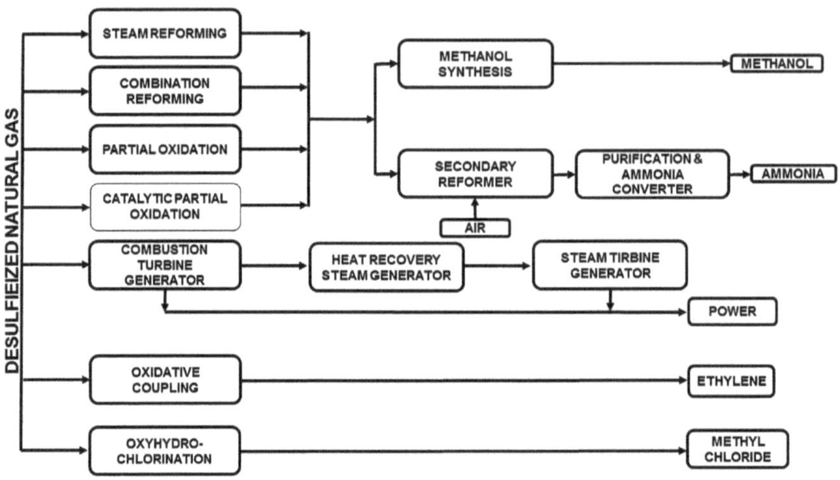

SCHEME 3.1 Conversion of natural gas to chemicals and power. *Reproduced from http://www.fischer-tropsch.org/DOE/DOE_reports/27346/27346-5098/de96000631_sec3.pdf.*

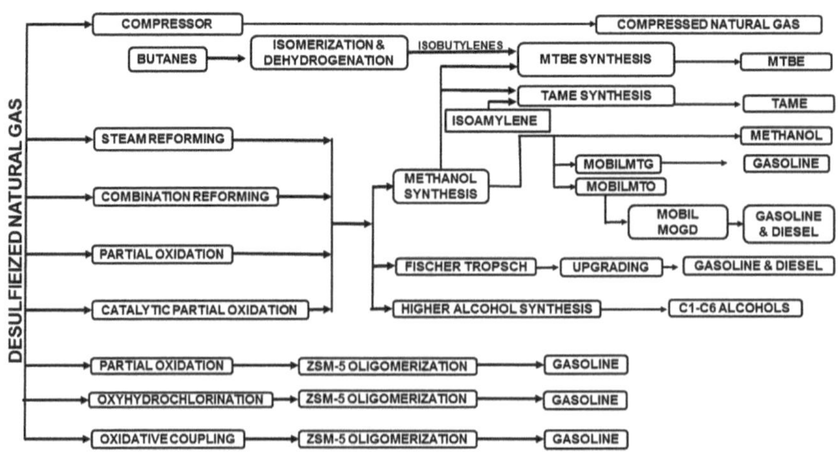

SCHEME 3.2 Flow sheet for the conversion of natural gas to transportation fuel. *Reproduced from http://www.fischer-tropsch.org/DOE/DOE_reports/27346/27346-5098/de96000631_sec3.pdf.*

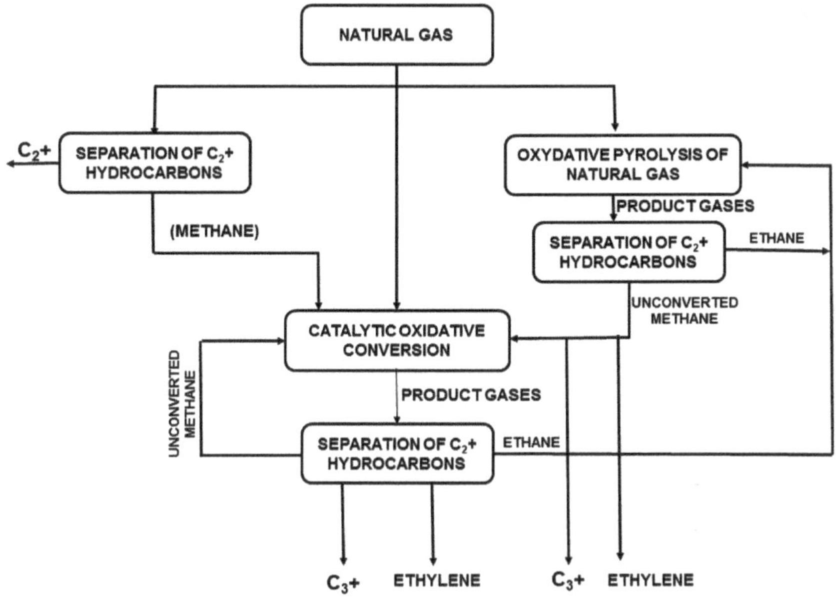

FIGURE 3.5 Schemes for the conversion of natural gas to ethylene.

is also an important feedstock for petrochemicals such as ethylene and propylene, which are key starting materials for the petrochemical industry. A simple flow diagram for the conversion of natural gas to ethylene is given in Fig. 3.5. Chloromethane, carbon black, and proteins are also derived from natural gas. Hydrogen cyanide and proteins for animal feed are also commercially produced from natural gas or methane. Details of the chemicals that can be derived from methane and the conditions employed in their manufacture are summarized in Table 3.6. The data given in this table are only representative and not a full listing.

The use of methane derived from natural gas as a chemical feedstock is demonstrated by data given in Table 3.6.

Hydrogen Cyanide

Methane can be converted to hydrocyanic acid (HCN) with ammonia in the presence or absence of oxygen. In the catalytic Andrussow process in the presence of a Pt or Pt/Rh catalyst at about 1273−1473K, hydrogen that is produced during the reaction is converted by oxygen to water to provide the necessary heat of reaction and to produce the required temperatures. Contact times must be short to avoid decomposition of the HCN produced. Unconverted ammonia is absorbed in an aqueous $NH_4H_2PO_4$ solution. In the Degussa process, which does not use oxygen, external heating is required. Points in

TABLE 3.6 Methane as a Chemical Feedstock for the Manufacture of Value-Added Chemicals

Product	Reaction	Conditions
Synthetic gas	$CH_4 + H_2O \rightarrow CO + 3H_2$	P: 30–50 bar T: 1123K Ni-supported catalyst
Hydrocyanic acid (HCN)	$CH_4 + NH_3 \rightarrow HCN + 3H_2$ $CH_4 + NH_3 + 1.5O_2 \rightarrow HCN + 3H_2O$	Degusaa process P: 1 bar T: 1273–1573K, Pt catalyst Andrussow process P: 1 bar T: 1273–1473K; Pt catalyst
Chloromethane CH_3Cl, CH_2Cl_2 $CHCl_3$, CCl_4	$CH_4 + xCl_2 \rightarrow CH_{4-x}Cl_x + xHCl$; $x = 0–4$	T: 673K; noncatalytic gas-phase reaction
Carbon disulfide (CS_2)	$CH_4 + 2S_2 \rightarrow CS_2 + H_2S$	P; 2.5 bar, T: 873K
Acetylene ethylene C_2H_2, C_2H_4	$2CH_4 \rightarrow C_2H_2, C_2H_4, H_2$	**1.** Electric arc process **2.** Partial combustion process
Ethylene and propylene	Oxidative methane coupling reaction	
Methanol		T: 633–666K P: 50–150 atm Catalyst: MoO_3 ZnO Fe_2O_3
Chloromethane	$CH_4 \rightarrow CH_3Cl$	T: 523K P: 230 psig Catalyst: Cu_2Cl_2, KCl and $LaCl_3$
Aromatics		H-ZSM-5,Ga-ZSM-5 Al-ZSM-5

Details with respect to the type of reaction and necessary experimental conditions are included in the table.

favor of this process are the high concentrations of HCN and the absence of oxygen-containing products.

Chloromethane

The exothermic chain reaction of methane with chlorine is carried out in the temperature range of 573–673K in the presence of a catalyst. This reaction can also be carried out photochemically, but in this case the product is a mixture of the four chloromethanes. After the removal of HCl, the chloromethanes are separated by distillation.

Acetylene

The production of acetylene from methane is not of importance currently because ethylene has replaced acetylene in many types of organic synthesis. The endothermic reaction requires temperatures above 1673K to avoid carbon formation. Short contact times, low acetylene partial pressures, and fast quenching of the product gas are essential for this process. Acetylene is separated from its by-products by absorption and distillation.

Carbon Disulfide

Carbon disulfide is produced by the reaction of methane with gaseous sulfur at temperatures of 923–948K using alumina or clay as a catalyst. After the precipitation of excess sulfur, carbon disulfide is separated by condensation. The product may contain small amounts of mercaptans and di- and polysulfides. The hydrogen sulfide formed can be reconverted back to sulfur.

Carbon Black

Natural gas is also a source for carbon black. It is produced by the oxidative decomposition of methane under oxygen-deficient conditions. Hydrogen is an important by-product obtained in the production of carbon black.

Proteins From Natural Gas

Since1952 the use of aerobic fermentation for the production of microbial biomass known as single-cell protein from hydrocarbon feedstock for incorporation into compounded animal feeds has received considerable attention throughout the world. A limited number of highly specialized species of bacteria are able to use methane for their growth. Methane is oxidized by the microorganism to form biomass and CO_2 through the following route:

$$CH_4 \rightarrow CH_3OH \rightarrow HCHO \rightarrow HCCOH \rightarrow CO_2$$

Methane is also used as a reducing agent, partly through synthetic gas for iron ore reduction [4]. In the Midrex process of the Midland Ross Corporation, methane is oxidized by carbon dioxide to give carbon monoxide and hydrogen. The iron ore is mainly reduced by hydrogen [5]. Synthetic gas in the NSC-DR process by the Nippon Steel Corporation is generated by the steam reforming of natural gas [6].

It is suggested that pure hydrogen may be produced by the catalytic decomposition of methane at 1273K using nickel catalysts. An autothermal operation is possible because the heat of combustion can be used for the endothermic decomposition of methane [7].

Chemicals From Methane by Synthetic Gas Route

The steam reforming of methane produces syngas as per the equation on a nickel catalyst at an elevated temperature:

$$CH_4 + H_2O \rightarrow CO + 3H_2$$

Syngas to Methanol

Methanol is produced by the reaction of syngas over a catalyst containing Cu and ZnO and alumina at temperatures of around 523K and at 50–70 atm pressure. Methanol can be converted to a host of other chemicals such as methyl chloride, methyl formate, methylmethacrylate, and methylamine, and also to lower olefins (methanol to olefins process) and gasoline (methanol to gasoline process). Methanol can be used to produce MTBE, which is used as a gasoline blend or an octane booster.

Fischer–Tropsch Synthesis

The Fischer–Tropsch (FT) conversion of syngas is essentially a catalytic route involving an oligomerization/polymerization process involving the formation of carbon–carbon bonds with the general basic reaction sequence:

$$nCO + 2nH_2 \rightarrow -(CH_2)- + nH_2O$$

FT synthesis is an extensively studied process for the conversion of syngas to higher hydrocarbons and oxygenates, such as lower olefins, polyolefins, aliphatic hydrocarbons, higher alcohols, and other oxygen-containing products. Depending on the catalyst (iron, cobalt, or other supported catalysts) and the process conditions, the product of the FT process may be predominantly lower olefins (C_2–C_5), olefins that are a suitable feedstock for petrochemicals or for blending with gasoline, or paraffinic hydrocarbons or oxygenated hydrocarbons.

A generalized scheme for the production of liquid hydrocarbons from natural gas is shown in Fig. 3.6.

The components of syngas can also be used in a variety of reactions. For example, CoO can be used in carbonylation for the production of acetic acid or propionic acid. CO is also used in an oxo process for the production of alcohols, in an oxidative carbonylation process for the formation of carbamates, and in the production of phosgene gas and a variety of other organic chemicals. Similarly, hydrogen is used for the synthesis of ammonia and the hydrogenation of olefins (Tables 3.7 and 3.8).

Oxidative Coupling of Methane

The oxidative coupling of methane is a reaction that has received considerable attention because it can give higher hydrocarbons. However, the oxygen

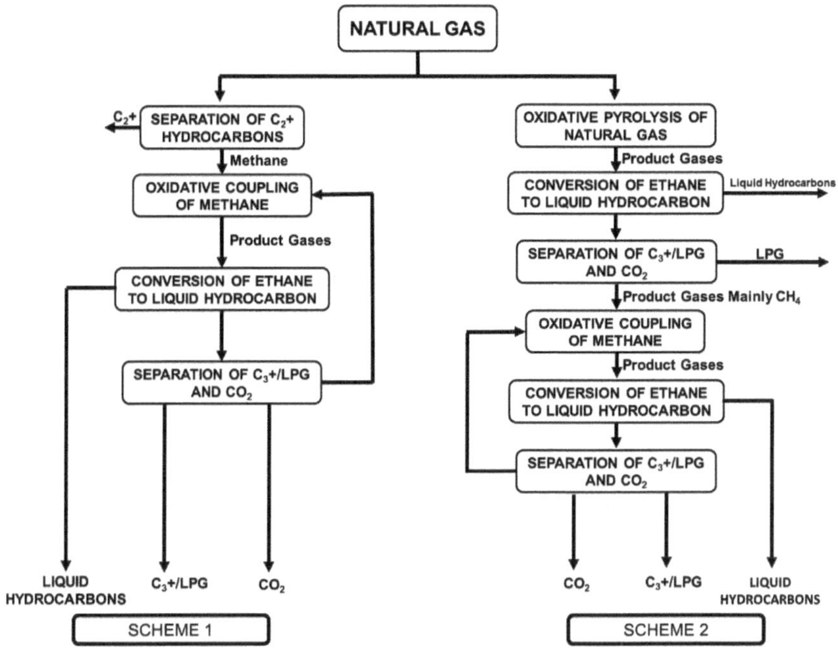

FIGURE 3.6 Schemes for the conversion of natural gas to liquid hydrocarbons.

TABLE 3.7 Production of Petrochemicals From Propylene and Ethylene That Are Produced From Methane—Natural Gas as Feedstock for Petrochemicals

Propylene Based	Butene Based	Natural Gas Liquid as Feedstock	Ethylene Based
Polypropylene Isopropyl alcohol Acrylonitrile Acrylonitrile copolymers Acrolein	Secondary butyl alcohol Butadiene Isobutene Tertiary butyl alcohol Butyl rubber Rubber	Maleic anhydride Synthetic gas Synthetic natural gas	Low-density polyethylene High-density polyethylene Ethylene oxide Ethylene glycol Ethanol Acetaldehyde Dichloromethane Vinyl chloride Polyvinyl alcohol Polyvinyl chloride Ethyl benzene Styrene Polystyrene

TABLE 3.8 Changes in Crystal Phase of Different Catalysts During Catalytic Reaction

Catalyst Composition	Crystalline Phase (Fresh)	Crystalline Phase (After Reaction)
18 mol% PbO/CaO	CaO, α-PbO, β-PbO	CaO, $Ca_3(CO_3)_2(OH)_2Ca(OH)_2$, α-PbO, β-PbO
31 mol% PbO/MgO	MgO, α-PbO, β-PbO	MgO, α-PbO, β-PbO, β-Pb
9 mol% PbO/CaO	CaO, α-PbO	CaO $Ca(OH)_2$
5 mol% PbO/γ-Al_2O_3	–	α-Al_2O_3, β-PbO, α-PbO,
14 mol% PbO/Al_2O_3	α-Al_2O_3, PbO	β-PbO, α-Pb, α-Al_2O_3
31 mol% PbO/Al_2O_3	γ-Al_2O_3, β-PbO, α-PbO	β-PbO, α-Pb
$PbAl_2O_4$	$PbAl_2O_4$	β-PbO, α-Pb, $PbAl_2O_4$
1% Na_2O/1% Li_2O/CaO	CaO	CaO
18 mol% SrO/CaO	CaO, SrO	CaO, $Ca(OH)_2$
$Pb_3(PO_4)_2$	$Pb_3(PO_4)_2$	$Pb_3(PO_4)_2$ $Pb_5(PO_4)_3OH$
Ca_2PbO_3	Ca_2PbO_3	$Ca_xPb_{10-x}(OH)_6(O)_{1+x}(CO_3)_{6-x}$
19% $PbSO_4/Ca_3(PO_4)_2$	$Pb(SO)_4$, $Ca_3(PO_4)_2$	$Ca_5(PO_4)_3(OH)$, $CaSO_4$, $PbSO_4$, Pb_2OSO_4, $Pb_3(PO_4)_2$

environment will result in the thermodynamically favorable products of complete combustion; the oxidative coupling of methane can be made to be selective to the desired products in the presence of a suitable catalyst. Direct conversion of methane into hydrocarbons is a thermodynamically unfavorable reaction since $2CH_4 \rightarrow C_2H_6 + 2H_2$ $\Delta G^0(1000K)$ 71 kJ/mol and $2CH_4 + \frac{1}{2} O_2 \rightarrow C_2H_6 + H_2O$ $\Delta G^0(1000K) = -159.7$ kJ/mol.

The literature shows that widely used catalyst materials for the oxidative coupling of methane can be classified as:

1. reducible metal oxides
2. alkaline earth oxides
3. rare earth oxides
4. supported lead oxides
5. alkali-promoted metal oxides or mixed oxides
6. LiCl-added transitional metal oxides
7. alkali carbonate melts
8. mixed oxides, especially molybdates

Essential results reported in the literature are:

1. The pulsed sequential mode of operation was shown to be favorable for C_2 product formation on oxides of Pb, Sb, Bi, Ti, Cd, and Mn, and lattice oxygen participation in the reaction has been shown to be responsible for selective product formation [8].
2. For steady-state operation, alkaline earth oxide doping with Li^+ or Na^+ seems to be suitable.
3. Lanthanides promote this reaction and the reaction results from a secondary reaction between methyl radicals and metal oxides, especially those capable of existing in multiple cationic oxidation states. Surfaces that can sustain methyl radicals give rise to a selective C_2 product whereas surfaces that form a methoxide species yield CO_2.
4. Lead oxide supported on MgO has shown methane conversion levels with good C_2 selectivity, and the facile mobility of lead between bulk and surface accounts for the nonsevere deactivation of the catalyst.

However the conversion levels are still low for commercial exploitation, possibly because of the slow regeneration process. The reaction is connected to the generation of surface oxygen species (possibly O^-), which will facilitate the dissociative adsorption of methane.

The economic promise of OCM has been the motivation for concentrating on this reaction. In the 1980s and 1990s, research efforts were pursued by both academic investigators and petrochemical companies. Hundreds of substances were tested as catalysts. However, so far the required level of selectivity for ethylene could not be achieved for economic operation. Mostly complete combustion predominated over the selective formation of ethylene. The stumbling steps are poor C—H activation and the requirement for high temperatures in the range of 750—950°C. The reaction appears to proceed in both heterogeneous mode and a homogeneous medium. The lack of necessary engineering skills for operation at high temperatures is another reason for limited success in developing this reaction on an industrial scale. These limitations have contributed to reduced interest in developing this reaction [9].

Shale Gas

Shale gas is a natural gas produced from shale. The matrix permeability of shale is too low, and hence one has to depend on fractures to tap the shale gas. Today all shale gas wells depend on artificially created fractures by hydraulic drilling. Since 2000, shale gas has become popular, and hence exploration has been attempted in Europe, China, and South Africa.

Liquefied Natural Gas

Natural gas at −161°C transforms into liquid. This is done for easy storage and transportation because it reduces the volume occupied by gas by a factor of

600. LNG is transported in specially built ships with cryogenic tanks. It is received at LNG receiving terminals and is regassified to be supplied as natural gas to consumers. Dedicated gas field development and production, liquefaction plants, transportation in special vessels, regassification plants, and transportation and distribution to gas consumers are various steps involved in the production and distribution of LNG.

Natural Gas and the Environment

All fossil fuels, such as coal, petroleum, and natural gas, release pollutants into the atmosphere when burned to provide the energy required. The list of pollutants they release reads like a chemical cornucopia: carbon monoxides, reactive hydrocarbons, nitrogen oxides, sulfur oxides, and solid particulates (ash or soot). The good news is that natural gas (among these three energy sources) is the most environmentally friendly fossil fuel. It is cleaner burning than coal or petroleum because it contains a lower percentage of carbon than its fossil fuel cousins. Natural gas also has less sulfur and nitrogen compounds and it emits less ash particulates into the air when burnt than do coal or petroleum fuels [10].

The conversion of coal into other chemicals (especially olefins and other higher hydrocarbons) is still not economically attractive. Thus research should be directed to converting available natural gas into value-added chemicals. Natural gas can be considered as an alternative source of chemical feedstock for petrochemicals to reduce dependence on oil. The development of active and selective catalysts is necessary to make the process of conversion of natural gas into olefins and liquid fuel economically viable. The oxidative coupling of methane into higher hydrocarbons shows promise in the near future. Natural gas is a viable short- and middle-term energy source for transport along with its industrial and residential applications [11].

REFERENCES

[1] B. Viswanathan (Ed.), Natural Gas Prospects and Possibilities, The Catalysis Society of India, 1992.

[2] R. Narayan, B. Viswanathan, Chemical and Electrochemical Energy System, Universities Press, 1998, pp. 28–35.

[3] A. Janssen, S.F. Lienin, F. Gassmann, W. Alexander, Model aided policy development for the market penetration of natural gas vehicles in Switzerland, Transport. Res. A 40 (2006) 316–333.

[4] P. Ruprecht, M. Baerns, Chem. Ing. Tech. 43 (1971) 894.

[5] Winnacker, Kiichler, Chemische Technologie, fourth ed., vol. 4, Carl Hanser Verlag, Munchen, 1986, p. 129.

[6] J.Muraki, N.Otsuki, Nippon Steel Technical report, No 12, (December 1978).

[7] K. Hedden, S. Ruch, Chem. Ing. Tech. 39 (1967) 1017.

[8] G.E. Keller, M.M. Bhasin, J. Catal. 73 (1982) 9.
[9] Oxidative coupling of methane, in: E.V. Kondratenko, M. Baerns (Eds.), Handbook of Heterogeneous Catalysis, vol. 6, 2008, pp. 3010–3023.
[10] Role of Alternative Energy Sources: Natural Gas Technology Assessment, DOE/NETL-2012/1539, June 30, 2012.
[11] https://mitei.mit.edu/system/files/NaturalGas_Report.pdf, The Future of Natural Gas, An Interdisciplinary MIT Study.

Chapter 4

Coal

INTRODUCTION

Energy security and energy independence are the two desires of any nation. The global energy requirement is great and always recurring. One has to look for changes beyond incremental measures. Energy sources are inevitable for progress and prosperity. Chemical sciences and technology hold an answer to the challenges ahead because most of industrial society is based on the following two reactions:

$$C + O_2 \leftrightarrow CO_2$$

$$H_2 + {}^1\!/_2\, O_2 \leftrightarrow H_2O$$

Despite their inherent differences, all chemical energy systems are related by the fact that they must involve, in some fashion, the making and breaking of chemical bonds and the transformation of chemical substances. A chemist with mastery over chemical structures (an understanding of the nature of the bonds involved between chemical entities and their relative strengths, and knowledge of activating C=C, C–C, C–H, C–O, C–N, C–S, H–H, and a few other bonds) can help in energy conversion processes and possibly provide a means to contain environmental pollution within limits.

For the sustainable development of a society, energy conversion processes occupy an important position. What chemists do to address this challenge will have an impact reaching far beyond our laboratories and institutions, because all human activities such as agriculture, transportation, construction, entertainment, and communication are energy driven. Food, clothing, and shelter are the basic amenities of life. This century has seemingly added another necessity: namely, energy. Any small interruption in the availability of required energy will have serious implications for the whole of our complex way of living. Global energy consumption and the living standards of the increasing population are interdependent. It is predicted that by 2050 there will be a twofold increase in the energy demand; at the current level of energy demand it is 12.8 terawatts (TW) and is predicted to increase to 28-35 TW [1,2].

Energy Sources. http://dx.doi.org/10.1016/B978-0-444-56353-8.00004-6

COAL: AN AGE-OLD ENERGY SOURCE

Probably coal is the one energy source whose utility is devoid of its physical form in the sense that it can cater to our energy needs in all three forms (solid, liquid, or gas) as the situation demands. No doubt the heating value changes depending on the amount of hydrogen present per unit weight, but this energy source is unique in a way that it can be molded in the hands of a chemist in accordance with the need. The heating value can also be tuned.

It is not well documented when exactly the use of coal started, but it is believed that coal was used for the first time in Europe during the Middle Ages.

Just as colors can be classified into primary (red, yellow, and blue) and secondary (a suitable combination of primary colors yielding green, purple, and orange), fuels can also be classified as primary and secondary, depending on the readiness of their utility. The major primary fuels are coal, crude petroleum oil, and natural gas (largely methane). These are naturally available. Coal and petroleum are sometimes referred to as fossil fuels, meaning they were once living matter. Secondary fuels are those derived from naturally occurring materials by some treatment resulting in drastic and significant alterations in physical and chemical properties, such as those of coal gas made from solid coal.

Coal is the most abundant fossil fuel available today worldwide. Except for coal, fossil fuel resources are limited. Coal is the most abundant fossil fuel on the planet. It is estimated that coal can last for more than another 500 years; this estimation is based on coal use for nearly 220 years. However, the global distribution of coal is nonuniform like any other mineral deposit or, as stated earlier, like petroleum. For instance, one-half of the world's known reserves of coal are in the United States.

THE GENESIS OF COAL

Several stages in the conversion of wood to coal are shown schematically in Fig. 4.1. These processes took several million years.

METAMORPHOSIS OF PEAT TO COAL

Coal is formed by the partial decomposition of vegetable matter and is primarily organic in nature. It is well studied as a sedimentary rock. Coal is a complex organic natural product that has evolved from precursor materials over millions of years. It is believed that the formation of coal occurred over geological times in the absence of oxygen, thereby promoting the formation of a highly carbonaceous product through the loss of oxygen and hydrogen from the original precursor species. A simplified representation of coal maturation by inspecting its elemental composition is presented in Table 4.1.

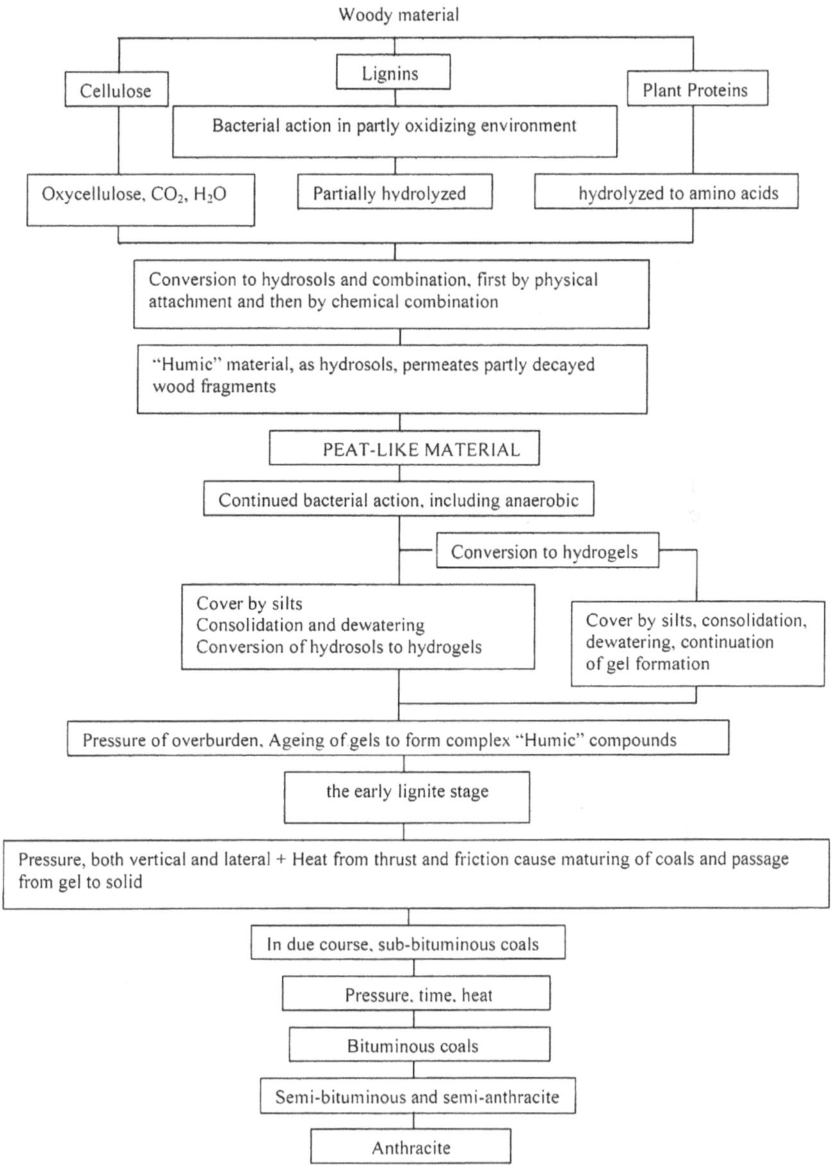

FIGURE 4.1 Schematic flow diagram of coal genesis. *Reproduced from J.C. Macrae, An Introduction to the Study of Fuel, Elsevier Publishing Company, 1966.*

TABLE 4.1 Maturation of Coal (Process Occurred With Increasing Pressure, Temperature, and Time)

	Composition (wt%)					
Type ↓	C %	Hydrogen %	Oxygen %	C/H Ratio	Heat Value (MJ/kg)	% Moisture
Wood	49	7	55	1.7	–	–
Peat	60	6	34	1.2	7–9.3	50–95
Lignite coal	70	5	25	0.9	14.6–19.3	45–60
Subbituminous coal	75	5	20	0.8	19.3–26.7	16–36
Bituminous coal	85	5	10	0.7	26.7–32.5	1–25
Anthracite coal	94	3	3	0.4	>32.5	0.5–3

Each class contains higher carbon content than the preceding one; eg, bituminous coal has greater carbon content than subbituminous coal. As shown, coal is generally composed of C, H, O, N, and S. A progressive change in composition is found through the *coal rank series* as given in Table 4.1. Typical representative structures of various grades of coal are given in Fig. 4.2.

The concept of a coal rank series is a largely undefined term that is often misused. A coal of a certain level of maturity, or degree of metamorphosis from the peat, is said to be of a certain rank. In the United States, coal is classified not on the basis of carbon but according to its property. The different types of coal that are clearly recognizable by their different properties and appearance can be arranged in the order of their increasing metamorphosis from the original peat material:

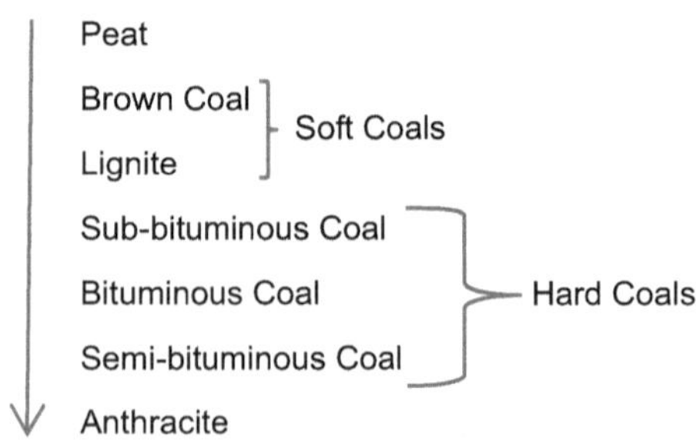

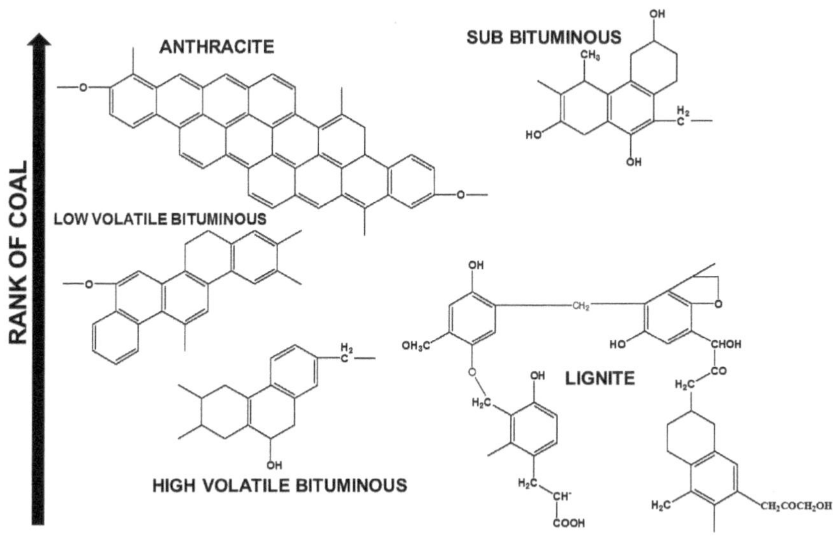

FIGURE 4.2 **Representative partial structures of different marks of coal.**

The most highly changed material, which is the final member of the series of coals formed from peat, is anthracite. Each member of the series represents a greater degree of maturity than the preceding one. The whole sequence is known as the "peat-to-anthracite series."

MOLECULAR STRUCTURE OF COAL

Peat

Peat is considered the precursor of coal. In some regions, this has some fuel value for industrial practice. When completely dehydrated, peat is employed as an effective adsorbent for oil spills and water. It is also used to condition soil, which facilitates the slow release of water.

Lignite (or Brown Coal)

Lignin structure is preserved in lignites. This means that the macromolecular structure of lignites consists of small aromatic units (mainly single rings) joined by cross-links of aliphatic (methylene) chains or aliphatic ethers. If the polymerization is random with cross-links heading off in all directions, the structure can be represented as seen in Fig. 4.3. Lignite is used as a fuel for power generation.

FIGURE 4.3 Sketch of the "open structure" with extensive cross-linking and small aromatic ring systems. *Reproduced from J.C. Macrae, An Introduction to the Study of Fuel, Elsevier Publishing Company, 1966.*

Subbituminous Coal

Subbituminous coal is primarily used as fuel for steam-electric power generation and is an important source for light aromatic hydrocarbons that are used in synthesis in the chemical industry.

Bituminous Coal

Compared with lignites, bituminous coals have a higher carbon content and lower oxygen content. The progression of changes that occur in the structure leads to an increase in the coal rank. The structure evolves toward graphite. It is used as a fuel in steam-electric power generation but also to make coke.

Viewed edge-on, graphite is represented as shown in Fig. 4.4, in which the hexagonal layers are perfectly stacked and aligned.

The structure of graphite is represented in Fig. 4.5.

Graphite is a crystalline solid. It gives a characteristic X-ray diffraction pattern that represents interplanar distances in the structure. Most coals, in contrast, are nearly amorphous and do not give the sharply defined X-ray diffraction patterns of graphite. When bituminous coals are examined by X-ray diffraction, it is possible to detect weak graphite-like signals emerging from the amorphous background. This observation indicates that in bituminous

FIGURE 4.4 Sideways view of graphite showing perfectly stacked aromatic planes. *Reproduced from J.C. Macrae, An Introduction to the Study of Fuel, Elsevier Publishing Company, 1966.*

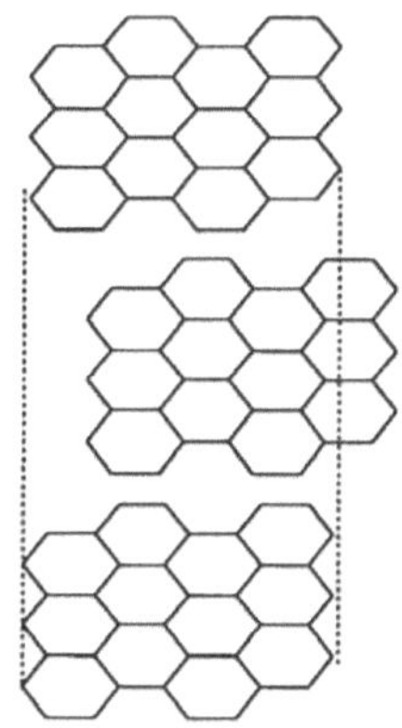

FIGURE 4.5 **Layered structure of graphite.**

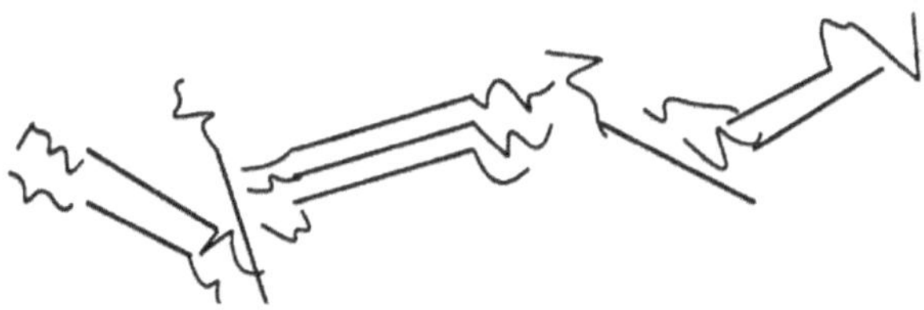

FIGURE 4.6 **"Liquid structure" of bituminous coals, with reduced cross-linking but increased size of aromatic units relative to the open structure.** *Reproduced from J.C. Macrae, An Introduction to the Study of Fuel, Elsevier Publishing Company, 1966.*

coals, the aromatic ring systems are beginning to grow and to become aligned. The structure of bituminous coal with a carbon content in the range of 85–91% is represented in Fig. 4.6.

Compared with the open structure shown earlier, aromatic units are larger, the cross-linkings are both shorter and fewer in number, and some vertical stacking of the aromatic units is present.

Some configurations understood to exist in coal, giving consideration to aromatic carbon, hydroaromatic carbon, and the kinds of structures and connecting bridges that join these structures are presented in Fig. 4.7. Bituminous coals consist of layers of condensed aromatic and hydroaromatic clusters ranging in size from one to several rings per cluster, with an average of three rings per condensed configuration. The principal types of links or bridges joining these clusters seem to be short aliphatic chains, some ether linkages, some sulfur linkages, and perhaps some biphenyl linkages.

Anthracites

Anthracites have a carbon content over 91%. The structure of anthracite approaches that of graphite, as represented in Fig. 4.8. X-ray diffraction data

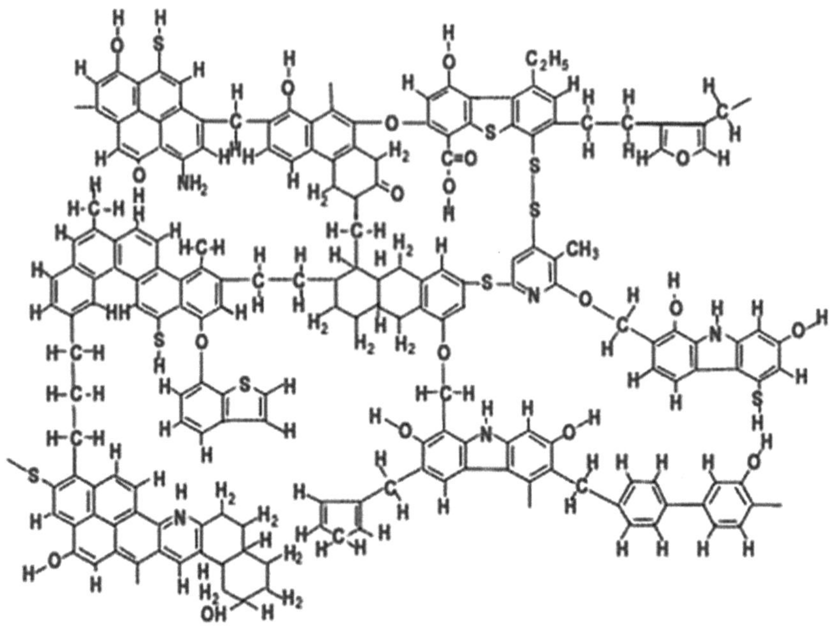

FIGURE 4.7 **Schematic representation of structural groups and connecting bridges in bituminous coal.**

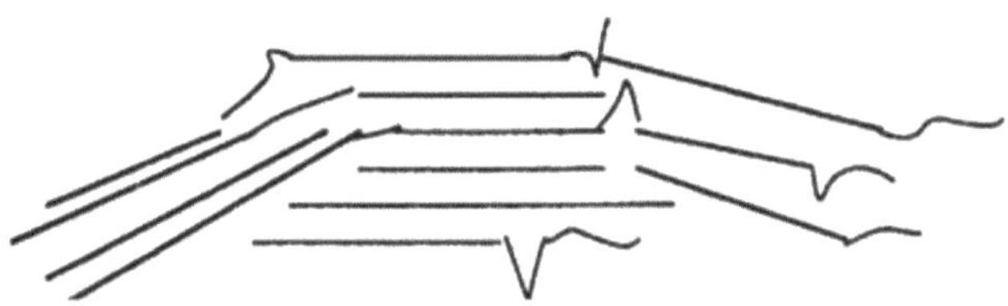

FIGURE 4.8 **"Anthracite structure" with large, fairly well-aligned aromatic units and minimal cross-linking.** *Reproduced from J.C. Macrae, An Introduction to the Study of Fuel, Elsevier Publishing Company, 1966.*

show increased alignment of the aromatic rings with little contribution from aliphatic carbon.

COAL PETROGRAPHY: THE STUDY OF MACERALS

The branch of science concerned with the visible structure of coal is coal petrology or petrography. The structure may be examined visually by the unaided eye or by an optical microscope. Marie Stopes, a British scientist, founded the discipline of coal petrography. In other words, petrography can be defined as the study of coal macerals. Analogous to the minerals of inorganic rock, the components of organic rock, ie, coal, are termed "macerals."

One needs to ask the following questions: What is the use of coal petrography? What are the importance and use of petrography in coal research?

The chemical behavior and reactivity of coal can be postulated with the knowledge of relative proportions of the different macerals contained in a coal sample. Different macerals come from different components of the original plant material that eventually resulted in the coal. Different plant components have different molecular structures. Substances with different molecular structures undergo different kinds of chemical reactions under a given set of conditions. Although plant components are altered chemically during coalification, the macerals should still reflect some of the chemical differences inherent in the original plant components. Consequently, one can expect various macerals to show differences in their chemical behavior. Thus by knowing the relative proportions of the different macerals in a coal sample, it should be possible to predict something about the chemical behavior and reactivity of the sample. There are four types of macro-components in coal: vitrain, clarain, durain, and fussian. These components correspond to anthraxylon, translucent attritus, opaque attritus, and fussain, respectively. In a typical coal seam, 50% of the seam may be clarain, 15–30% durain, 10–15% vitrain, and 1–2% fussain.

Vitrain: Vitrain is a bright black brittle coal normally occurring in very thin bands. It is generally translucent and amber red in color. A typical thin section of vitrain is shown in Fig. 4.9.

The cells of vitrain consist of complete pieces of bark. Bark tissues are more resistant to decay. As a result, they form a larger proportion of coal than expected.

Clarain: Clarain is bright black but less bright than vitrain. It is often finely banded so that it tends to break irregularly. In thin sections it shows partly the same appearance as vitrain in thin bands, but these are interbanded with more opaque bands consisting largely of fragmented plant remains among which can be identified cellular material, spore exines, and cuticle. A typical clarain structure is shown in Fig. 4.8. It contains more plant remains than vitrian and is the most common of the four types of coal substance.

Durain: Durain is a dull grayish black coal that is hard and tough and breaks irregularly. It is fairly opaque in thin sections and has large and small pore exines and woody fragments in a matrix of opaque grains. A typical durain structure with large flattened macrospores is shown in Fig. 4.9.

In the coal seam, durain bands are often thick and can be followed throughout the area of the seam. It is highly charged with durable plant remains and is supposed to be formed from silts or muds of small particles of vegetable matter.

Fussain: Fussain is a soft powdery form occurring in thin seams between the bands of other types. It is a friable, charcoal-like substance that dirties the hand when coal is touched. It is noncoking, but when fines are present at a small percentage in coal charge, they help increase the strength of the coke

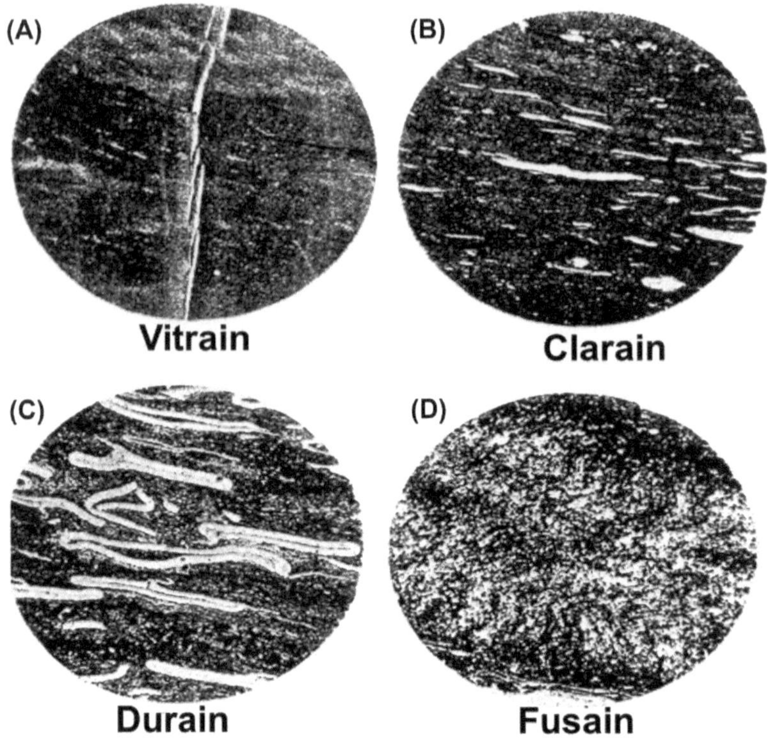

FIGURE 4.9 **Sections of bituminous coal taken perpendicular to the bedding plane.**

produced. The fixed carbon content is higher and the volatile matter is lower in fussain than in other banded ingredients.

CONSTITUTION OF COAL

Variation of Oxygen Content With Rank

There is a large variation in oxygen content as a function of carbon content, from nearly 30% in brown coals to ≈2% in anthracites. The variation in oxygen content with rank is illustrated in Fig. 4.10. The oxygen content and quality of coal (rank) are related; as the ranking increases, the oxygen content decreases. On a weight basis, oxygen is the second most important element in coal. The oxygen content of coal has several practical implications. The presence of oxygen reduces the calorific value. As a rule, for a given amount of carbon, as the oxygen content increases and hydrogen content decreases, the calorific value will decrease. This is seen from the values given in Table 4.2.

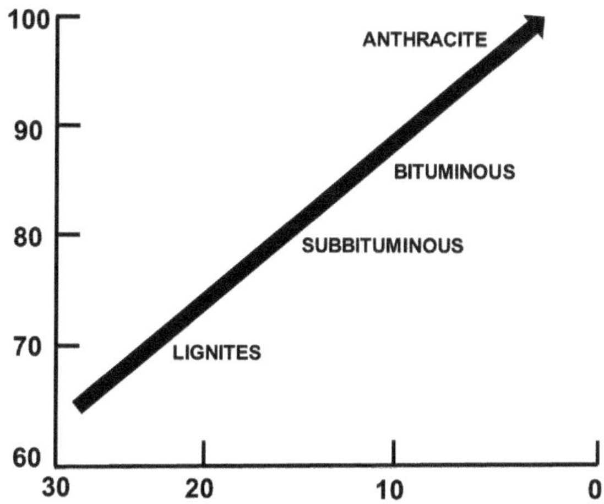

FIGURE 4.10 Variation of oxygen content with rank. *Reproduced from J.C. Macrae, An Introduction to the Study of Fuel, Elsevier Publishing Company, 1966.*

TABLE 4.2 Effect of Oxygen Content on Calorific Value

Compound	Hydrogen/Carbon	Oxygen/Carbon	Heat of Combustion (kJ/mol)
Methane	4	0	88
Formaldehyde	2	1	543
Butane	2.5	0	2880
1-Butanol	2.5	0.25	2675
2-Butanone	2.0	0.25	2436
Butanoic acid	2.0	0.5	2193

Variation of the Principal Oxygen Functional Groups With Carbon Content

The oxygen-containing structures represent functional groups, the sites where chemical reactions occur. The principal oxygen functional groups in coals are carboxylic acids, phenols, ketones or quinones, and ethers (Fig. 4.11).

Other oxygen functional groups of less importance are esters, aliphatic alcohols, aldehydes, and peroxides.

A variation of oxygen functional groups as a function of carbon content in vitrinites (the most common component of coal) is shown in Fig. 4.12.

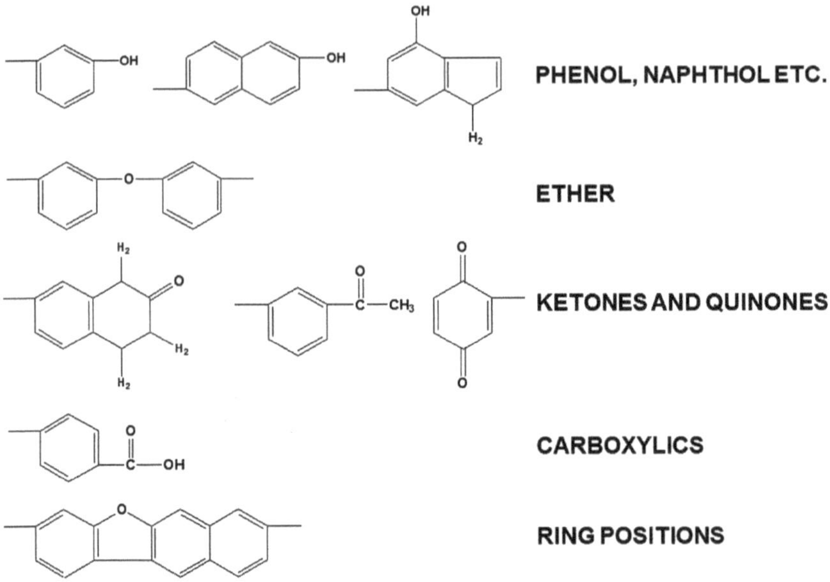

FIGURE 4.11 Principal oxygen functional groups in coal.

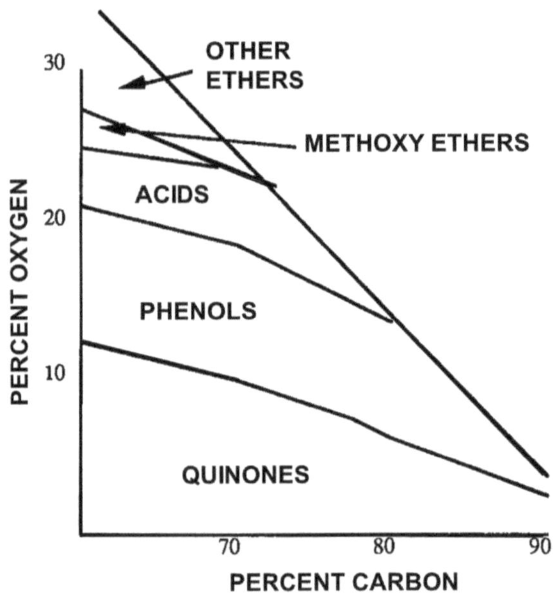

FIGURE 4.12 Variation of principal oxygen functional groups with carbon content. *Reproduced from H.H. Schobert, The Chemistry of Hydrocarbon Fuels, Butterworths & Co (Publishers), 1990.*

Methoxy groups are important only in coals with a carbon content less than 72 wt%. These groups are derived from lignin and their loss with increasing coalification suggests that lignin structures have been completely coalified by the time the coal has reached subbituminous rank. Phenols and quinones are the main oxygen groups in high-rank coals, although some ether may persist in high-rank coals.

Determination of Fixed Carbon Content

When coal is heated in an inert atmosphere to about 105°C, weight loss occurs. An inert atmosphere can be nitrogen or preferably argon. Weight loss is the result of water being driven off; this weight loss is used to calculate the moisture content of the sample. If the temperature is substantially increased, to 950°C under an inert atmosphere, a second weight loss is observed. Under these conditions a variety of materials, including carbon dioxide and carbon monoxide, and a mixture of hydrocarbons are evolved. The components emitted during this heating experiment are never determined individually. Rather they are lumped together under the term "volatile matter." Thus the weight loss observed at 1223 K provides a measure of the total volatile matter associated with the coal sample. At the end of the volatile matter test, a black carbonaceous solid remains containing carbon that was not removed during the evolution of volatile matter. If this carbon material remaining from the removal of volatile matter is heated in air, it burns, leaving behind an incombustible inorganic, residue, and ash. The ash is collected and weighed [7].

Three components, moisture, volatile matter, and ash, are directly determined. The fixed carbon is calculated indirectly as:

$$\%\ FC = 100 - (\%\ M + \%\ VM + \%\ A)$$

where FC = fixed carbon, M = moisture, VM = volatile matter, and A = ash.

Proximate Analysis

A proximate analysis is not approximate analysis! It is unfortunate that the name "proximate" sounds much like the word "approximate." The procedures for proximate analysis are rigorously established by the American Society for Testing and Materials along with the standards for acceptable levels of error within a laboratory and among different laboratories.

In this described method of analysis of percent moisture, percent volatile matter, percent ash, and therefore percent fixed carbon, the actual components of the volatile matter, ash, or fixed carbon are never determined. That is, the carbon dioxide evolved in the volatile matter is never collected, determined, and reported as some percentage of carbon dioxide. In analytical chemistry the practice of lumping together a variety of components and reporting them as a single entity is called proximate analysis. The proximate analysis of coal is therefore the determination of moisture of volatile matter, fixed carbon, and ash.

Ash Content

Strictly speaking, there is no ash in coal. The incombustible residue, ash, remaining after the combustion of coal is actually the product of high-temperature reactions of inorganic components, termed "mineral matter," originally present in the coal. The amount of ash is not necessarily equal to the amount of original mineral matter.

Fuel Ratio of Coal

The fuel ratio of coal is defined as the ratio of fixed carbon to volatile matter. Coal with a high-volatile matter is usually easy to ignite, burn with a large, often smoky flame, and burn quickly. Coal with high-fixed carbon is hard to ignite but it burns slowly with a short, clean flame [7].

Ultimate Analysis

Determination of the principal elements of coal, namely, carbon, hydrogen, oxygen, nitrogen, and sulfur, is called the ultimate analysis. It is not "ultimate" in the sense of completely determining the elemental composition of coal. It is because the careful analysis of a coal sample, including the mineral matter, to the trace level would show that coal contains virtually every element in the periodic table except for the rare gases and the man-made highly, unstable elements, and so the complete determination of composition and ultimate analysis are not one and the same, which should be remembered.

Mineral Matter of Coal

All elements in coal except C, H, N, O, and S are termed "mineral matter" even if they are present as organometallics, chelates, or absorbed species. Inorganic sulfur and inorganic oxygen (sulfur and oxygen not present in the heteroaromatic structure) are considered mineral matter. Iron, silica, and alumina constitute the major portion of mineral matter in coal. The four major species of mineral matter include aluminosilicates (clay minerals), carbonates, sulfides, and silicates (mainly quartz). Clay materials are dominant in most coals whereas pyrite is dominant among sulfides (Table 4.3).

CARBONIZATION

The process of heating coals in the absence of air to temperatures at which the coal decomposes is known as carbonization. Some coals when carbonized fuse and form a coal; others do not. Instead they leave a solid residue upon carbonization of the charred material that is of the same shape as the original coal (it has not fused) and is soft and easily crushed. Even those coals that form fused residues (fused cokes) differ in the extent to which they do so. They form cokes of different strengths. Coals are commonly referred to in terms of

TABLE 4.3 Mineral Matter Present in Coal

Normal State of Combination	% of Total Mineral Matter
Silica and silicates of aluminum	50–90
Pyrite (FeS_2), ferrous carbonate, ferrous and ferric sulfate, and silicates	0–20
Carbonate, sulfate, and silicates of calcium	0–20
Carbonate and silicate of magnesium	0–8
Chloride, carbonate, and silicate of sodium and potassium	0–4
Titanium oxide	0–2
Carbonate and silicate of manganese	Traces
Sulfur as sulfides and sulfates	0.5–10
Oxides of phosphorous, gallium, germanium, nickel, beryllium, etc.	<1

Reproduced from O.P. Gupta, Elements of Fuels, Furnaces and Refractories, Khanna Publishers, 1997.

their ability to form strong coke when carbonized. They are described as noncaking, weakly caking, or strongly caking. Alternately, they are said to have poor, weak, or strong caking properties [8].

Caking Property

One should know whether the caking property is an advantageous or disadvantageous property and to what extent. What does coal have to do with caking? What does gasoline have to do with sweetening? What does "approximate" have to do with proximate analysis? What does "complete" need have to do with ultimate analysis? What does a bomb have to do with calorific value? Does petrography have anything to do with petrol? Does the concept of caking imply that coal can be converted to a delicious, delightful, and flexible cake?

In science, the use of certain terms is inappropriate with the onslaught of time the original thoughts or views of the inventor are misinterpreted or manipulated or misunderstood in many instances since we have no time to go into details and facts and also short routes are always available in all walks of life whether good or bad. It appears that there is a mismatch between the terminology and the property which the terminology is addressing. The reasons for this in general are not known. But even now, enough care can be taken by researchers to look into facts as to what they mean exactly and as they implied when they were proposed by the scientists since all the proposals withstood the test of time, well documented and made available to one who examines for the same.

Knowledge of caking properties is essential because these decide the end application of coals such as in the manufacture of metallurgical coke, in coal-fired furnaces, or in any other specific application. A measure of caking property is necessary for the classification and selection of coal for different uses. Caking properties influence the production of coke and the performance of the coal in combustion and gasification.

As a rule, caking coals are coals of high-volatile, medium-volatile, and low-volatile bituminous rank. Medium-volatile bituminous coals are particularly good caking coals. Caking coals form an integral part of the bituminous family of coal. Peat, lignite, subbituminous, semianthracite, and anthracite are all noncaking coals.

When heated in an inert atmosphere, coal decomposes with the evolution of a variety of volatile materials, as stated previously. Some coals pass through a plastic state in which the coal appears to soften, swell, and then resolidify into a porous mass. Coals that exhibit this behavior are called caking coals. When the plastic behavior is such that the resulting solid product is a strong, hard, porous mass, the product is called coke and the original coal is called a coking coal. Coke is an important fuel in the metallurgical industry, particularly in the iron and steel industry. Supplies of coking coals are valuable. All coking coals are caking coals, but not all caking coals yield a commercially desirable coke. Thus, not all caking coals are coking coals.

Free Swelling Index

The caking behavior of a coal is measured in the United States by the free swelling index (FSI). It denotes the caking capacity of coal. In this test a known quantity of coal is heated in a standard-sized crucible under standardized conditions (1093 K for 2.5 min). The FSI is determined by comparing the size and shape of the resulting solid "button" with a series of standard outlines, as depicted in Fig. 4.13 and assigning a value from 1 to 9 at an interval of 0.5. A standard profile of coke buttons numbered from 1 to 9 (at an interval of 0.5 units) is shown in Fig. 4.12. An FSI of 0 is assigned to a material that does not form a coherent mass but rather falls apart when removed from the crucible.

In the United Kingdom, the caking property is evaluated in terms of the British Standard Swelling Number (BSS No.) Each of these numbers from 1 to 9 at an interval of 0.5 has great significance as far as the properties and end use of coal are concerned. They indicate the industrial purpose for which a coal is suitable.

For a BSS No. less than 2.5, these materials have very weak caking properties, or are noncaking. These coals are suitable for steam raising and other furnaces but are unsuitable for any form of carbonization.

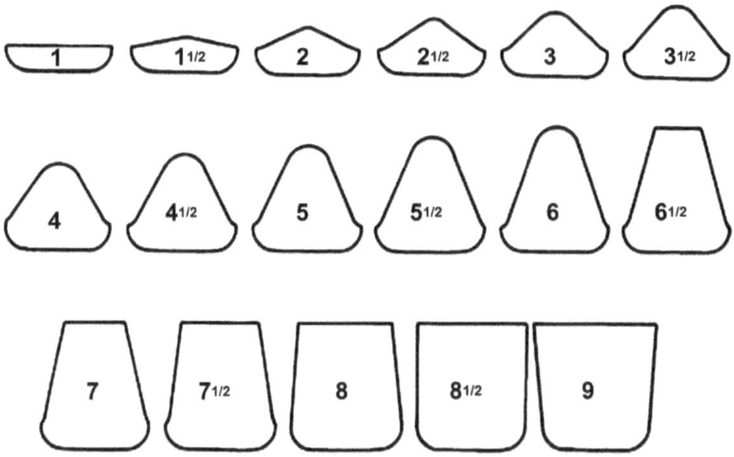

FIGURE 4.13 Standard profiles of the British Standard Swelling Number Test. *Reproduced from J.C. Macrae, An Introduction to the Study of Fuel, Elsevier Publishing Company, 1966.*

Materials with a BSS. No. of 3 or 3.5 are coals of weak to moderate caking power. These are suitable for all combustion purposes. They are suitable for carbonization in gasworks but are unsuitable for making metallurgical coke.

Materials with a BSS. No. of 4–6.5 are coals of moderate caking power. These are suitable for combustion purposes but tend to be too strongly caking for some forms of mechanical operations. They are suitable for gasworks and for making second-grade metallurgical coke.

BSS No. 7–9 are strongly caking coals. These coals are too strongly caking to be suitable for combustion in furnaces or for use in gasworks. This range includes coals that are best for making metallurgical cokes in coke ovens.

COAL FOR GENERATION OF ELECTRICITY

From coal to electricity generation is not a single-step conversion, as we see in fuel cells where chemical energy from any of the hydrocarbon fuels or hydrogen is directly converted to electrical energy. The main purpose of coal is only to get heat energy. This heat energy in turn is used to convert water to steam. The steam makes the propeller-like blades of the turbine rotate at high speeds. A generator connected to the turbine converts mechanical energy to electrical energy [9].

Zero-Emission Power Plants: The Need of the Hour

Zero-emission power generation refers to a range of technologies that are capable of producing electricity with carbon dioxide emissions that are at least 10% lower than those from a conventional plant. The significance of zero-emission power plants is well recognized from the statement of the US President George W. Bush: "I am pleased to announce that the United States will sponsor a $1 billion, 10-year demonstration project to create the world's first coal-based, zero-emission electricity and hydrogen power plant..." (Statement on February 27, 2003).

Hydrogen Content and Heating Value

Hydrogen is a well-known energy carrier like electricity. Hydrogen content has a role in deciding the heating value per unit of any fossil fuel such as coal, oil, or natural gas. Oil and natural gas are considered to be more versatile than coal because of the higher hydrogen content they possess compared with coal. The hydrogen content of coal, oil, and natural gas is 7.0, 13, and 25 wt% respectively.

The question is whether we can increase the hydrogen content per unit weight of coal. The answer lies in the processes of gasification and liquefaction of coal, in which the prime mode of operation is to add hydrogen. The processes of gasification and liquefaction of coal differ primarily in terms of operating conditions. Although the operative chemical reactions are different, they are only of secondary importance.

When starting with coal, the material possesses around 0.8–0.9 hydrogen atoms per carbon atom (Fig. 4.14). In a raw primary liquid fuel produced from coal, the hydrogen–carbon ratio on an atomic basis is about 1.1–1.2. If a distillate-type fuel is needed, a ratio of at least 1.6 is required.

If the optimized production of liquid is desired, the production of gases must be minimized because four hydrogen atoms per carbon atom are

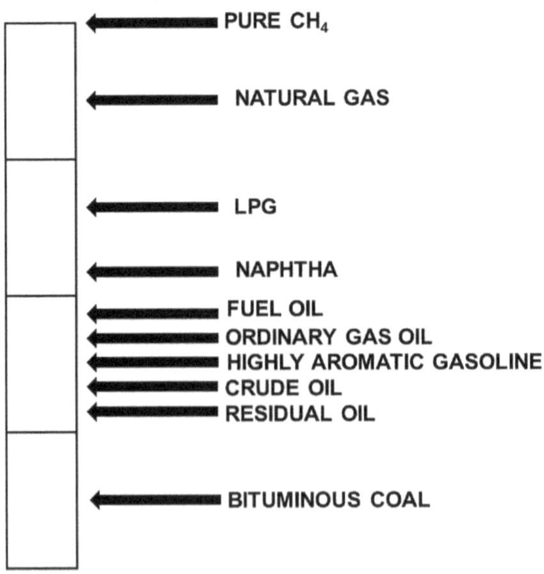

FIGURE 4.14 **Atomic hydrogen–carbon ratios for bituminous coal and other fuels.**

needed to produce methane. The cost of hydrogen is significantly expensive in coal liquefaction processes.

COAL LIQUEFACTION

Historical Background

It may seem surprising to convert a rock into liquid. Coal is often termed a heterogeneous rock of organic origin containing significant inorganic mineral matter. The idea of transforming coal into liquid began in 1869, when Bertholet demonstrated that coal could be hydrogenated. However, it was only in 1913 when Bergius began high-pressure hydrogenation and hydrodesulfurization of coal on a practical scale. Bergius employed the high-pressure technology of the Haber ammonia process to hydrogenate coal at elevated temperatures and pressure. The major commercial developers of hydrogenation were I. G. Farben Industrie A. G. (I. G. Farben) and BASF. The Initial goals of research at I. G. Farben were to design an effective slurry-phase hydrogenation catalyst and a sulfur-resistant vapor-phase catalyst that would survive the sulfur inherent in the coals. The development of a supported catalyst composed on pelleted Mo and Zn oxides was a breakthrough by Pier in 1926. In 1934, I. G. Farben switched from molybdenum to iron oxide as a slurry-phase catalyst. In slurry-phase mode, disposable catalysts such as inexpensive ores, coal minerals, and metallic wastes were employed. Catalyst self-sufficiency and no need for concern regarding catalyst deactivation or recovery are the advantages of this mode. Slurry-phase catalysts are coal minerals and iron sulfides.

Although the use of "a disposable" or a slurry-phase, catalyst in direct coal liquefaction was initiated in the Bergius process, currently SRC-II technology is being employed in which the presence of recycled coal mineral matter considerably enhances coal liquefaction.

The Process of Liquefaction of Coal

Liquid fuel from coal could be a future alternative to conventional petroleum. From each ton of coal, 1 to 4 barrels of oil can be produced. Despite the vast technology base for coal liquefaction, efforts to build commercial plants have been subsidized. The high-capital costs of synthetic oil plants and declining oil prices are major obstacles. If costs become competitive, coal could be the future source of liquid fuels.

A simple flowchart is given in Fig. 4.15 for the two indirect coal liquefaction processes.

Major issues of concern in the liquefaction of coal are:

1. process thermal efficiency
2. hydrogen use
3. materials and components reliability
4. solid–liquid separation

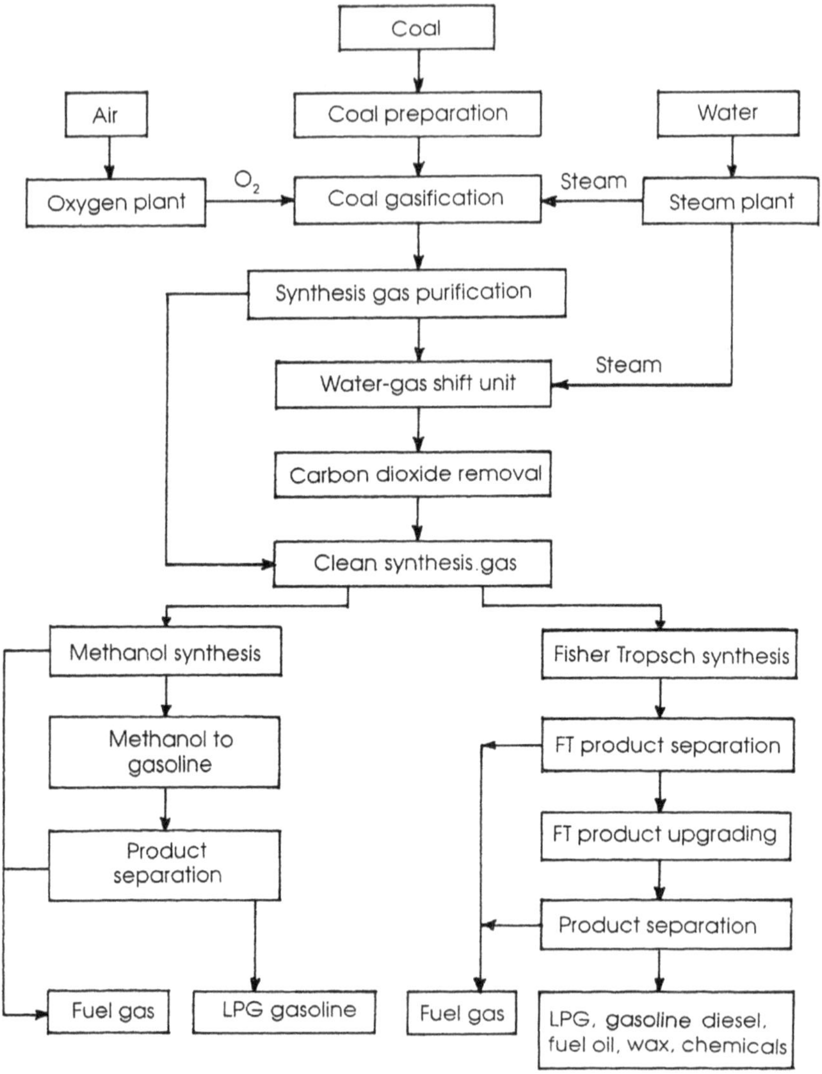

FIGURE 4.15 **Schematic simplified flowchart for two indirect coal liquefaction processes.**

5. product quality and flexibility
6. feed coal flexibility
7. process severity.

Among these major areas of concern, the efficient use of hydrogen is a critical aspect in direct liquefaction. The effective use of hydrogen is important from the standpoints of reaction chemistry, the reaction mechanism, and the economics of the process. The cost of hydrogen production itself is 20–33%

of the direct capital expenditure of coal liquefaction plants. Hydrogen stabilizes the reactive and unstable free radicals formed by thermal rupture of chemical bonds of the coal macromolecules. This prevents the occurrence of retrogressive reactions that produce undesirable high–molecular weight products that reduce the yield of desired liquid products.

A number of competing parallel reactions may be seen:

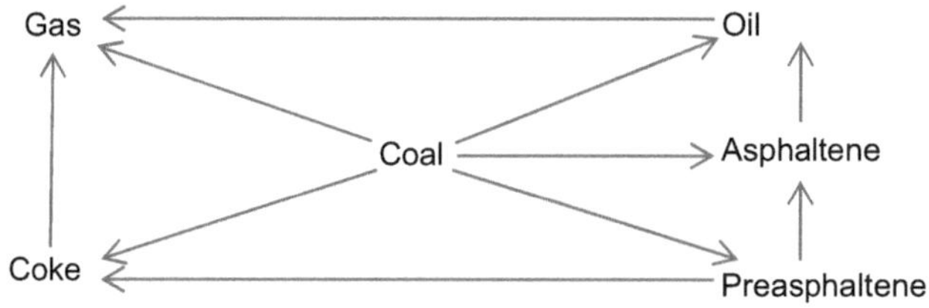

The overall conversion of coal can be regarded as a series of thermal decomposition–hydrogenation reactions such as those discussed previously. Preasphaltene, asphaltene, and oil are defined as tetrahydrofuran-soluble–toluene-insoluble, toluene-soluble–pentane-insoluble, and pentane-soluble coal liquids, respectively.

Catalysis in the Liquefaction of Coal

One important strategy in the process of the liquefaction of coal is hydrogen economy, or the effective use of hydrogen. The question is, what is the most efficient method of hydrogen transfer? This is where the role of catalyst becomes vital in the liquefaction of coal.

A catalyst is needed to transfer and distribute hydrogen efficiently.

An ideal catalyst should possess the following properties:

1. high activities for hydrogenation, cracking, and heteroatom removal;
2. selectivity for maximum liquid yield with minimum gas production;
3. adequate physical strength to overcome mechanical degradation with aging and regeneration; and
4. resistance to deactivation caused by coke and metal deposition, poisoning, sintering, and pore-mouth blockage.

An effective catalyst should help liquefy coal at lower severity conditions and improve liquid yield and its quality. Catalysts in direct coal liquefaction process, lower the severity of the operations essentially by activating both the molecular hydrogen and the coal molecules that too at lower reaction temperatures. They improve product quality by cracking and reforming heavy molecules to desired products, and removing heteroatoms (atoms other than C), such as N, O, and S from coal and coal liquids.

There are many advantages offered by catalysts in this regard. However, it is not simple and hurdles arise because of the complex nature of coals. Coal, frequently referred as "nature's dump," tends to deactivate the catalyst. Most vital reasons for catalyst deactivation are coke formation and metal deposition, in addition to poisoning and sintering, which are not as pernicious as the first two.

Studies reveal that the addition of iron pyrite to coal/oil slurry produces higher conversion and improved product quality. In quantitative terms, addition of 5% iron pyrite produces the same increase in conversion as that produced by a 20°C increase in reaction temperature. Unfortunately, the catalytically active chemical state of the iron sulfide is not definitely known. Most studies strongly suggest that pyrite is not the active catalyst in coal liquefaction but it has the role of a precursor. Pyrrholite ($Fe_{1-x}S$, in which $0 < x < 0.125$), a nonstoichiometric sulfur-rich crystalline structure, has been suggested as the catalytically active ingredient in liquefaction. Thus the catalytic action of pyrite is attributed to its transformation products (pyrrholite and H_2S). The critical parameter in the formation of catalytically active pyrrholite is the partial pressure of H_2S. In addition, hydrogen sulfide by itself is catalytically active in the liquefaction of coal.

Hydrogen Sources

The hydroaromatic portion of coal itself can act as a source for hydrogen. Apart from this gaseous hydrogen, hydrogen-donor solvents can also serve that purpose. The process of adding hydrogen or hydrogenation can be accomplished by catalytic or noncatalytic means. However, the noncatalytic addition of hydrogen to coal molecules requires the employment of drastic conditions of temperature and pressure. Use of a catalyst reduces the severity of the process by bringing down the temperature and pressure. Good hydrogen-donor solvents such as 1,2,3,4-tetra-hydronaphthalene, generally known as tetralin, can serve the purpose of supplying hydrogen and stabilizing the free radicals generated by coal molecules. The reaction can be envisioned as:

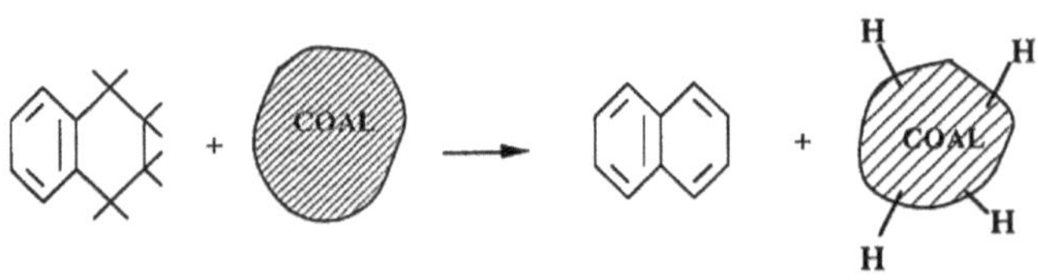

which may proceed sequentially through 1,2-dihydronaphthalene. The hydrogen donated by the tetralin is used to cap radicals, thus facilitating coal depolymerization and liquids production.

However, once the solvent is depleted of hydrogen, it needs to become rehydrogenated through catalytic means. Tetralin is regenerated as:

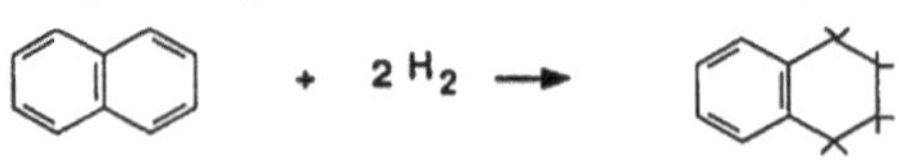

The net process of hydrogenation is represented pictorially as:

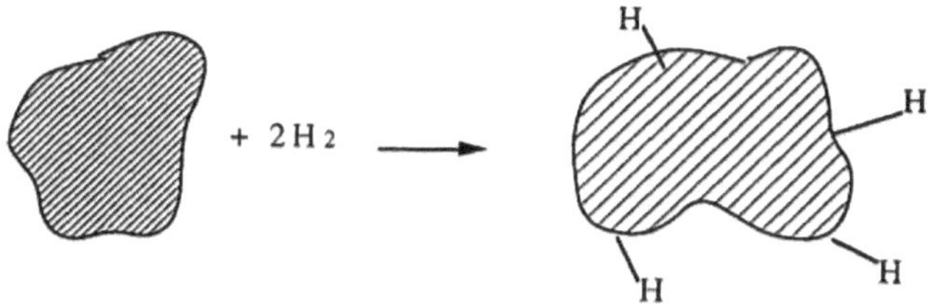

COAL BLENDING

Coal blending is a process of mixing two coals of different reactivity. Alternately, coal-derived ash and/or mineral matter from a reactive coal is added to a less reactive coal. Coals differ in their liquefaction reactivity owing to rank, petrography, and the concentration of mineral matter. The addition of a highly reactive coal to a low-reactive coal can increase conversion and lower operating conditions in a liquefaction plant. Coal blending is beneficial that this decreases plant operating conditions.

Coal Gasification [10]

The steps involved in coal gasification and the products thus obtained are shown schematically in Fig. 4.16.

Ideally it is desirable to gasify coal according to the reaction:

$$2C + 2H_2O \rightarrow CH_4 + CO_2 \quad \Delta H = 2.8 \text{ cal/g mol}$$

Methane, the principal component of natural gas, is produced directly by this reaction. Although a small amount of heat is required for the reaction to sustain, as indicated by the small positive ΔH (endothermic reaction), several favorable reactions compete with this reaction, making the production of methane indirect. In addition to the small amount of methane produced directly in the gasifier, the exit gas will contain CO, CO_2, H_2, H_2S, and H_2O, as shown in Fig. 4.16. The presence of ammonia depends on whether air or oxygen is used in the process and the presence of nitrogen in air obviously leads to the evolution of ammonia.

The gasification of the coal takes place in stage 1. The remaining stages represent processes for the removal of carbon dioxide and water and the removal of sulfur, generally as H_2S and also upgrading the heating value of the gas by methanation. CO and H_2O evolved in stage 1 have no heating value. They undergo water gas shift reaction

$$2CO + H_2O \rightarrow CO_2 + H_2$$

The objective of this operation is to bring the molar ratio of hydrogen to CO to a value of 3:1. Once this molar ratio of hydrogen and CO is achieved,

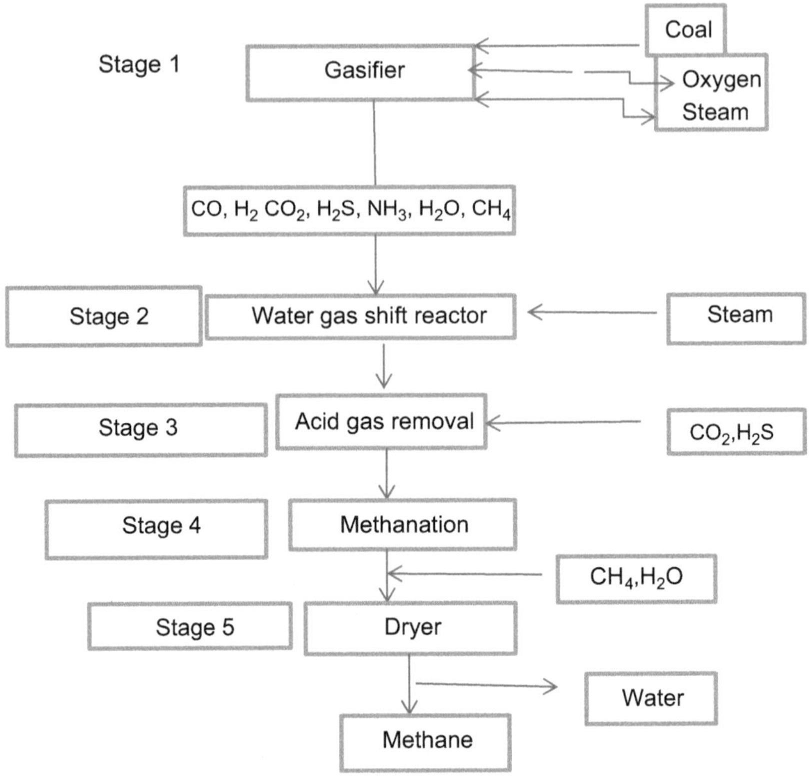

FIGURE 4.16 Schematic representation of steps in the gasification of coal. Normally three kinds of gasifier operations are employed; (1) moving bed gasifier, (2) fixed fluidized bed, and (3) entrained fluidized bed.

CO can be converted to methane and other hydrocarbons (by Fischer–Tropsch synthesis, a process in which CO and hydrogen are converted to higher hydrocarbons and alcohols on supported cobalt or iron catalyst) (stage 4).

The acid gases, hydrogen sulfide, and carbon dioxide are removed in stage 3. This is normally done in a wet collector called a scrubber. The proper choice of a scrubbing liquid allows the selective absorption of hydrogen sulfide and carbon dioxide.

Methanation is accomplished in stage 4 by the conversion of CO and hydrogen according to the reaction:

$$CO + 3H_2 \rightarrow CH_4 + H_2O \quad \Delta H = -49{,}071 \text{ cal/g mol}$$

This reaction has to be catalyzed to proceed at an economical rate. The large, negative ΔH indicates that the reaction is strongly exothermic and hence cooling must be provided.

The methane produced in stage 4 is referred to as synthetic natural gas. The process of drying is carried out in stage 5, in which water is eliminated with the objective of increasing the heating value of the gas. The drying process is carried out by passing it through an agent such as sulfuric acid, calcium chloride, or silica gel. The water vapor is absorbed by these dehydrating agents [11].

In Situ Gasification of Underground Coal

The in situ combustion of coal is yet another option to extract heat energy from coal. This is employed where coal cannot be mined economically or for coal deposits remaining after deep-mining operations. Oxygen and steam are piped down to the deposit and the gasified products are brought to the surface through the wells [12,13].

The following are some advantages with this form of gasification: (1) It can be used for unminable coal deposits, (2) miners' health and safety problems are eliminated, (3) the cost of mining is eliminated, and (4) air pollution and solid waste disposal normally encountered in above-ground gasifications are eliminated.

This process is has disadvantages: (1) there can be groundwater contamination and gas leakage problems, (2) it consumes more oxygen per energy content in the produced gas than in the case of conventional gasifiers, (3) it can generate only low-pressure synthesis gas, and (4) the underground coal gasification can be used only for reactive and permeable coals.

This method consists of several steps. First, chemical explosives are used in an array of drilled holes to fracture a coal bed. Explosions create passages for the flow of gasification fluids as well as gaseous products. Collection wells would be drilled to the bottom of the fractured zone. The top would be ignited and a steam–oxygen mixture would be pumped into the coal vein. The product gases would contain CH_4, CO, CO_2, H_2S, and H_2 similar to surface gasification operations. This can be burned for electric power generation after cleaning or further processed by methanation to upgrade its heating value [14–16].

Chemicals From Coal

Coal has significance in terms of producing industrially important chemicals, apart from being used as a source of energy. Table 4.4 lists major organic and inorganic chemicals that can be made from coal. Half of the chemicals that can be obtained are olefins and aromatics [14].

The current production of chemicals from coal is based on the Eastman process. The Eastman process involves the production of methanol from synthesis gas:

$$C + H_2O \rightarrow CO + H_2$$

$$CO + H_2O \rightarrow CO_2 + H_2$$

$$CO + 2H_2 \rightarrow CH_3OH$$

TABLE 4.4 Some Major Chemicals Derived From Coal

Organic chemicals	Acetic acid Acetone Acrylonitrile Benzene Carbon tetrachloride Cumene Cyclohexane Ethylene Ethylene oxide Formaldehyde Isopropyl alcohol Methanol Perchloroethylene Phenol Phthalic anhydride Propylene Propylene oxide
Inorganic chemicals	Ammonia Sulfuric acid

Reproduced from R. Narayan, B. Viswanathan, Chemical and Electrochemical Energy Systems, University Press, 1998.

Methanol is reacted with acetic acid to form methyl acetate:

$$CH_3COOH + CH_3OH \rightarrow CH_3COOCH_3$$

Methyl acetate can be further reacted with carbon monoxide to form acetic anhydride:

$$CH_3COOCH_3 + CO \rightarrow (CH_3CO)_2O$$

Acetic acid can be reacted with ethane to make vinyl acetate:

$$CH_2 = CH_2 + CH_3COOH \rightarrow CH_2 = CHOOCCH_3$$

Vinyl acetate can be polymerized to poly(vinyl acetate).

CALORIFIC VALUE AND ITS DETERMINATION

The calorific value is defined as the quantity of heat liberated by the combustion of a unit quantity of fuel.

Determination of Calorific Value

The accurate determination of a calorific value requires highly specialized apparatus and facilities. There is no acceptable, simple means of making this determination. The only worthwhile determination is that using a bomb calorimeter.

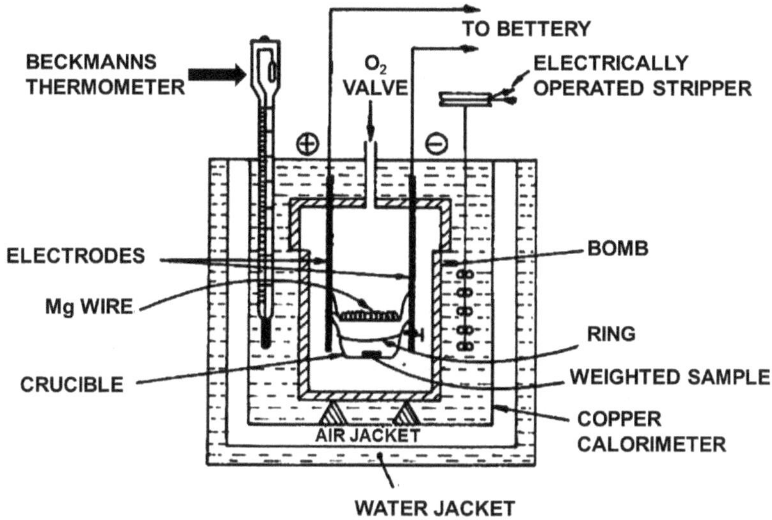

FIGURE 4.17 Typical design of a bomb calorimeter for the determination of heat of combustion.

A bomb calorimeter consists of a strong stainless-steel vessel called a bomb, which is capable of withstanding high pressures. The bomb is provided with a lid that can be screwed firmly onto the bomb. In turn, the lid is provided with two electrodes and an oxygen inlet valve. One of the electrodes is provided with a ring to accommodate the silica crucible.

The bomb is placed in a copper calorimeter with a known weight of water. The copper calorimeter is surrounded by an air jacket and water jacket to prevent loss of heat owing to radiation. The calorimeter is provided with an electrical stirrer for stirring water and a Beckmann thermometer (Fig. 4.17).

Working

A weighted amount of the fuel is placed in the silica crucible. The crucible is supported over the ring. A fine magnesium wire touching the fuel sample is stretched across the electrodes. Oxygen supply is forced into the bomb until a pressure of 25–30 atm is reached. The initial temperature of the water in the calorimeter is noted after thorough stirring. The current is switched on and the fuel in the crucible burns with the evolution of heat. The heat produced by burning of the fuel is transferred to water, which is stirred throughout the experiment by the electric stirrer. The maximum temperature shown by the thermometer is recorded.

To calculate the heat liberated by the combustion of the fuel, the weight of water in the whole apparatus must be known, as well as the amount of heat required to raise all the parts (metal, glass, etc.) in contact with water through the observed rise in temperature. This is evaluated in a calibrating experiment. In this, a known weight of a pure hydrocarbon of known calorific value is burned in the apparatus and the rise in temperature is noted.

Gross and Net Calorific Value

The heat evolved by burning the pure substance minus the heat required to raise the temperature of the water by the observed amount is the heat required to raise the temperature of the various parts of the apparatus by the same amount. A constant for the apparatus is derived by dividing this amount of heat by the observed rise in temperature. This constant is known as the water equivalent of the apparatus. It is expressed as the number of grams of water that are equivalent in heat capacity to the various parts of the apparatus that are heated by the calorimeter water.

Gross calorific value = the heat evolved (in calories) by a known weight of fuel = the rise in temperature in degrees centigrade × [water in calorimeter vessel + water in bomb + water equivalent of the apparatus (all in grams)].

Need for Net Calorific Value

Calorific values as determined with the bomb calorimeter represent the heat produced by a unit weight of coal when completely oxidized, when the products of the combustion are cooled to room temperature. This value is not realized in practice because the products of combustion are not cooled to room temperature before being discharged to waste.

Sensible heat is lost in the hot waste products. Apart from this, further heat loss occurs in practice as the latent heat of steam in the hot waste gases. Water is present as such because moisture in the air-dried coal and a further amount are formed by the combustion of the hydrogen combined with carbon in the coal. In the bomb calorimeter the moisture is first evaporated and then condensed to liquid water. Similarly the water formed as steam by combustion is condensed to liquid water; the latent heat of condensation of the steam is recovered. In industrial practice water from both sources is discharged as steam, so that both latent heat and sensible heat are lost. It is therefore useful to distinguish the calorific value as determined with the bomb calorimeter by calling it the gross calorific value.

A lower value can be derived, which is the gross calorific value minus the latent heat of condensation at 15.5°C of all of the water involved. This is named the net calorific value. The net calorific value is a more realistic statement of realizable potential heat than the gross value.

The correction to the gross calorific value is 586 cal/g water (latent heat of steam = 586 cal/g). The water referred to is the weight of water produced by the complete combustion of unit weight of coal plus the water existing as moisture in the coal. The former is calculated from a known hydrogen content of the coal:

$$\text{Net calorific value} = \text{gross calorific value} - 586\,(\text{water as moisture} + \text{water formed from } H_2)\ \text{cal/g}.$$

The calorific value of coal has been used to exemplify gross and net calorific value. The same correction can be applied to any fuel of any physical state if care is taken with the units of weight or volume.

COAL BURNING: ENVIRONMENTAL HAZARDS AND MEASURES

Coal is the least hydrogen rich of fossil fuels, which means that more carbon is oxidized to carbon dioxide per gram of fuel. Even though coal reserves guarantee energy for over a century, the path is not free from a host of environmental problems in addition to large-scale carbon dioxide generation that must be addressed. Emissions of NO_x from power generation using coal are an important environmental problem. NO_x contributes to the formation of photochemical smog and acid rain. HCN and ammonia, which are formed during pyrolysis of coal, are assumed to be two important precursors for NO_x [17].

Carbon Sequestration

Carbon sequestration is a new way to manage carbon. It is a provision for the long-term storage of carbon in the terrestrial biosphere, underground, or in the oceans so that the buildup of carbon dioxide concentration in the atmosphere can be reduced or slowed.

Atmospheric levels of carbon dioxide have risen from preindustrial levels of 280 parts per million (ppm) to the current levels of nearly 400 ppm. This is because of the expanding use of fossil fuels for energy. Continued increases in carbon emissions are suggested because the world cannot afford to abandon fossil fuels. However, the technology that ultimately succeeds may be radically different from today's fossil fuels. New technologies that could capture carbon dioxide either at the point of emission from power plants or from the air and dispose it of safely and permanently are under development. These approaches would make fossil fuel energy use sustainable for at least another century.

Carbon Dioxide Capture From the Air

This can be accomplished by letting wind carry air over an absorber that pulls off carbon dioxide. This could be achieved through a variety of methods, including blowing air over lime water, which will remove the carbon dioxide and produce limestone. The volume of air that needs to be processed is surprisingly little compared with the volumes required to harness wind energy.

Carbon Dioxide Capture From Power Plants

The capture of carbon dioxide directly from power plants could be accomplished through new plant designs. Plant designs that could generate hydrogen or electricity from coal with no emissions to the atmosphere are under development. In these methods, lime captures waste carbon dioxide and generates heat for additional hydrogen production. Captured carbon dioxide can be disposed by underground injection, a low-cost procedure whose effectiveness has been proven in enhanced oil recovery. Although underground injection will be more difficult because the demand for repositories increases, this alternate method of disposal is available. Magnesium silicates can be used to react with carbon dioxide to form magnesium carbonate and silica (quartz), which can be disposed of safely and permanently [17].

Norway became the first country to impose a federal tax on atmospheric carbon dioxide emissions from combustion-based point sources such as coal-fired power plants. The day is not far off when other countries follow Norway's lead to protect the environment and curb rising temperatures of the planet.

The development of strategies for the safe disposal of carbon dioxide waste streams is the need of the hour.

CONCLUSION

The time will inevitably come when there will be no more coal or petroleum for the rate at which the reserves are being consumed. Before the disappearance of coal and petroleum from everyday life, mankind must develop a new source of power or perish. Nuclear energy appears to be a ray of hope even though it requires mutual cooperation among people worldwide in terms of fuel supply and distribution apart from the intricacies of technological advancement. It is evident that uranium and thorium reserves worldwide can match humanity's never-quenched thirst for energy. Apart from nuclear energy, renewable sources such as water, wind, and sunlight can be thought of as energy sources for the future even though they can never serve our common needs 100 years from now despite the toiling efforts of scientists in this sector.

REFERENCES

[1] R. Eisenberg, D.G. Nocera, Inorg. Chem. 44 (2005) 6799.
[2] J.L. Dempsey, A.J. Esswein, D.R. Manke, J. Rosenthal, J.D. Soper, D.G. Nocera, Inorg. Chem. 44 (2005) 6879.
[3] R. Narayan, B. Viswanathan, Chemical and Electrochemical Energy Systems, University Press, 1998.
[4] H.H. Schobert, The Chemistry of Hydrocarbon Fuels, Butterworths & Co (Publishers), 1990.
[5] O.P. Gupta, Elements of Fuels, Furnaces and Refractories, Khanna Publishers, 1997.

[6] J.C. Macrae, An Introduction to the Study of Fuel, Elsevier Publishing Company, 1966.
[7] J.S.S. Brame, J.G. King, Fuel – Solid, Liquid and Gaseous, fifth ed., Rewritten by J.G. King, fifth ed., Edward Arnold (Publishers) Ltd, 1955.
[8] B.R. Cooper, Scientific problems of coal utilization, in: Proceedings of a Conference at West Virginia University, May 23–25, 1977, Technical Information Centre, U. S. Department of Energy, 1978.
[9] M.L. Shepard, J.B. Chaddock, F.H. Cocks, C.M. Harman, Introduction to Energy Technology, Ann Arbor Science Publishers Inc., 1976.
[10] J.L. Johnson, Kinetics of Coal Gasification, John Wiley & Sons, 1979.
[11] H.H. Lowry, Chemistry of Coal Utilization, Supplementary Volume, John Wiley & Sons, Inc., 1963.
[12] G.R. Gavalas, Coal Pyrolysis, Coal Science and Technology, vol. 4, Elsevier Scientific Publishing Company, 1982.
[13] E.J. Badin, Coal Combustion Chemistry – Correlation Aspects, Coal Science and Technology, vol. 6, Elsevier Scientific Publishing Company, 1984.
[14] A. Volborth, Coal Science and Chemistry, Coal Science and Technology, vol. 10, Elsevier Scientific Publishing Company, 1987.
[15] T.R. Martha, A. Bhattacharya, K. Vinod Kumar, Curr. Sci. 88 (1) (2005) 21.
[16] E. Shoko, B. McLellan, A.L. Dicks, J.C. Diniz da Costa, Int. J. Coal Geol. 65 (2006) 213.
[17] A. Verma, A.D. Rao, G.S. Samuelsen, J. Power Sources 158 (2006) 417.

Chapter 5

Nuclear Fission

INTRODUCTION

Some well-known scientists are of the opinion that nuclear power is the greenest option even though this source of energy has received considerable criticism and opinions regarding the possible dangers associated with nuclear reactors. Especially after disasters such as the Japanese Fukushima-Daiichi, Russia's Chernobyl, and Three Mile Island, emphasis has focused on safety in nuclear operations. Without entering into this controversial debate [1], in this chapter, only the scientific background behind this source of energy will be considered. In simple terms, an atomic nucleus consists of neutrons (zero charge) and protons (positively charged) and every atomic nucleus contains Z protons and N neutrons, which give the atomic weight and $(Z + N)$. Because the atom is neutral, it should have Z electrons (negatively charged) for a nuclear particle.

Fission takes place as a result of net electrostatic repulsion in the high amount of positively charged protons in the heavy nucleus. The net internal electrostatic repulsion between positively charge protons will be less in smaller nuclei than in the heavy nucleus because the number of positively charged protons is higher in the heavy nucleus. Therefore in the heavy nucleus, when the strong nuclear attractive force is overcome by the net repulsive force of the protons, the nucleus undergoes fission to generate two or more lighter nuclei in which the electrostatic force is less than the attractive nuclear forces that holds the atom intact. It is necessary to get an idea of the energy that can be harnessed in a nuclear fission reaction. This energy is manifested in the form of the mass difference between the heavier nucleus undergoing fission and the sum of the atomic masses of light product nuclei that are produced as a result of the fission reaction. The net energy that is generated from the fission reaction results from the mass defect that is experienced in the fission reaction; this can be assessed by the mass–energy relationship. One uranium fuel pellet has as much energy available as nearly 42 gallons of oil, 1 ton of coal, 2.5 tons of wood, or 17,000 cubic feet of natural gas. A heavy nucleus such as uranium has essentially no other significant positive application except for energy conversion. Observation of the fission process in a heavier nucleus (such as

Energy Sources. http://dx.doi.org/10.1016/B978-0-444-56353-8.00005-8

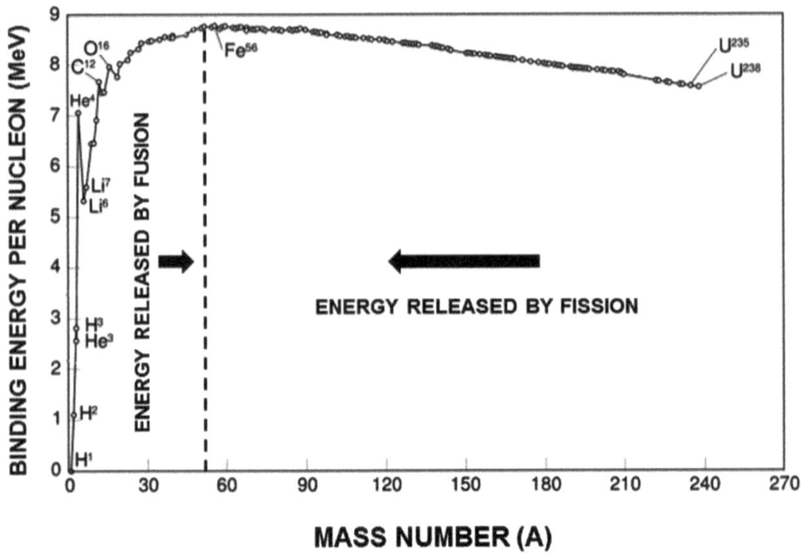

FIGURE 5.1 Binding energy per nucleon as a function of the mass number of the atom.

uranium) implies that the net binding energy per nucleon is lower than the binding energy in lighter nuclei. This concept is shown graphically in Fig. 5.1.

Consider a typical fission reaction:

$$^{235}_{92}U_{143} + ^{1}_{0}n_{1} \rightarrow ^{134}_{54}Xe_{80} + ^{100}_{38}Sr_{62} + 2^{1}_{0}n_{1}$$

The products shown in this equation are only one set of possible product nuclei that can be formed by the fission of ^{235}U. The only condition to be satisfied in a fission reaction is that the total number of protons and neutrons in the product nuclei must be equal to that present in the fissionable nucleus and nuclear particle inducing the fission.

A typical set of 10 nuclear fission reactions involving ^{235}U is given in Table 5.1. The energy distribution in the various processes in a typical fission of ^{235}U is given in Table 5.2.

Of this 202 MeV nearly 191 MeV of energy is in a usable form. Therefore if one were to use 1 g ^{235}U, the total power available would be:

$$[1\ \text{g} \times 6.023 \times 10^{23}\ \text{atom/mol}/235.04\ \text{g/mol}] \times 191\ \text{MeV} \times 1.602 \times 10^{-13}\ \text{J/MeV} = 78.4\ \text{GJ}$$

It is seen that the energy yield for every kilogram of nuclear fuel will be approximately 2.66×10^6 times that obtainable from mined coal. Even if one were to use 1 kg of natural uranium (containing 0.71% of ^{235}U), the total heat energy available is approximately 556 GJ, equivalent to that available from 19 tons of mined coal. A similar comparison of energy equivalents in terms of other fuel sources was given earlier.

TABLE 5.1 Set of 10 Possible Processes (Among Nearly 30 Possible) in the Fission of ^{235}U

$^{235}U + {}^{1}n \rightarrow {}^{134}Xe + {}^{100}Sr + 2{}^{1}n$
$^{235}U + {}^{1}n \rightarrow {}^{89}Kr + {}^{144}Ba + 3{}^{1}n$
$^{235}U + {}^{1}n \rightarrow {}^{96}Zr + {}^{136}Xe + 4{}^{1}n$
$^{235}U + {}^{1}n \rightarrow {}^{99}Nb + {}^{133}Sb + 4{}^{1}n$
$^{235}U + {}^{1}n \rightarrow {}^{145}La + {}^{88}Br + 3{}^{1}n$
$^{235}U + {}^{1}n \rightarrow {}^{95}Mo + {}^{139}La + 2{}^{1}n$
$^{235}U + {}^{1}n \rightarrow {}^{144}Ba + {}^{90}Kr + 2{}^{1}n$
$^{235}U + {}^{1}n \rightarrow {}^{141}Ba + {}^{92}Kr + 3{}^{1}n$
$^{235}U + {}^{1}n \rightarrow {}^{94}Zr + {}^{139}Te + 3{}^{1}n$
$^{235}U + {}^{1}n \rightarrow {}^{144}Cs + {}^{90}Rb + 2{}^{1}n$

TABLE 5.2 Typical Energy Distribution in Various Processes in Fission of ^{235}U

Process	Energy (MeV)
1.Kinetic energy of fission fragments	168
2. Kinetic energy of fissioned neutrons	5
3. Fission gamma radiation (prompt)	5
4. Beta emission from fission products	7
5. Gamma emission from fission products	6
6. Neutrinos (not available as heat source)	11
7. Total	202

Despite all of these enormous energy equivalents in nuclear energy from a small mass of nuclear fuel, there are advantages and disadvantages to harnessing this source of energy; this information is given in Table 5.3.

To understand the concepts of nuclear reactions, some basic terminology has to be introduced.

THE NUCLEUS AND ITS CONSTITUENTS

An atom consists of a centrally located nucleus surrounded by electrons revolving in certain physically permitted orbitals. The nucleus itself is made

TABLE 5.3 Typical Advantages and Disadvantages of Nuclear Energy

Advantages	Disadvantages
Nuclear power costs about the same as coal, so it is inexpensive.	Although not much waste is produced, the waste as such is dangerous. It must be properly sealed and buried for many years for all of the radioactivity to die away. Nuclear power is reliable but a lot of effort has to be made for safety. If it does go wrong, a nuclear accident can be a major disaster.
The energy conversion process does not produce smoke or carbon dioxide and does not contribute to the greenhouse effect.	
It produces a large amount of energy from a small amount of fuel.	
It produces a small amount of waste.	
This form of power is reliable.	

TABLE 5.4 Characteristics of the Constituents of an Atom

Fundamental Nuclear Particle	Charge	Mass (U)
Proton	+e	1.007276
Neutron	O	1.008665
Electron	−e	0.000549

up of neutrons and protons, collectively called nucleons. The number of protons (Z) is called the atomic number and the total number (A) of nucleons in a nucleus is called the atomic (or nuclear) mass number. The number of neutrons (A−Z) is represented as N. The basic properties of the atomic constituents are summarized in Table 5.4.

Charge

Protons have a positive charge of magnitude $e = 1.6022 \times 10^{-19}$ C (Coulombs) equal to and opposite that of the electron. Neutrons do not possess a charge. Thus, a neutral atom (A, Z) contains Z electrons and can be written as $^{A}X_{N}$.

Mass

Nuclear and atomic masses are expressed in atomic mass units (u), based on the definition that the mass of a neutral atom of $^{12}C_6$ is exactly 12.000 u ($1\ u = 1.6605 \times 10^{-27}$ kg).

Isotopes, Isotones, and Isobars

Atoms are classified as isotopes, isotones, and isobars based on their nuclear content. Isotopes of an element are atoms whose nuclei have the same Z but different N. They have similar electronic structure and therefore similar chemical properties. For example, hydrogen has three isotopes: $^1H_0$, $^2H_1$, $^3H_2$ whose nuclei are, respectively, the proton p, the deuteron d, and the triton t. Nuclei with the same N and different Z are called isotones, and nuclides with the same mass number A are known as isobars. In a symbolic representation of a nuclear species, or nuclide, it is usual to omit the N and Z subscripts and include only the mass number as a superscript, because $A = N + Z$ and the symbol representing the chemical element uniquely specifies Z.

Mass Defect and Binding Energy

Careful measurements have shown that the mass of a particular atom is always slightly less than the sum of the masses of the individual neutrons, protons, and electrons of which the atom consists. The difference between the mass of the atom and the sum of the masses of its parts is called the mass defect (Δm). The mass defect can be calculated using equation $\Delta m = [Z(mp + me) + (A - Z)m_n] - m_{atom}$, where: Δm = mass defect [atomic mass unit (amu)]; m_p = mass of a proton (1.007277 amu); m_n = mass of a neutron (1.008665 amu); m_e = mass of an electron (0.000548597 amu); m_{atom} = mass of nuclide $^A_Z X$ (amu); Z = atomic number (number of protons); and A = mass number (number of nucleons).

In calculating the mass defect it is important to use the full accuracy of mass measurements because the difference in mass is small compared with the mass of the atom. Rounding off the masses of atoms and particles to three or four significant digits before the calculation will result in a calculated mass defect of zero.

Binding Energy

The binding energy is the amount of energy that must be supplied to a nucleus to separate its nuclear particles completely. The binding energy is the energy equivalent of the mass defect. The binding energy can be calculated by multiplying the mass defect by the factor of 931.5 MeV/amu.

Binding Energy and Nuclear Stability

As the number of particles in a nucleus increases, so does the total binding energy. The rate of increase, however, is not uniform. This lack of uniformity results in variation in the amount of binding energy associated with each nucleon within the nucleus. This variation in the binding energy per nucleon (BE/A) is easily seen when the average BE/A is plotted versus atomic mass number (A), as shown in Fig. 5.1.

Fig. 5.1 illustrates that as the atomic mass number increases, the binding energy per nucleon decreases for elements with $A > 60$. The BE/A curve reaches a maximum value of 8.79 MeV at $A = 56$ and decreases to about 7.6 MeV for $A = 238$. The general shape of the BE/A curve can be explained using the general properties of nuclear forces. The nucleus is held together by very short-range attractive forces that exist between nucleons. On the other hand, the nucleus is being forced apart by long-range repulsive electrostatic (Coulomb) forces that exist among all the protons in the nucleus.

As the atomic number and atomic mass number increase, the repulsive electrostatic forces within the nucleus increase owing to the greater number of protons in the heavy elements. To overcome this increased repulsion, the proportion of neutrons in the nucleus must increase to maintain stability. This increase in the neutron-to-proton ratio only partially compensates for the growing proton–proton repulsive force in the heavier, naturally occurring elements. Because the repulsive forces are increasing, less energy must be supplied, on average, to remove a nucleon from the nucleus. The BE/A has decreased. The BE/A of a nucleus is an indication of its degree of stability. Generally, stable nuclides have higher BE/A than do the less stable ones. The increase in the BE/A as the atomic mass number decreases from 260 to 60 is the primary reason for the energy liberation in the fission process. In addition, the increase in the BE/A as the atomic mass number increases from 1 to 60 is the reason for the energy liberation in the fusion process, which is the opposite reaction of fission.

The heaviest nuclei require only a small distortion from a spherical shape (small energy addition) for the relatively large Coulomb forces forcing the two halves of the nucleus apart to overcome the attractive nuclear forces holding the two halves together. Consequently, the heaviest nuclei are easily fissionable compared with lighter nuclei.

Normally, as a result of the neutron bombardment that the ^{235}U or ^{236}U that is formed undergoes in addition to fission, it emits gamma rays and decays to the ground state. Therefore one has to take into account all of the reaction cross-sections (some kind of probability) while selecting a fissile material. For information, relevant data for typical fissile nuclei are assembled in Table 5.5.

RADIATION AND NUCLEAR REACTIONS

Traditional chemical reactions occur as a result of the interaction between valence electrons around an atom's nucleus. In 1896, Henri Becquerel expanded the field of chemistry to include nuclear changes when he discovered that uranium emitted radiation. Soon after Becquerel's discovery, Marie Sklodowska Curie began studying radioactivity and carried out much of the pioneering work on nuclear changes. Curie found that radiation was proportional to the amount of radioactive element present, and she proposed that radiation was a property of atoms (as opposed to a chemical property of a compound).

TABLE 5.5 Values of Cross-Sections for Typical Fissile Nuclei Bombarded by Thermal Neutrons

Fissile Nuclei	Fission (barns)	Radiative Capture (barns)	Absorption (barns)	Scattering (barns)
^{233}U	527	54	581	–
^{235}U	580	107	687	10
^{238}U	$<0.5 \times 10^{-3}$	2.71	2.71	8.3
^{239}Pi	746	280	1026	9.6
Natural U	4.18	3.5	7.68	8.3

In 1902, Frederick Soddy proposed the theory that "radioactivity is the result of a natural change of an isotope of one element into an isotope of a different element." Nuclear reactions involve changes in particles in an atom's nucleus and thus cause a change in the atom itself. All elements heavier than bismuth (Bi) (and some that are lighter) exhibit natural radioactivity and thus can "decay" into lighter elements. Unlike normal chemical reactions that form molecules, nuclear reactions result in the transmutation of one element into a different isotope or a different element altogether (the number of protons in an atom defines the element, so a change in protons results in a change in the atom). There are three common types of radiation and nuclear changes.

Alpha Radiation

Alpha radiation (α) is the emission of an alpha particle from an atom's nucleus. An α particle contains two protons and two neutrons (and is similar to a He nucleus: ${}^{4}_{2}He$). When an atom emits an α particle, the atom's atomic mass will decrease by four units (because two protons and two neutrons are lost) and the atomic number (z) will decrease by two units. The element is said to "transmute" into another element that is two units of z smaller. An example of an α transmutation takes place when uranium decays into the element thorium (Th) by emitting an α particle, as depicted in the equation:

$$^{238}U_{92} \rightarrow {}^{4}He_{2} + {}^{234}Th_{90}$$

Beta Radiation

Beta radiation (β) is the transmutation of a neutron into a proton and an electron (followed by the emission of the electron from the atom's nucleus: ${}^{0}_{-1}e$). When an atom emits a β particle, the atom's mass will not change (because there is no change in the total number of nuclear particles). However the atomic number will increase by 1 (because the neutron is transmuted into an additional proton). An example of this is the decay of the isotope of carbon, named carbon-14, into the element nitrogen:

$$^{14}C_6 \rightarrow {}^{0}e_{-1} + {}^{14}N_7$$

Gamma Radiation

Gamma radiation (γ) involves the emission of electromagnetic energy (similar to light energy) from an atom's nucleus. No particles are emitted during γ radiation, and thus γ radiation does not itself cause the transmutation of atoms. However γ radiation is often emitted during, and simultaneous with, α or β radioactive decay. X-rays, emitted during the β decay of cobalt-60, are a common example of γ radiation.

Half-Life

Radioactive decay proceeds according to a principle called the half-life ($T_{1/2}$). The half-life is the amount of time necessary for half of the radioactive material to decay. For example, the radioactive element bismuth (^{210}Bi) can undergo α decay to form the element thallium (^{206}Tl) with a reaction half-life equal to 5 days. If we begin an experiment starting with 100 g of bismuth in a sealed lead container, after 5 days we will have 50 g of bismuth and 50 g of thallium in the jar. After another 5 days (10 from the starting point), half of the remaining bismuth will decay and we will be left with 25 g of bismuth and 75 g of thallium in the jar.

The fraction of parent material that remains after radioactive decay can be calculated using the equation:

$$\text{Fraction remaining} = 1/2^n, \text{ where } n = \text{half-lives elapsed.}$$

The amount of a radioactive material that remains after a given number of half-lives is therefore:

$$\text{Amount remaining} = \text{original amount} \times \text{fraction remaining.}$$

The decay reaction and $T_{1/2}$ of a substance are specific to the isotope of the element undergoing radioactive decay. For example, ^{210}Bi can undergo α decay to ^{206}Tl with a $T_{1/2}$ of 5 days. ^{215}Bi, by comparison, undergoes β decay to ^{215}Po with a $T_{1/2}$ of 7.6 minutes, and ^{208}Bi undergoes yet another mode of radioactive decay (called electron capture) with a $T_{1/2}$ of 368,000 years!

NUCLEAR FISSION

Although many elements undergo radioactive decay naturally, nuclear reactions can also be stimulated artificially. Although these reactions occur naturally, we are most familiar with them as stimulated reactions. There are

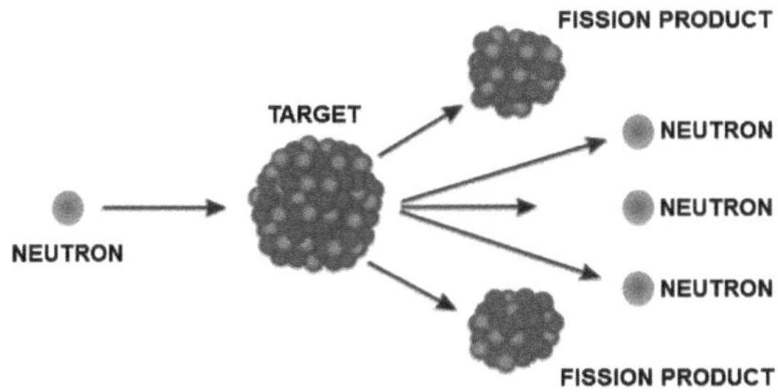

FIGURE 5.2 Pictorial model of a typical fission reaction.

two such types of nuclear reactions: nuclear fission and nuclear fusion. This chapter deals exclusively with nuclear fission reactions.

As stated earlier, nuclear fission denotes reactions in which an atom's nucleus splits into smaller parts, releasing a large amount of energy in the process (Fig. 5.2). Most commonly, this is done by "bombarding with" a neutron at the nucleus of an atom. The energy of the neutron causes the target element to split into two (or more) elements that are lighter than the parent atom. When a nucleus undergoes fission, it splits into several smaller fragments. These fragments, or fission products, are about equal to half the original mass. Two or three neutrons are also emitted. The sum of the masses of these fragments is less than the original mass. This "missing" mass (about 0.1% of the original mass) has been converted into energy according to Einstein's equation.

Fission can occur when a nucleus of a heavy atom captures a neutron, or it can happen spontaneously.

Chain Reaction

A chain reaction refers to a process in which neutrons released in fission produce an additional fission in at least one further nucleus. This nucleus in turn produces neutrons, and the process repeats (Fig. 5.3). The process may be controlled (nuclear power) or uncontrolled (nuclear weapons).

$$^{235}\text{U} + \text{n} \rightarrow \text{fission} + 2 \text{ or } 3\ \text{n} + 200\ \text{MeV}$$

If each neutron releases two more neutrons, the number of fission doubles each generation. In that case, in 10 generations there are 1024 fissions, and in 80 generations about 6×10^{23} (1 mol) fissions.

Critical Mass

Although two or three neutrons are released for each fission reaction, not all of these neutrons are available for continuing the fission reaction. If the conditions are such that the neutrons are lost at a faster rate than they are formed by

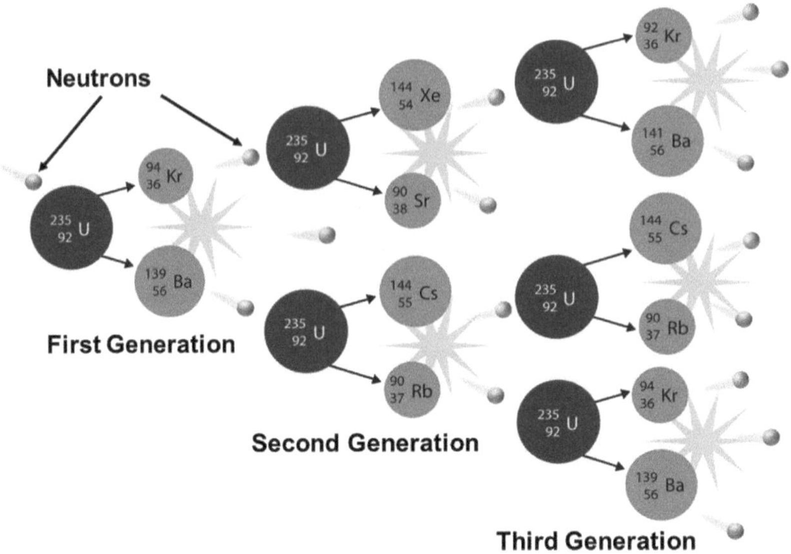

FIGURE 5.3 Model of a nuclear chain reaction.

fission, the chain reaction will not be self-sustaining. At the point where the chain reaction can become self-sustaining, this is referred to as critical mass. In an atomic bomb, a mass of fissile material greater than the critical mass must be assembled instantaneously and held together for about a millionth of a second to permit the chain reaction to propagate before the bomb explodes. The amount of a fissionable material's critical mass depends on several factors: the shape of the material, its composition and density, and the level of purity. A sphere has the minimum possible surface area for a given mass, and hence minimizes the leakage of neutrons. By surrounding the fissionable material with a suitable neutron "reflector," the loss of neutrons and the critical mass can be reduced. By using a neutron reflector, only about 11 lb (5 kg) of nearly pure or weapon-grade ^{239}Pu or about 33 lb (15 kg) ^{235}U is needed to achieve critical mass.

^{235}U and ^{238}U

Uranium, which is used in nuclear power generation, includes ^{235}U and ^{238}U. Almost like twins, these two isotopes of uranium differ only in the number of their neutrons. When a ^{235}U isotope absorbs a neutron, it loses its stability, which causes nuclear fission. Nuclear power generation uses thermal energy emitted at the time of nuclear fission. A ^{238}U nucleus, on the other hand, does not split when a neutron is absorbed; instead ^{238}U changes into ^{239}Pu.

^{232}Th

One can also consider a thermal reactor that can breed a fuel ^{235}Th, and β emission of the product nuclei results in ^{233}U, which is the fissile fuel used to generate the necessary neutrons. The nuclear reaction sequence is:

$$^{232}Th_{90} + {}^{1}n_{0} \rightarrow {}^{233}Th_{90} \rightarrow {}^{0}e_{1} + {}^{233}Pa_{91} \rightarrow {}^{0}e_{1} + {}^{233}U_{92}$$

This U–Th cycle is at best at the demonstration stage.

Uranium Enrichment

The concentration of ^{235}U, with which nuclear fission occurs, is increased from approximately 0.7% to 3–5%. Enrichment methods include the gaseous diffusion process, the laser enrichment method, and the process using centrifuge.

Controlled Nuclear Fission and Nuclear Reactors

To maintain a sustained controlled nuclear reaction, for every two or three neutrons released, only one must be allowed to strike another uranium nucleus (Fig. 5.4). If this ratio is <1, the reaction will die out; if it is >1, it will grow uncontrolled (an atomic explosion). A neutron-absorbing element must be present to control the amount of free neutrons in the reaction space. Most reactors are controlled by means of control rods that are made of a strongly neutron-absorbent material such as boron or cadmium.

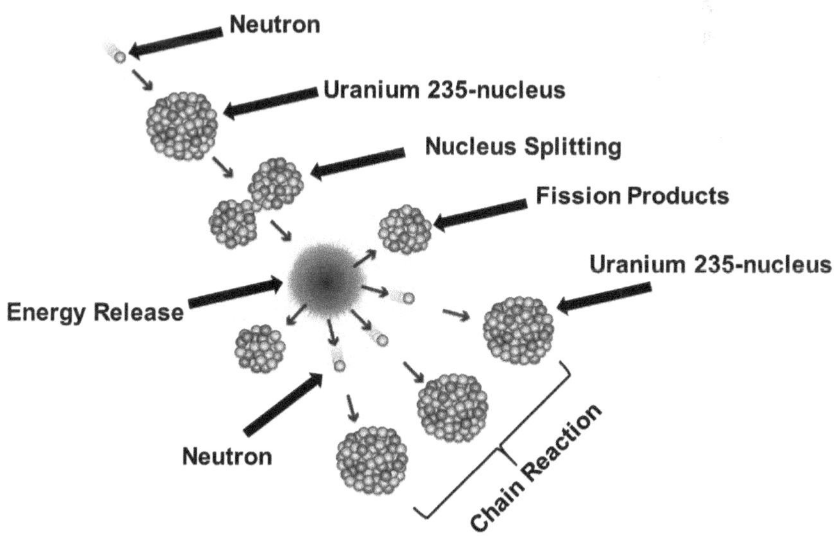

FIGURE 5.4 Model for uncontrolled and controlled nuclear fission.

There are different types of nuclear reactors, such as pressurized water, boiling water, gas-cooled, pressurized heavy water, light-water graphite, and so on. These reactors are mostly used for power generation but some can also produce plutonium for weapons and fuel. Two components are common to all reactors: control rods and a coolant. **Control rods** determine the rate of fission by regulating the number of neutrons. These rods are made with neutron-absorbing material such as cadmium, hafnium, or boron and are inserted or withdrawn from the core to control the rate of reaction or to halt it. (Secondary shutdown systems involve adding other neutron absorbers, usually as a fluid, to the system.) The **coolant** removes the heat generated by fission reactions. Water is the most commonly used coolant, but pressurized water, helium gas, and liquid sodium have also been used. In light-water reactors the moderator also functions as coolant.

In addition to the need to capture neutrons, the neutrons often have too much kinetic energy. These fast neutrons are slowed through the use of a moderator such as heavy water and ordinary water. Some reactors use graphite as a moderator, but this design has several problems. Once the fast neutrons have been slowed, they are more likely to produce further nuclear fissions or be absorbed by the control rod.

Slow-neutron reactors operate on the principle that ^{235}U undergoes fission more readily with thermal or slow neutrons. Therefore, these reactors require a **moderator** to slow the neutrons from high speeds upon emerging from fission reactions. The most common moderators are graphite (carbon), light water (H_2O), and heavy water (D_2O). Because slow reactors are highly efficient in producing fission in ^{235}U, slow-neutron reactors operate with natural or slightly enriched uranium. Light-water reactors are classified as either pressurized-water or boiling-water reactors, depending on whether the coolant water is kept under pressure. The long periods, typically 12–18 months, between refueling of light-water reactors make it difficult to use them as a source of plutonium.

FAST BREEDER REACTOR

The fast breeder reactor (FBR) is a fast neutron reactor designed to breed fuel by producing more fissile material than it consumes. It is supposed to minimize nuclear waste. FBRs usually use a mixed-oxide fuel core of up to 20% plutonium dioxide (PuO_2) and at least 80% uranium dioxide (UO_2). The plutonium used can be from reprocessed civil or dismantled nuclear weapons sources. Surrounding the reactor core is a blanket of tubes containing non-fissile ^{238}U which, by capturing fast neutrons from the reaction in the core, is partially converted to fissile ^{239}Pu, which can then be reprocessed for use as nuclear fuel. No moderator is required because the reactions proceed well with fast neutrons. Early FBRs used metallic fuel, either highly enriched uranium or plutonium.

Fast reactors typically use liquid metal as the primary coolant to cool the core or heat the water used to power the electricity-generating turbines. Sodium is the normal coolant for large power stations, but lead and Na-K have both been used successfully for smaller generating rigs. Some early FBRs used mercury. One advantage of mercury and Na-K is that they are both liquids at room temperature, which is convenient for experimental rigs but less important for pilot or full-scale power stations. At its best, the breeder reactor system produces no nuclear waste at all: literally everything eventually gets used. In the real world, there actually may be some residual material that could be considered waste, but its half-life, or the period of time it takes for half the radioactivity to dissipate, is on the order of 30–40 years.

FISSION TO ELECTRICITY

Nuclear power is the controlled use of nuclear reactions (currently limited to nuclear fission and radioactive decay) to do useful work including propulsion, heat, and the generation of electricity. Nuclear energy is produced when a fissile material, such as ^{235}U, is concentrated such that the natural rate of radioactive decay is accelerated in a controlled chain reaction and creates heat that is used to boil water, produce steam, and drive a steam turbine. The turbine can be used for mechanical work and to generate electricity.

During the fission of ^{235}U, three neutrons are released in addition to the two daughter atoms. If these released neutrons collide with nearby ^{235}U nuclei, they can stimulate the fission of these atoms and start a self-sustaining nuclear chain reaction. This chain reaction is the basis of nuclear power. As uranium atoms continue to split, a significant amount of energy is released from the reaction. The heat released during this reaction is harvested and used to generate electrical energy. A nuclear power plant produces electricity in almost exactly the same way that a conventional (fossil fuel) power plant does. A conventional power plant burns fuel to create heat. The fuel is generally coal, but oil is also used sometimes. The heat is used to raise the temperature of water, thus causing it to boil. The high temperature and intense pressure steam that result from the boiling water turn a turbine that then generates electricity. A nuclear power plant works the same way, except that the heat used to boil the water is produced by a nuclear fission reaction using ^{235}U as fuel, not the combustion of fossil fuels. A nuclear power plant uses less fuel than a comparable fossil fuel plant does. A rough estimate is that it takes 17,000 kg of coal to produce the same amount of electricity as 1 kg of nuclear uranium fuel.

A WORD OF CAUTION

In this chapter, only the possible energy output itself may be on the right order of magnitude; all other aspects regarding the magnitude and design of the reactor and other subsidiary units have been mentioned in passing. Only the scientific side of nuclear energy has been dealt with. How it is actually harnessed is beyond the scope of this presentation, and hence the readers must refer to authoritative treatments on these aspects from other sources.

REFERENCE

[1] R.A. Chaplin, Nuclear Energy Materials and Reactors, 1, Fundamentals of nuclear energy; http://www.eolss.net/Eolss-sampleAllChapter.aspx.

FURTHER READING

[1] J. Lilley, Nuclear Physics, John Wiley and Sons, Chichester, 2001.
[2] K.S. Krane, Introductory Nuclear Physics, John Wiley and Sons, New York, 1998.
[3] M.N. Sastri, Introduction to Nuclear Science, Affiliated East West Press Private Limited, New Delhi, 1983.
[4] http:/en.wikipedia.org/wiki/nuclear-power.
[5] http://www.visionlearning.com/library/module_viewer.php?mid=59.
[6] http://www.barc.ernet.in/webpages/about/anu1.htm.
[7] Larry Foulke, A Look at Nuclear Science and Technology, University of Pittsburgh https://dl.dropboxusercontent.com/content_link/d4MaW7NtOimMmu6FZU0amZJjs2q36ARAFkSLWQBhTCxAUf9xuPwVcM4gw55vbKKY?dl=1.

Chapter 6

Nuclear Fusion

INTRODUCTION

In simple terms nuclear fusion is a process in which one or more light nuclei fuse together to generate a relatively heavier nucleus in which in there is some mass deficiency that is released as energy, and the quantity of energy released follows Einstein's formula: $E = mc^2$, in which E is the energy in joules, m is the mass difference in kilograms, and c is the speed of light (approximately 300,000,000 or 3×10^8 m per second). If one were to look carefully at Fig. 5.1, one would notice that iron and nickel possess the highest binding energy whereas elements on either side have lower binding energy and will tend toward a more stable form. Nuclei with lesser mass will fuse together to generate stable configurations. Heavier nuclei may also fuse but these are astrophysical events that can lead to short periods of fusion, and this process gives rise to nucleosynthesis, which is the creation of heavy elements. This is not our concern in this chapter.

Let us consider a short history of nuclear fusion. The concept of quantum tunneling was first proposed by Friedrich Hund. Robert Atkinson and Fritz Houtermans in 1929 measured the masses of light elements to predict the considerable amount of energy that could be released during the fusion of light nuclei. Nuclear transmutation experiments were carried out earlier than this by Ernest Rutherford. Combing these concepts, the laboratory fusion of hydrogen isotopes was first achieved by Mark Oliphant in 1932. For nearly a decade, Hans Bethe worked out the cycle of nuclear fusion in stars. Essentially first nuclear fusion on a large scale was carried out on November, 1, 1952 in a hydrogen bomb test. This exercise is still continuing and various ways to contain the fusing nuclei in a specific volume are being attempted by various means even today.

The energy release in a fusion reaction is also the result of the interplay of the very short-range attractive force that keeps the protons and neutrons together in the nucleus and the Coulomb force that results in the repulsion of the protons. Because the nuclear forces are stronger than the Coulomb forces for atomic nuclei lighter than iron and nickel, building these nuclei from the

Energy Sources. http://dx.doi.org/10.1016/B978-0-444-56353-8.00006-X

lightest nuclei releases energy and is called the **fusion energy**. It is understood that a fusion reaction will be possible only when the nuclei fusing together are brought together in a confined volume (recollect the dimensions of a nucleus).

METHODS FOR CARRYING OUT FUSION

Thermonuclear fusion: When matter is heated to a high degree (similar to the plasma state of matter), fusion may take place as a result of collisions with high kinetic energy of the particles. However, harnessing fusion energy in a controlled fashion by this process has not yet been achieved, although the energy in stars is obtained only through this process.
Inertial confinement fusion: In this process the reaction is carried out by heating and simultaneously compressing a fuel target.
Inertial electrostatic confinement: This method employs a device with which electric field is used to heat and confine the nuclei. Various models of these have been tried since 1999.
Beam–beam or beam–target fusion: In this method the fusion reaction takes place by accelerating one or both the nuclei. However, the fusion cross-sections are several orders of magnitude lower than that of Coulomb interaction cross-sections. Therefore, a major fraction of the energy is dissipated out in Bremsstrahlung radiation or ionization of the atoms of the target.

For a reaction to be considered a potential fusion reaction, several criteria have to be fulfilled:

1. The reaction has to be exothermic. Upon inspection of the binding energy curve (Fig. 5.1), it is seen that the limits of the reactants are to low values of the Z (number of protons) side of the curve of binding energy. Naturally, ^{4}He, the nucleus with extraordinary tight binding, appears to be the predominant product.
2. The process involves low Z nuclei; it is known that electrostatic repulsion has to overcome before the nuclei can be brought close enough to fuse.
3. The process involves two reactants: three body collisions are mostly improbable except in stellar densities. In inertial confinement both densities and temperature are exceeded but the confinement times are very short.
4. The products are two or more species. This satisfies the conservation of energy and momentum without relying on electromagnetic force.
5. Proton and neutron conservation: This situation will harness required energy and also make appropriate cross-sections for the reaction.

Possible nuclear fusion reactions are:

$$^{2}D_{1} + {}^{3}T_{1} \rightarrow {}^{4}He_{2}\ (3.5\ MeV) + {}^{1}n_{0}\ (14.1\ MeV)$$
$$^{2}D_{1} + {}^{2}D_{1} \rightarrow {}^{3}T_{1}\ (1.01\ MeV) + {}^{1}H_{1}\ (3.02\ MeV)$$
$$^{2}D_{1} + {}^{2}D_{1} \rightarrow {}^{3}He_{2}\ (0.82\ MeV) + {}^{1}n_{0}\ (2.45\ MeV)$$
$$^{2}D_{1} + {}^{3}He_{2} \rightarrow {}^{4}He_{2}\ (3.6\ MeV) + {}^{1}H_{1}\ (14.7\ MeV)$$
$$^{3}T_{1} + {}^{3}T_{1} \rightarrow {}^{4}He_{2} + 2{}^{1}n_{0} + 11.3\ MeV$$
$$^{3}He_{2} + {}^{3}He_{2} \rightarrow {}^{4}He_{2} + 2{}^{1}P^{+} + 12.9\ MeV$$
$$^{3}He_{2} + {}^{3}T_{1} \rightarrow {}^{4}He_{2} + P^{+} + n_{0} + 12.1\ MeV$$
$$^{3}He_{2} + {}^{3}T_{1} \rightarrow {}^{4}He_{2}\ (4.8\ MeV) + {}^{2}D_{1}\ (9.5\ MeV)$$
$$^{2}D_{1} + {}^{6}Li_{3} \rightarrow 2{}^{4}He_{2} + 22.4\ MeV$$
$$^{2}D_{1} + {}^{6}Li_{3} \rightarrow {}^{3}He_{2} + {}^{4}He_{2} + {}^{1}n_{0} + 2.56\ MeV$$
$$^{2}D_{1} + {}^{6}Li_{3} \rightarrow {}^{7}Li_{3} + {}_{1}P^{+} + 5.0\ (MeV)$$
$$^{2}D_{1} + {}^{6}Li_{3} \rightarrow {}^{7}Be_{4} + {}^{1}n_{0} + 3.4\ MeV$$
$$^{1}p^{+} + {}^{6}Li^{3} \rightarrow {}^{4}He_{2}\ (1.7\ MeV) + {}^{3}He_{2}\ (2.3\ MeV)$$

Like nuclear fission reactions, fusion reactions are governed by parameters such as the reaction cross-section and energy distribution in the nuclei involved in the reaction. For a fusion reaction to occur, the Coulomb barrier has to be surmounted between two positively charge nuclei. This energy barrier is given by:

$$\left[\{Z_1 Z_2 e^2\}/[4\pi\varepsilon_0 R_0]\right]$$

where Z_1 and Z_2 are the atomic numbers of the two reacting nuclei and R_0 is the distance between two nuclei centers at which attractive forces become dominant. For a typical value of R_0 in the range of 5×10^{-15} m, the barrier to be surmounted is approximately $10^{-14}Z_1Z_2$. This may amount to 5×10^4 K (if the energy is set equal to $3/2kT$) required for the fusion of hydrogen isotopes to fuse. If fusion were to take place between two accelerated nuclei, the barrier to be surmounted may be smaller. The cross-section for the fusion as a function of total energy is given by:

$$\text{ie,}\quad \sigma(E) = \{\text{constant}/E\}\ \exp\left\{-\ 2^{3/2}\pi^2 M^{1/2} Z_1 Z_2 e_2 \Big/ 4\pi\varepsilon_0 E^{1/2}\right\}$$

where M is the reduced mass of the interacting nuclei. Because the temperatures involved are high and the reacting materials have to be in a state of plasma, the energy distribution of the particles can be described by a Maxwellian function in the following manner:

$$dn = N(E)d = \text{constant} \times \left(E^{1/2}\Big/T^{3/2}\right)\exp(-E/kT)dE$$

where T is the kinetic temperature related to the energy distribution among the particles. The probability function of particles in the energy range between E and $E + dE$ is given by:

$$P(E) = (dn/dE)\sigma(E)$$
$$= \left[C\Big/E^{1/2}T^{2/3}\right]\exp\left[-2^{3/2}\pi^2 M^{1/2} Z_1 Z_2 e^2\Big/\ 4\pi\varepsilon_0 hE^{1/2} - E/kT\right]$$

where C is a constant. It is therefore possible to compute both the probability function and $\sigma(E)$ as a function of E. These functional dependences obtained are shown in Fig. 6.1.

The reaction rate of a fusion reaction generally depends on:

1. the number densities (n_1 and n_2) of the two reacting species,
2. the reaction cross-sections of the respective species, and
3. the speed (ϑ) of the particles, because the reaction cross-section varies with energy.

Therefore the reaction rate $R_{12} = n_1 n_2 \sigma\vartheta$ reactions per unit volume per second.

The reaction cross-section is measurable quantities in the accelerator experiments and the data generated for typical deuterium–deuterium (D–D), D–tritium (D–T), and D–He^3 reactions are shown in Fig. 6.2. Similarly one can define the reaction mean free path as equal to $1/\sigma n$. The reaction free path depends not only on the number density but also on the average value of σ or the average kinetic temperature.

A fusion reaction can occur only if the following conditions are fulfilled:

1. The temperature of the reacting species should be high.
2. The plasma must be contained within a limited volume.

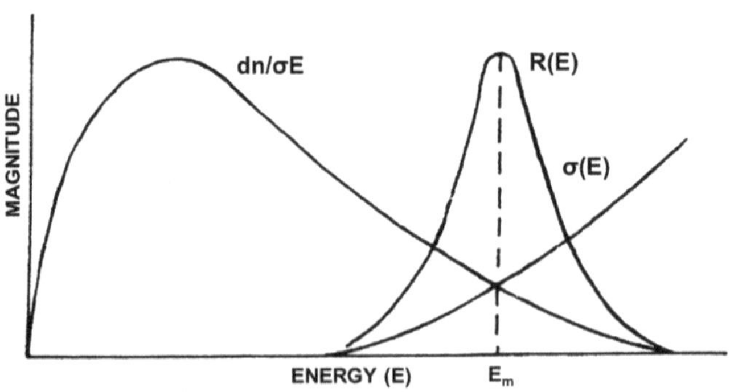

FIGURE 6.1 Maxwellian distribution fusion reaction cross-section and probability R of a fusion reaction.

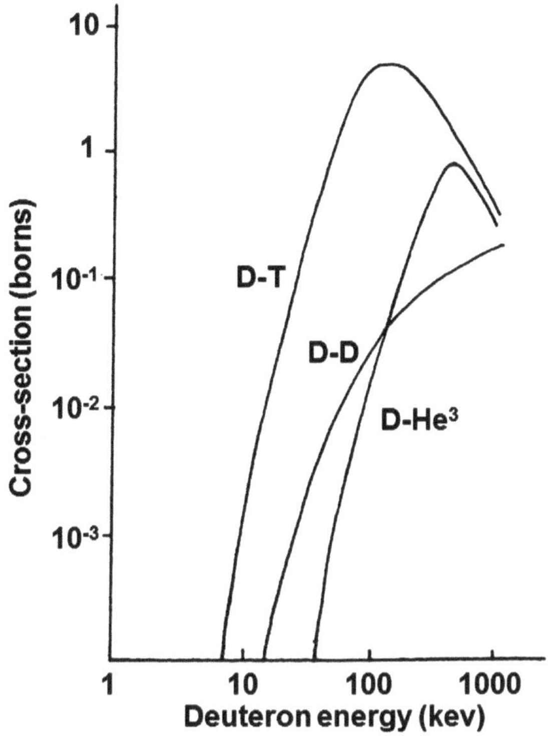

FIGURE 6.2 Variation with energy of fusion reaction cross-section.

The rate of energy release in a fusion reaction is given by:

$$R_{12} = n_1 n_2 \sigma\vartheta Q \text{ watts/unit volume}$$

where Q is the energy in joules released per fusion reaction. Let us consider a typical D–T reaction with $n_D = n_T = n$, $Q = 3.5\text{ MeV} = 1.6 \times 10^{-13}$ J, and a temperature as high as 10 keV with $\sigma\vartheta = 10^{-16}\text{ cm}^3/\text{s}$. Then the power density is given by:

$$R_{DT}\ (10\text{ keV}) = 6.2 \times 10^{-29} n^2\ \text{W/cm}^3$$

If one wants to operate at 100 W/cm^3, the number of nuclei of each kind required is of the order of 10^{15} to 10^{16} per cubic centimeter.

However there can be some competing reactions in which the energy will be lost:

1. Power loss caused by instability in the plasma. This loss results from a reduction in the temperature of the plasma or by the dissipation of plasma.
2. Ohmic losses in the magnetic field coils and associated circuitry.

3. Absorption or emission of radiation (Bremsstrahlung) as a result of excitation or deceleration of charges.
4. Electron–electron collisions as well as impurities present which may lead to loss due to radiation. This loss can be captured in the form of heat.
5. The charged particles in the magnetic field used for confinement move in curved paths and give rise to radiation. This radiation is in the infrared or microwave region of the electromagnetic radiation and is termed "cyclotron radiation." This loss has to be appropriately contained.

To harness fusion on earth, different, more efficient fusion reactions than those at work in the sun are considered.

For the successful operation of a fusion reactor, the temperature of the plasma should be as high as possible and the number density should also be high. This can be achieved only if the particles are confined in a limited volume. Confining the plasma and intermittently extracting power is difficult. The continuous mode of operation is still in development and only the pulsed mode of operation is a viable option. The number density for a D–D reaction needs to be greater than 10^{16} atoms s/cm^3, and for a D–T reaction this has to be greater than 10^{14} atoms s/cm^3. This is termed the Lawson criterion.

For power generation in the pulsed mode of operation the following are necessary conditions:

1. The temperature of the plasma must be above the ignition temperature, which is about 8×10^7 keV.
2. The confinement time must be such that $n\iota$ has a value of 10^{14} atoms s/cm^3. This will correspond to about 0.1 s.
3. The number density of the nuclei should be around 10^{15} cm^{-3}.
4. The plasma should not be made to come in contact with the walls of the confinement vessel. If this is not fulfilled the plasma will subjected to a cooling effect and to sputtering of the vessel, which will contaminate the plasma. This will cause power loss by Bremsstrahlung radiations.

Plasma confinement has to be carried out in magnetic fields. These involve the application of an extreme magnetic field, the stellerator, astron, the magnetic mirror, and the Tokomak mode of operation. Magnetic mirror systems have a stronger magnetic field at the ends compared with the center. Thus the magnetic field strength constrains the charged particles of plasma within a certain volume. In the Tokomak operation an external solenoid provides the confining magnetic field and the reaction vessel is designed in such a manner that the field is around it in the form of a torus.

COMPARISON OF ENERGIES RELEASED FROM VARIOUS PROCESSES

A fusion reaction can take place at a sufficient rate only at high temperatures greater than 100 million Kelvin. At this condition the reacting nuclei exists in a plasma state and this kind of energy is required for positively charged nuclei to overcome their natural charge repulsion.

	Chemical	Fission	Fusion
Reaction	$C + O_2 \rightarrow CO_2$	U-235	${}^2{}_1H + {}^2{}_1H + {}^3{}_2He + {}^1{}_0n$
Starting material	Coal	UO_2 ore	^2H and ^3H isotopes
Temperature needed (K)	700	1000	$1 \times e^{+8}$
Energy J/kg of fuel	33 MJ	2000 GJ	3,400,000 GJ

CONDITIONS FOR A FUSION REACTION

Successful fusion requires certain conditions to be fulfilled; these will be briefly considered. Statements have already been made about some of these conditions. Three parameters (plasma temperature, density, and confinement time) need to be achieved simultaneously for sustained fusion to occur in plasma. The product of these is called a fusion (or triple) product. For D—T fusion to occur, this product has to exceed a certain quantity, derived from the so-called **Lawson Criterion**, named for British scientist John Lawson, who formulated it in 1955. These criteria have been already stated: that is, for a fusion reaction to occur, $n\iota \geq 10^{14}$ s/cm^3 for a D—T reaction, and $n\iota \geq 10^{16}$ s/cm^3 for a D—D fusion reaction. The closest approach to Lawson's criterion has been achieved at the Tokamak Fusion Test Reactor at Princeton. It reached ignition temperature and got close to Lawson's criterion, although not at the same time.

Attaining conditions to satisfy the Lawson criterion ensures that the plasma exceeds the **break-even** point at which the fusion power exceeds the power required to heat and sustain the plasma.

TEMPERATURE

Fusion reactions occur at a sufficient rate only at very high temperatures, at which the positively charged plasma can overcome their natural repulsive forces. Typically, over 100 million Kelvin is needed for the D—T reaction to occur; other fusion reactions (eg, D—D, D—He^3) require even higher temperatures.

DENSITY

The density of fuel ions (the number per cubic meter) must be sufficiently large for fusion reactions to take place at the required rate. The fusion power generated is reduced if the fuel is diluted by impure atoms or by the accumulation of helium ions from the fusion reaction itself. As fuel ions are burned in the fusion process they must be replaced by new fuel and the helium products must be removed.

ENERGY CONFINEMENT

The energy confinement time is a measure of how long the energy in the plasma is retained before being lost. It is officially defined as the ratio of the thermal energy contained in the plasma and the power input required to maintain these conditions. Magnetic fields are used to isolate the very hot plasma from the relatively cold vessel walls to retain the energy for as long as possible. Losses in magnetically confined plasma are mainly the result of radiation. The confinement time increases dramatically with plasma size (large volumes retain heat better than do small volumes). The ultimate example is the sun, whose energy confinement time is massive.

For sustained fusion to occur, the following plasma conditions need to be maintained (simultaneously): plasma temperature: (T) 100–200 million Kelvin; energy confinement time: (t) 1–2 s; central density in plasma: (n) 2–3 $\times$ 10^{20} particles/m^3 (approximately 1/1000 g/m^3).

MAGNETIC PLASMA CONFINEMENT

Because plasma is composed of charged particles, ions (positive) and electrons (negative), powerful magnetic fields can be used to isolate the plasma from the walls of the containment vessel, thus enabling the plasma to be heated to temperatures in excess of 100 million Kelvin. This isolation of the plasma reduces the conductive heat loss through the vessel and minimizes the release of impurities from the vessel walls into the plasma that would contaminate and further cool the plasma by radiation.

In a magnetic field, charged plasma particles are forced to spiral along the magnetic field lines. The most promising magnetic confinement systems are toroidal (from "torus" = ring-shaped); of these, the most advanced is the Tokamak.

Other, nonmagnetic plasma confinement systems are being investigated, notably inertial confinement or laser-induced fusion systems.

The plasma is heated in a ring-shaped vessel (or torus) and kept away from the vessel walls by the applied magnetic fields. Basic components of magnetic confinement system include the toroidal field, which produces a field around the torus. This is maintained by magnetic field coils surrounding the vacuum vessel. The toroidal field provides the primary mechanism of confinement of the plasma particles. The poloidal field produces a field around the plasma cross-section. It pinches the plasma away from the walls and maintains the plasma's shape and stability. The poloidal field is induced both internally by the current driven in the plasma (one of the plasma-heating mechanisms) and externally by coils that are positioned around the perimeter of the vessel.

The main plasma current is induced in the plasma by the action of a transformer. A changing current in the primary winding or solenoid (a multiturn coil wound onto a large iron core) induces a current [up to 5 million amperes (5 MA)] in the plasma, which acts as the transformer secondary circuit.

A main requirement for fusion is to heat the plasma particles to very high temperatures or energies. The next section describes methods typically used to heat the plasma.

PRINCIPLE METHODS OF HEATING PLASMA: OHMIC HEATING AND CURRENT DRIVE

Currents up to 5 MA are induced in the plasma, typically via the transformer or solenoid. In addition to providing natural pinching of the plasma column away from the walls, the current inherently heats the plasma by energizing plasma electrons and ions in a particular toroidal direction. A few megawatts of heating power are provided in this way.

NEUTRAL BEAM HEATING

Beams of high-energy neutral deuterium or tritium atoms are injected into the plasma, transferring their energy to the plasma via collisions with the plasma ions. The neutral beams are produced in two distinct phases. First, a beam of energetic ions is produced by the application of an accelerating voltage of up to 140,000 V. However, a beam of charged ions will not be able to penetrate the confining magnetic field in the Tokamak. Thus, the second stage ensures the accelerated beams are neutralized (ie, the ions are turned into neutral atoms) before injection into the plasma.

RADIO-FREQUENCY HEATING

As the plasma ions and electrons are confined to rotation around the magnetic field lines in the Tokamak, electromagnetic waves of a frequency matched to the ions or electrons are able to resonate or damp its wave power into the plasma particles. As energy is transferred to the plasma at the precise location where the radio waves resonate with the ion–electron rotation, such a wave-heating scheme has the advantage of being localized at a particular location in the plasma.

A number of antennae in the vacuum vessel propagate waves in the frequency range of 25–55 MHz into the core of the plasma. These waves are tuned to resonate with particular ions in the plasma, thus heating them. This method can inject up to 20 MW of heating power.

Waves can also be used to drive current in the plasma by providing a "push" to electrons traveling in one particular direction. Ten megawatts of these so-called lower hybrid microwaves (at 3.7 GHz) accelerate the plasma electrons to generate a plasma current of up to 3 MA in a typical unit.

SELF-HEATING OF PLASMA

Helium ions (or so-called alpha particles) produced when deuterium and tritium fuse remain within the plasma's magnetic trap for a time before they are pumped away through the diverter. The neutrons (which are of course neutral) escape the magnetic field and their capture in a future fusion power plant will be the source of fusion power to produce electricity.

Fusion energy contained within the helium ions heats the deuterium and tritium fuel ions (by collisions) to keep the fusion reaction going. When this self-heating mechanism is sufficient to maintain the required plasma temperature for fusion, the reaction becomes self-sustaining (ie, no external plasma heating is required). This condition is referred to as ignition.

MEASURING THE PLASMA

Measuring the key plasma properties is one of the most challenging aspects of fusion research. Knowledge of the important plasma parameters (temperature, density, and radiation losses) is very important for understanding plasma behavior and design with confidence. However, because the plasma is contained in a vacuum vessel and its properties are extreme (extremely low density and extremely high temperature), conventional methods of measurement are not appropriate. Thus, plasma diagnostics are normally innovative and often measure a physical process from which information about a particular parameter can be deduced or postulated.

Measurement techniques can be categorized as active or passive. In active plasma diagnostics, the plasma is probed (via laser beams or microwaves) to see how the plasma responds. For instance, in interferometers, the passage of a microwave beam through the plasma will be slowed by the presence of the plasma (compared with passage through vacuum). This measures the refractive index of the plasma from which the density of plasma ions and electrons can be interpreted. With all active diagnostics, it must be ensured that the probing mechanism does not significantly affect the behavior of the plasma.

With passive plasma diagnostics, radiation and particles leaving the plasma are measured; this knowledge is used to deduce how the plasma behaves under certain conditions. For instance, during D–T operation, neutron detectors measure the flux of neutrons emitted from the plasma. All wavelengths of radiated waves (visible, UV waves, X-rays, etc.) are also measured, often from many locations in the plasma. Then detailed knowledge of the process that created the waves can enable a key plasma parameter to be deduced.

Other aspects of plasma, such as its stability and the methods to confine it, are not considered in this presentation.

COLD FUSION

Another exciting development in this field was introduced in 1989 by Stanley Pons and Martin Fleichmann, which they termed "cold fusion." We have already seen that a normal fusion reaction can take place at high temperatures, but they harnessed excess heat energy at around room temperature by confining D—D or D—T in Pd—Ti electrodes. Atomic confinement fusing nuclei were supposed to have occurred while electrolysis was generating hydrogen atoms (here deuterium) from electrolysis of heavy water. **Cold fusion** is therefore a hypothetical type of nuclear reaction (?) that might occur at or near room temperature if the nuclei undergoing fusion are confined in a small volume, especially in so-called hydrogen-absorbing materials such as Pd and other intermetallics such as FeTi, $LaNi_5$, and Mg_2Ni. However, in real plasma where the hot fusion takes place, temperatures on the order of millions of degrees are obtained, and hence so-called "hot" **fusion** takes place naturally within stars.

Cold fusion, possibly a revolutionary source of energy, postulated about 25 years ago was cold-shouldered by mainstream scientists throughout that period. The concept is currently staging a comeback thanks to researchers who doggedly pursued the science behind it all these years. There is optimism in the statement that experimental evidence shows that "Low Energy Nuclear Reaction (LENR) can have large energy gains, but the engineering of commercial prototypes is still relatively crude" and possibly elusive, according to David Nagel of George Washington University.

Perhaps it is too soon to bring this exciting development to a textbook level of knowledge.

SUGGESTED READING

[1] https://en.wikipedia.org/wiki/Cold_fusion.
[2] http://zeenews.india.com/news/sci-tech/cold-fusion-is-real-claim-scientists_1559235.html.
[3] http://undsci.berkeley.edu/article/cold_fusion_01.
[4] http://www.extremetech.com/extreme/191754-cold-fusion-reactor-verified-by-third-party-researchers-seems-to-have-1-million-times-the-energy-density-of-gasoline.
[5] https://en.wikipedia.org/wiki/Nuclear_fusion.
[6] H.J. Arnikar, Essentials of Nuclear Chemistry, fourth ed., New Age International (P) Limited, Publishers, 1995.
[7] J.K. Long, G. Dawson, Chemistry of Nuclear Power, 1949.
[8] R.I. Murry, Nuclear Energy. Pergamon Press Inc., 1975.
[9] http://hyperphysics.phy-astr.gsu.edu/HBASE/nucene/fusion.htm/.
[10] http://fusedweb.pppl.gov.

Chapter 7

Solar Energy: Fundamentals

Solar energy is radiant light and heat received from the sun harnessed on earth using technologies such as solar heating, photovoltaics (PV), solar thermal energy, and photosynthesis. It is philosophically considered that the energy for life on earth originates from the sun in some form or another. It is recognized today that it is the most important renewable energy source. Energy from the sun is usually harnessed on earth employing active devices such as PV cells and passive methodologies such as appropriate constructions to absorb solar radiation. It is recognized that "the development of affordable, inexhaustible and clean solar energy technologies will have huge long-term benefits. It will increase countries' energy security through reliance on an indigenous, inexhaustible and mostly import independent resource, enhance sustainability, reduce pollution lower the cost of mitigating so-called global warming" and possibly control the price of fossil fuels. In Fig. 7.1 the solar spectra in the outer atmosphere and on the earth's surface are shown compared with black body radiation.

Corresponding data at various wavelengths in terms of the percent contribution are given in Table 7.1.

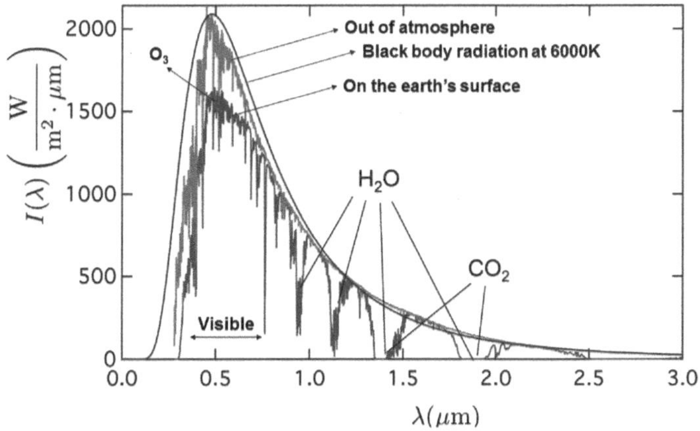

FIGURE 7.1 Spectrum of solar radiation solar energy density at each wavelength.

Energy Sources. http://dx.doi.org/10.1016/B978-0-444-56353-8.00007-1

TABLE 7.1 Energy Ratio of Each Wavelength Range (Color Ranges) in the Solar Spectrum

Wavelength (nm)	% Total Solar Spectrum	Spectral Range
0–310	0.019	Far UV
310–325	0.21	Far UV
325–350	1.0	Far UV
350–400	3.8	UV
400–760	43.9	Visible
760–	52.3	Infrared to far infrared

UV, ultra violet.

It is customary to start with a profound statement about the extent of solar energy available on earth. The statement is usually that "the amount of energy reaching the surface of the earth every hour is greater than the amount of energy used by the earth's population over an entire year." This implies that the energy scarcity we are often shouting about is only an illusion. If only this energy source could be harnessed and stored, humanity would not need to be worried about the shortage of energy sources or to fight for its share of energy sources.

There are various estimates of the amount of solar radiation received by earth. One document shows that the earth receives 174,000 terawatts (TW) of incoming solar radiation at the upper atmosphere. About 30% is reflected back to space while the rest is absorbed by clouds, oceans, and landmasses. The spectrum of solar light at the earth's surface is mostly spread in the visible and near-infrared ranges with a small part in the near-UV. The actual distribution is shown in Fig. 7.1. The radiation in solar spectrum results from thermonuclear reactions occurring in the sun. Every second, the sun converts about 650,000,000 tons of hydrogen to helium. It has been estimated that the sun has already used approximately 50% of its hydrogen source. This thermonuclear reaction generates heat and electromagnetic radiation. Heat is used to maintain the thermonuclear reaction. Electromagnetic radiation in the form of photons (mostly 45% visible light, 52% infrared light, and a small fraction of UV radiation) (Table 7.1) streams out into space in all directions. A typical flowchart for the energy distribution of solar radiation is given in Scheme 7.1.

The data shown in Table 7.1 indicate that more than half of solar radiation is in the infrared or far-infrared region, and hence it is necessary to use this heat component of the freely available radiation. When the higher energy component of solar radiation that is visible and UV radiation is used, as in photosynthesis and other photochemical routes for energy storage, the process

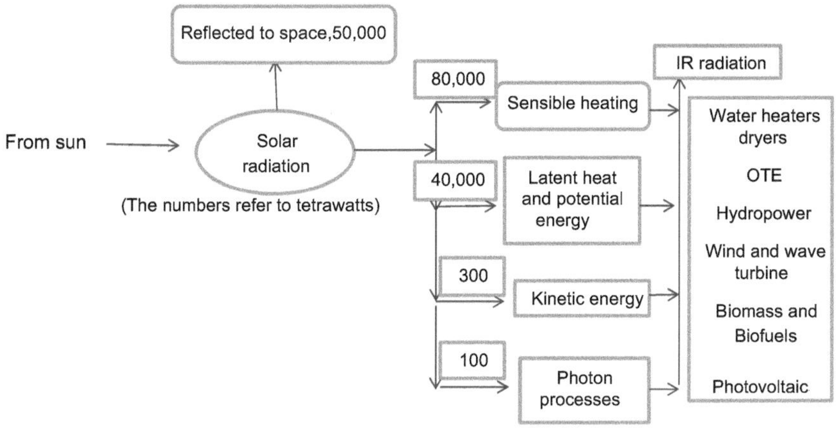

SCHEME 7.1 Flowchart of solar radiation.

is governed by the rules of photon energy and not the heat energy. This imposes a condition under which one should be capable of harnessing the heat energy component from solar radiation for effective use, and hence one has to employ appropriate solar collectors. There are optimized solar collectors with respect to design (parabolic, flat plate, or other configurations such as Fresnel lens and optical fiber tube leads), but these are not considered within the scope of this exercise. Interested readers can look up appropriate resource material on these topics.

After an appropriate projection for the increase in world population and after suitably normalizing for the gross domestic product (GDP) and energy consumption rate per unit of GDP, it is projected that the world energy consumption rate will more than double from 13.5 TW in 2001 to 27.6 TW by 2050, and then more than triple to 43.0 TW by 2100. These estimates have uncertainties but they represent a reasonable and reliable starting point for assessing the magnitude of other issues that follow this rate of energy consumption. Under this scene, the world carbon emission intensity may work out to be approximately 0.49 kg C/(W year) in 2001, which will have values of 0.4 by 2050 and 0.31 kg C/(W year) by 2100. These numbers will work out to be 6.6 Gt C/year in 2001, and 11 and 13.3 Gt C/year in 2050 and 2100, respectively. These are only projections and have built-in uncertainties. These situations can have adverse effects on atmospheric conditions. The so-called greenhouse effect and carbon dioxide concentrations in the atmosphere have increased from before1750 at 280 parts per million (ppm) to 371 ppm in 2001; they has now reached a value around 395 ppm, which near the limit. To control carbon dioxide emissions, one has to resort to C-neutral energy sources. In other words, any further delay in the development of C-neutral energy sources will only impose more C-neutral energy sources to offset the CO_2 emission in the interim because of the delay in introducing in C-Neutral energy source.

Almost all renewable energy sources considered, such as, hydropower, ocean, wind, geothermal, and solar, are C-neutral sources. These sources have differing potential to supply the projected power of about 15 TW. Among these, solar, wind, geothermal, and ocean waves have the theoretical potential to provide more than about 15 TW, and among these only two sources are amenable for extracting this kind of power: solar and wind. Of these two, only solar has the technical potential to safely exceed 15 TW. Order of magnitude estimates of some of the renewable energy sources are given in Table 7.2.

Among renewable energy sources, hydropower and solar fuels supply the most power followed by wind and geothermal. At this stage, it is necessary to evaluate the theoretical and harnessable potential of solar energy. The solar constant (the solar flux intercepted by the earth) is 1.37 kW/m^2. The time- and space-averaged solar flux striking the outer atmosphere of earth is 342.5 W/m^2. If one were to take into account the amount of flux that is scattered (30%) and is absorbed by clouds and atmosphere (19%), the average flux striking the earth surface is 174.7 W/m^2. This figure is averaged accounting for the different regions receiving different fluxes. Thus, the total solar power over the total area of the earth is $P = (174.7\ \text{W/m}^2) \times (4\pi r^2) = 89{,}300$ TW. If the required power is 15 TW, one can harness this energy even using a 10% efficient device using only 0.17% of the earth's surface (this is approximately

TABLE 7.2 Order of Magnitude Estimates of World Potentials of Renewable Energy Sources

	World			
Energy Source	Theoretical Potential (TW)	Extractable Potential ($TW_{chemical}$)	Technical Potential ($TW_{chemical}$)	2001 Supply ($TW_{chemical}$)
Hydropower	12 TW_m	3.5	1.2	0.23
Ocean tidal	2.4 TW_m	0.60	0.037	5×10^{-5}
Wind	1000 TW_m	250	14	5×10^{-3}
Geothermal	44 TW_t	2.8	1.9	5×10^{-3}
Solar electricity	89,000 TW_p	58,000	7500	15×10^{-5}
Solar fuels	89,000 TW_p	61,000	2500	0.19
Solar thermal	89,000 TW_p	19,000	5600	6×10^{-4}

Subscripts denote chemical, electrical, mechanical, photonic, and thermal.
Reproduced from working draft version, 2006 April 20, Solar FAQs.

the area of Venezuela). Of the 89,300 TW, how much power is extractable in the three forms: electricity, fuel, and heat? The conversion of solar power to electricity depends on technology, which depends on the device's efficiency (0.87 TW_c/TW_P) and also subsequent conversion efficiency (0.75 TW_c/TW_P), which works out to be 58,300 TW_c.

For the generation of solar fuels this figure will be a fraction (0.68 TW_c/TW_P) of the theoretical potential, and hence the potential for solar fuel conversion works out to be 60,700 TW_c. For the thermal route, the initial conversion has to be from heat to mechanical energy (87%) and for a subsequent mechanical to electrical to chemical conversion, this works out to be 89,300 TW × (0.87 TW_m/TW_P) × (0.25 TW_c/TW_m) = 19,400 TW_c. These figures are obtained assuming solar power is concentrated and the devices are working in good stipulated efficiencies. Of this extractable solar energy, one cannot do it effectively over the oceans (0.708) or in "frigid" zones (0.0345). Furthermore, current technologies are less efficient than thermodynamically limiting efficiencies. For example, a PV device that converts solar energy into electricity is about 40% and the subsequent conversion to fuel is only 75% efficient and hence the net efficiency will be only 30%. In short, the technical potential of solar energy can be summarized as:

Solar electricity	89,000 TW_C × 0.292 × (1−0.0345) × 0.3	~7500 TW_C
Solar fuels	89,000 TW_C × (1−0.708) × (1−0.0345) × 0.1	~2500 TW_C
Solar thermal	89,000 TW_C × (1−0.708) × (1−0.0345) × (0.225)	~5600 TW_C

A solar cell's (PV cell's) installed capacity is around 1.00 GW_{peak}. Of this, only one-fifth electrical power is obtained because this quantity represents the peak power harnessed. This works out to be 15×10^{-4}% of the world's electrical power requirement. Solar fuel in the form of biomass conversion works out to be 1.43 TW, but taking into account other ecosystem effects, the sustainable quantity works out to be 0.2 TW, which is about 1.4% of the world's power requirement. Solar thermal considering only the "active" systems supplies nearly 6.4 GW_t and another 0.1 GW_e is obtained from thermal to electricity-generating systems. This works out to be roughly 67×10^{-4} TW or 0.05% of the world's power. The numerical values given are subject to gross errors; these can be taken to be representative and actual figures may differ considerably from the values given. If all three forms of solar energy are put together, one can roughly assume that the harnessed amount can be on the order of 0.2 TW_c or 1.5% of the world's power. There is great efforts to make all PV systems considerably more efficient than they currently are and thus facilitate harnessing solar power to better satisfy better the world's power requirement. This will be taken up in the chapter on PV.

Solar energy has some advantages, the main one being that all chemical and radioactive polluting by-products of so-called thermonuclear reactions are retained in the sun itself and only pure radiant energy is transmitted to the earth. Under these conditions, solar energy can be considered nonpolluting. We have already seen that the total amount of solar energy reaching the earth is enormous. There are various estimates of this in quantitative terms. If one can harness 30 days of sunshine striking the earth, the energy equivalent is the total of all of the earth's fossil fuels, both used and in reserve. There is another way of comprehending the energy received by earth from solar radiation. If one could harness the solar energy for an hour, this will be on the order of the energy requirement of the earth for a year. Although these positive aspects of solar energy are encouraging, one should also keep in mind that solar energy can be inconsistent and diffuse in nature and hence requires concentration and storage in the form of heat and electricity.

The dynamics of incoming solar radiation can be rationalized as a fraction in backscattered and reflected components (31%); an absorption component by water vapor, dust, and ozone (19%); and by clouds (4%), and the net 46% only reaches the earth. Only this amount of energy is usable. This energy is now appears as long-wavelength radiation and sensible and latent heat fluxes. This situation is represented in pictorial form in Fig. 7.2.

One effective way of using solar energy has been to harness it in the other forms such as fuels, chemicals, and electricity. These solar fuel generator devices require electrical asymmetry to effectively generate, separate, and transport (possibly vectorially so that the desired selective chemical is formed) charge carriers generated by photon absorption. If these conditions are not satisfied, either the deleterious recombination of charge carrier will take place or a loss of chemical potential will result in undesired fuel−oxidant mixture. Essentially, this vectorial transport of the charge carriers is effected by formulating chemical or potential gradients that will naturally present at an

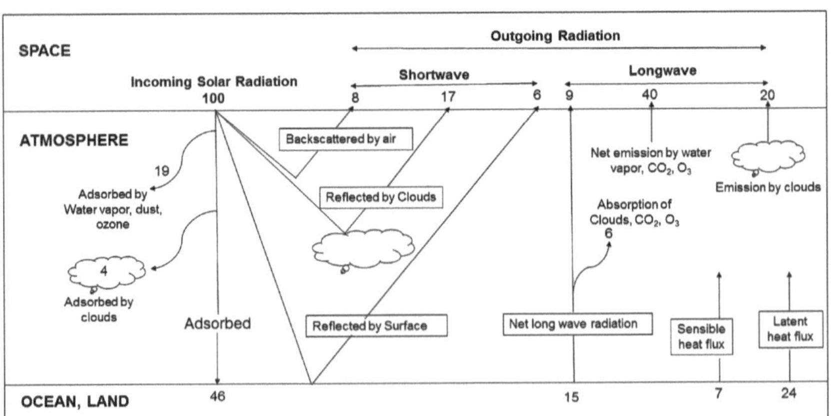

FIGURE 7.2 Dynamics of incoming total solar radiation and how are they distributed. *Reproduced from http://zebu.uoregon.edu/disted/ph162/l4.html.*

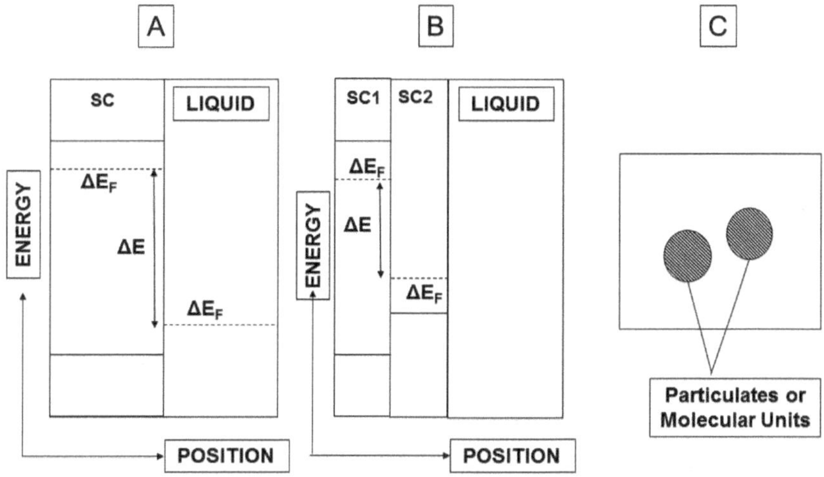

FIGURE 7.3 Three typical types of fuel generators from solar energy.

interface. The desired directional formation of the required storage of energy in the form of chemicals can also be created by the kinetic asymmetries that can occur at the interfaces (Figs. 7.3 and 7.4).

The three types of solar fuel generators: (Fig 7.3A) Semiconductor (SC)/electrolyte. The photovoltage and photocurrent depend on the Fermi level (E_F) of the semiconductor and the electrochemical potential of the electrolyte (E_{Redox}) denoted and ΔE in the figure. Two semiconductors, one buried into another in this case, the photopotential and photocurrent, depend on the difference between the Fermi levels of semiconductor 1 and that of semiconductor 2 (Fig 7.3B). Particulate photocatalysts suspended in solution.

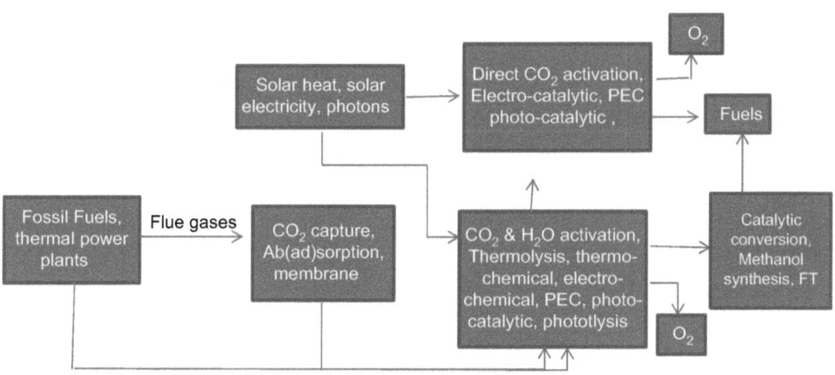

FIGURE 7.4 Simplified flowchart of solar refinery. *Adopted from J.A. Herron, et al., Energy and Environmental Science, 2015, pp. 8126–8157.*

In this case each unit individually absorbs light and facilitates the charge transfer process at the interface (Fig 7.3C).

The concept of solar refinery has been proposed. Although this concept is in the conceptual stage involving some developing areas such as the photoelectrochemical cell (PEC), which has not yet reached the desired level of efficiency despite persistent efforts, and other feasible concepts of thermolysis, thermochemical, and photocatalytic steps, all of them have yet to reach the maturity level of modern oil refineries in the world. It is true that it will take some time for any evolving technology to reach the maturity level of any existing technology.

In a typical solar refinery, solar radiation is used to derive chemical reactions in photocatalytic, PEC, and photolysis modes, which are totally new for a refinery concept, and also for solar heating and solar electricity generation. In addition, solar utilities can be used for process heating and process electricity. Solar to thermal energy is well-known for a number of applications and widely used. There are two main methodologies for converting solar energy to electricity: PV and solar thermal power. PV technology will

TABLE 7.3 Evaluation of Different Carbon Dioxide Capture Modes

Method/ Extent	Material	Conditions Pressure/ Temperature	Advantages	Disadvantages
Absorption (~>90)	Monoethanol amine	1/40–60	Known process, viable low partial pressures	High cost
				Degrading scale-up difficulty
Adsorption (~90%)	Zeolite metal-organic frameworks, Supported amines metal oxide	0.1–1/0–100 1/<100 0.05 –1/50–75 1/600–700	Mostly high surface area, stable, tunable	Impurity effect thermal degradation, stability
Membranes (~85%)	Polymer	15–20/100	No reagents, simple	High pressures, low-purity product, thermal degradation
Cryogenic (>95%)			No reagents	High capital cost
				High energy cost

be dealt with in a separate chapter. This is a device in which solar energy is used to generate direct-current electricity. Briefly, PV cells can be broadly classified as:

1. crystalline single crystal silicon solar cells
2. single-junction GaAs cells
3. multijunction cells
4. thin-film technologies
5. evolving PV technologies

Details of these technologies will be dealt with separately.

Solar thermal power (also referred to as concentrated solar power) was used to drive a heat engine at high temperatures to generate electricity. Because high temperatures are involved, the systems have to be large and hence there are higher investment costs.

The solar refinery feedstock is CO_2. Different methods are employed for capturing CO_2 from the diffuse levels (about 400 ppm) in the atmosphere (Table 7.3).

SUGGESTED READING

[1] E.W. Brown, An Introduction to Solar Energy. http://www.ccs.neu.edu/home/feneric/solar.html.

[2] Solar Energy. A Reference Guide for Users John Maina and Peter Gichohi. http://my.ewb-usa.org/theme/library/myewb-usa/project-resources/technical/solarenergyareferenceguide.pdf.

[3] http://www.sandia.gov/~jytsao/Solar%20FAQs.pdf.

[4] Basic Needs in Solar Energy Utilisation Report (BES 2005). http://science.energy.gov/~/media/bes/pdf/reports/files/seu_rpt.pdf.

Chapter 8

Photovoltaic Systems

There have been many attempts to meet the energy requirement to raise it from the current 13 terawatts (TW) per year (TWyr) to as great as 30 TWyr by the middle of this century. In this connection, roughly 125,000 TW of solar energy is striking the earth at any time, and hence solar energy may be the only possible renewable energy source that can meet a large fraction of the future needs of earth. Various options are available for humanity to exploit solar energy, among which photosynthetic solar energy conversion for biofuel production is one that has received consideration as a possible alternative. Solar thermal systems convert solar radiation to thermal energy for heating applications or in conjunction with conversion to electricity. Photovoltaic (PV) systems convert solar energy directly to electricity without the interface of a heat engine. PV systems are generally considered to be rugged and simple in design with few maintenance needs.

A PV power system consists of multiple components such as a solar cell, mechanical and electrical connections, and mounting and devices for regulation and converting electrical output from direct to alternate current for connectivity to the grid system. Strictly speaking, a solar array mainly consists of an assembly of solar panels; this does not include other hardware components often referred to as balance of system (BOS). As of 2015, the global market reached 200 GW about 40 times the installed capacity in 2006, thus showing that PV systems are already accounting for 1% of worldwide electricity generation. It is expected that about 16% of the world's electricity demand by 2050 will be met by PVs whereas about 11% of the global demand will be contributed by the solar thermal conversion process. PV power generation is already exploited extensively in many parts in the world including China, Japan, and the United States, and Europe, Germany, and Italy already contribute nearly 8% of their domestic electricity demand by the PV mode of electricity generation. There are various estimates reported in literature regarding the cost of electricity generation by PVs and its economic comparison with other sources of energy; this will not be considered in this chapter. Interested readers can refer to the original literature on this topic [1−12].

The main component of a PV system is the so-called solar array or solar panel. Other components are generally termed BOS. The term originates from the fact that these components balances the power-generating subsystem of the PV system with the power-using system, namely the load. BOS may include a

Energy Sources. http://dx.doi.org/10.1016/B978-0-444-56353-8.00008-3

maximum power point tracker, storage system, solar tracker, sensor to measure solar irradiance, and task-specific accessories to meet the demands of the owner of the system.

A solar cell is generally a semiconductor device that converts photons from the sun into electricity. The central component of a solar cell is a light-absorbing crystalline material, usually crystalline silicon (C-Si). However, other semiconducting materials have been employed and the current favorites are multi-crystalline silicon (~53%), Single-crystalline silicon (~38.5%), amorphous silicon (~5%), silicon ribbon (~3%), CdTe (~1.5%), and copper indium diselenide (~0.2%). The figures in parentheses are the current extent of exploitation [13]. As a result of photon absorption by the semiconducting material, the charge carriers generated are separated and collected at the contacts. There are essentially two modes of charge carrier separation: drift of carriers mainly driven by the built-in electrostatic field of a p–n junction and diffusion of carriers from regions of high to low concentration mainly promoted by the electrochemical potential gradient. When drift is the mode of charge transport, the structure of the cell and the depletion width will be used to compensate for poor diffusion. When diffusion is the predominant mode of transport, the material quality becomes important. This will be important in p–n junction devices and also for dye or organic solar cells (Fig. 8.1).

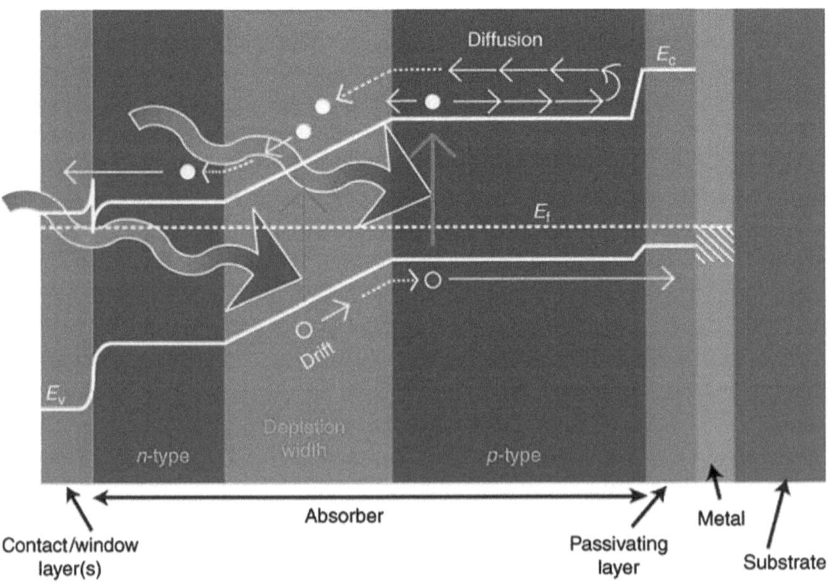

FIGURE 8.1 Representation of a p–n junction solar cell depicting the structure, carrier generation, and separation. *Arrows* indicate photons; E_v, E_c, and E_F are the valence band, conduction band, and Fermi energy, respectively. *Reproduced from A. Slaoui, R.T. Collins, MRS Bull. 32 (2007) 211.*

TABLE 8.1 Essential Properties of Four Typical Solar Cell Materials

Property	Perovskites	Silicon	CIGS	GaAs
Band gap eV	1.5 (tunable)	1.1	1.12	1.43
Absorption coefficient	10^{4-5}	10^3	10^{4-5}	10^{4-5}
Carrier mobility $cm^2/(V \cdot s)$	Up to 2000	1500	<10	8500
Carrier lifetime	>100 ns	ms	5–200 ns	<100 ns

CIGS, copper indium gallium selenide.

All PV devices require a light-absorbing material within the cell structure and produce electrons and holes as a result of photon absorption from the sun. Photon energy is imparted to the charge carriers generated, thereby raising their energy level and even freeing them. The built-in-potential in the cell manifests as voltage, providing current to flow in the circuit. The photon-absorbing materials have seen many developments. It is generally believed that silicon technology is the dominant one for the manufacture of PV devices. In Table 8.1 the essential characteristics of four typical solar cell materials are given for comparison.

The first-generation solar cells used mainly silicon but their best power conversion efficiency were only around 24%. Semiconductors employing III-V group elements are considered better because they are better absorbers of solar radiation than are those using silicon. Because these materials can be stacked into layers, they provide better a match between the absorber and solar spectrum and hence the loss of sun light as heat is reduced, whereas in silicon solar cells the loss of solar radiation as heat can be as high as 80–85%. A second advantage of these materials is that they can harness different parts (wavelengths) of the solar spectrum, thus accounting for an overall increase in efficiency. Nanomaterials and quantum dots are other classes of materials that have evolved in which the optical and electrical properties of these materials are being exploited [14].

Let us summarize at this stage the advantages of solar cells:

1. Solar energy is clean and inexhaustible whereas conventional energy sources will be exhausted or depleted within the next few decades.
2. Harvesting of solar energy does not involve exploration, which will have implications for the ecosystem.
3. Solar cells do not require periodic maintenance and can function for decades because no moving parts are involved.
4. Although solar cell materials are costly at present, it is expected that extensive deployment may reduce initial installation costs considerably.

5. Solar cells can be installed in remote places where the demand for electricity exists, thus avoiding the cost of transportation and other attendant losses.
6. Solar cells do not produce greenhouse gases and thus avoid global warming and other attendant changes such as climate change, the rise in sea level, and thinning of the ozone layer.
7. Charging of low-power gadgets such as calculators and mobile devices can be done easily and effectively.

Despite these advantages, solar cells have some specific disadvantages:

1. The main limitation is that solar radiation is available only a fraction of the day (daytime only) and that, too, depends on regional, seasonal, and daily fluctuations.
2. The land area required to harness solar energy is high and also depends on the intensity of available solar radiation.
3. There must be concurrent storage devices such as batteries to use electricity during the periods (typically nighttime) when solar radiation is not immediately available. These devices occupy space and require maintenance and periodic replacement.

Two new classes of solar cells have been introduced and have received considerable attention for various reasons including the efficiency that could be achieved:

1. dye-sensitized solar cells (DSSC)
2. perovskite-based solar cells

DYE-SENSITIZED SOLAR CELLS

It will be appropriate if some aspects of these solar cells are considered, because in the near future these two solar cells will be receiving considerable attention.

The basic operation principles of any solar cell consist of the absorption of radiation, charge carrier separation, and collection of these charge carriers at the appropriate carriers so that electricity can be used. It is necessary to optimize these three parameters to increase the efficiency of the device. In the case of DSSC [15−19], optimization of the first parameter is focused upon. Under exposure to radiation, the dye that performs as a sensitizer absorbs the photon, resulting in an excited state of the dye (D^*). The photoexcited dye facilitates the electron injection in the conduction band of the semiconductor, thus making the sensitizer reach an oxidized state, D^+. The oxidized species can return to the original state (D) by an electron donated from a redox species in the electrolyte. The usual couple used is iodide/triiodide $\left(I^-/I_3^-\right)$. Iodide regenerates the sensitizer and is regenerated by the reduction of

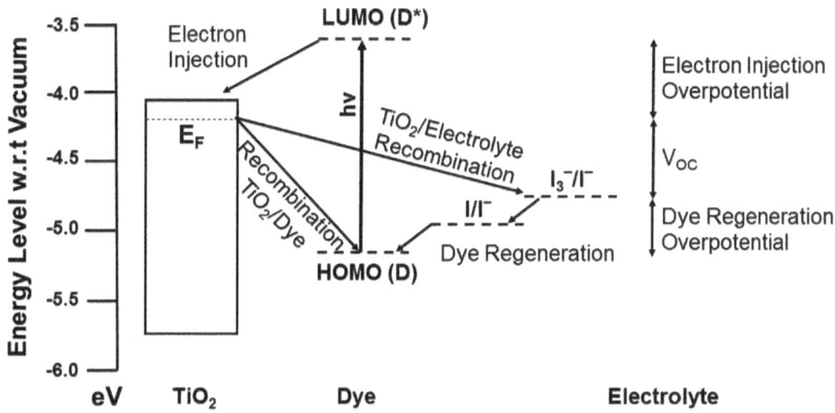

FIGURE 8.2 Schematic of a dye-sensitized solar cell device. Energy-level diagram and operation principle of a dye-sensitized cell: the sensing dye absorbs a photon (hν); from the excited dye the electron is injected into the conduction band of the semiconductor (TiO_2) and then transferred (not shown). The oxidized dye is reduced by the electrolyte, which is regenerated at the counter-electrode (not shown) to complete the circuit. V_{OC} is determined by the Fermi level (E_F) of the semiconductor (TiO_2) and the redox potential $\left(I_3^-/I^-\right)$ of the electrolyte.

triiodide (Fig. 8.2). Eqs. (8.1)–(8.4) show the working of a DSSC in a simple manner:

$$D + h\nu\ (\text{Photon}) \rightarrow D^*_{(\text{absorbed})} \tag{1}$$

$$D^* \rightarrow D^+ + e^-_{(\text{injected into SC})} \tag{2}$$

$$I_3^- + 2e_{(\text{cathode})} \rightarrow 3I^-_{(\text{cathode})} \tag{3}$$

$$D^+ + 3/2I^- \rightarrow D + 3/2I_3^- \tag{4}$$

A number of sensitizers have been tried and have been classified into organic and inorganic dyes. Inorganic dyes have given better results compared with organic dyes owing to the stability of these dyes to photodegradation. Typical inorganic dyes employed for this purpose are metal complex dyes such as complexes of ruthenium, osmium, or iridium. The choice of the dye is based on the fact that it should be capable of absorbing photons below a threshold wavelength and also capable of grafting to the semiconductor surface. Many different compounds have been studied for sensitization, including porphyrins, phthalocyanines, coumarins, carboxylated derivatives of anthracene, and polymeric films.

There are various attempts to explore various other types of solar cells, such as DSSC, quantum dot sensitized solar cells, organic PV, quantum dot–polymer hybrid solar cells, and perovskite solar cells. Among these, perovskite-based solar cells have been receiving considerable attention.

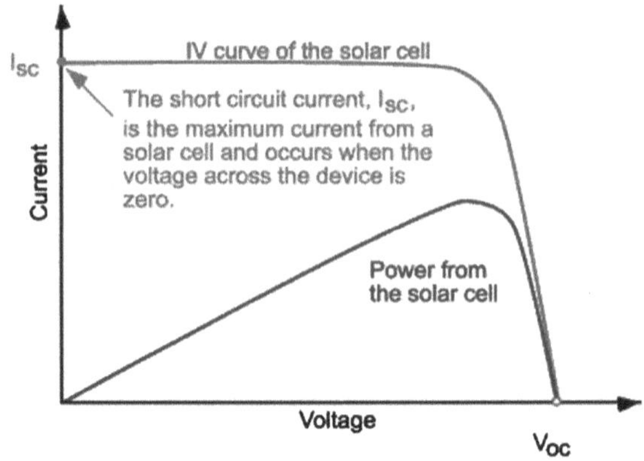

FIGURE 8.3 Typical I–V curve for a solar cell.

It is necessary to comprehend some terminology commonly employed in the solar cell literature.

Short circuit current is the current through the solar cell when the voltage across solar cell is zero (when the cell is short-circuited). The short circuit current results from the generation and collection of light-generated carriers. For an ideal solar cell, at most moderate resistive loss mechanisms, the short-circuit current and light-generated current are identical. Short-circuit current depends on: (1) the area of the solar cell, (2) the number of photons, (3) the spectrum of incident light, and (4) the optical properties of the material (Fig. 8.3).

The second term is open circuit voltage. This is the maximum voltage obtained from a solar cell; it occurs at zero current.

The third term is called the fill factor, which determines the maximum power that can be derived from a solar cell. It is defined as the ratio of maximum power from a solar cell to the product of V_{OC} and I_{SC}.

Efficiency is defined as the fraction of incident power that is converted to electricity; it is defined as:

$$P_{max} = V_{OC} I_{SC} FF$$

$$\eta = \frac{V_{OC} I_{SC} FF}{P_{in}}$$

The Shockley–Queisser limit refers to the maximum theoretical efficiency of a solar cell using a p–n junction to collect power from a cell. The limit places maximum solar efficiency around 33.7% assuming a single p–n junction with a band gap of 1.34 eV (using the AM 1.5 spectrum). (Terrestrial solar cells are measured under AM 1.5 and at 25°C and an input power of 1000 W/m^2.)

PEROVSKITE-BASED SOLAR CELLS

Perovskite-based solar cells represent most the significant breakthrough in solar cell technology since the 1970s [20–32]. Perovskites take solar technology to a whole new level, namely theoretical maximum efficiency (>30%) far exceeding silicon for single-junction cells. Extremely fast progress in R&D demonstrates a game-changing potential. This technology uses abundant, inexpensive raw materials with a simple cell structure, low waste, and low manufacturing cost. The current scene of perovskite-based technology compared with silicon or copper indium gallium selenide–based technologies is shown in Fig. 8.4. Although only 20.1% efficiency is shown, it may improve in coming years.

Progress in increasing the efficiency of perovskite-based solar cell technology has seen considerable in the past 10 years, as seen in Fig. 8.5.

A comparison of perovskite-based solar cells with respect to other solar cells is given in Table 8.2.

The band structure of hybrid perovskite based on inorganic-organic perovskite, namely $CH_3NH_3PbI_3$, has been computed; essential features of this study are that: (1) Energy levels in the conduction band minimum (CBM) are mainly contributed by nonbonding p orbitals of lead. (2) Energy levels in the valence band maximum (VBM) is contributed from antibonding hybridized lead 6s and iodine 5p orbitals. (3) Orbitals of organic cation are located at least 5 eV below the VBM and 2.5 eV above CBM; thus $CH_3NH_3^+$ functions as an electrostatic charge compensation species. (4) The effective mass of electrons (holes) around the bottom of the conduction band (the top of the valence band) can be approximately fitted by $m^* = h^2\,[\partial^2\varepsilon(k)/\partial^2 k]^{-1}$, where $\varepsilon(k)$ is the energy dispersion relation function. (5) Intuitively, the more dispersive (flat) the band is near the edges, the lighter (heavier) the effective

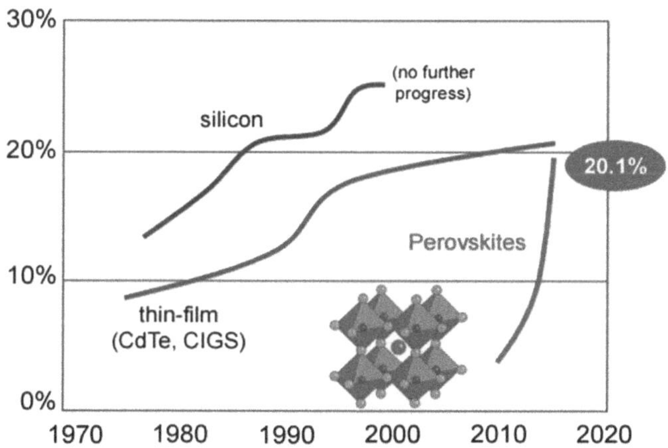

FIGURE 8.4 Comparison of progress of efficiency of silicon, thin-film, and perovskite solar cells.

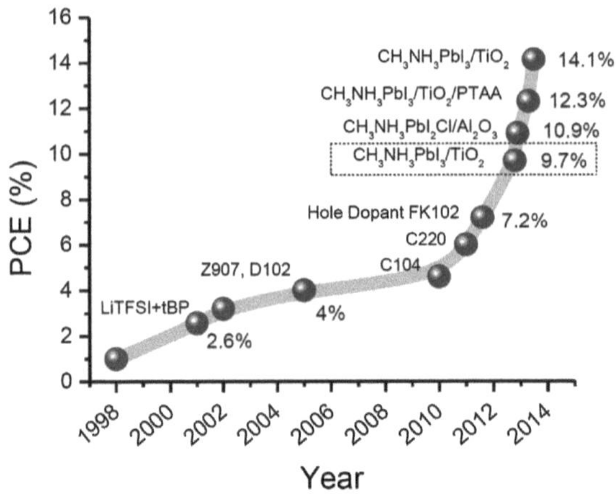

FIGURE 8.5 Efficiency of perovskite-based solar cell technology. *Reproduced from N.-G. Park, J. Phys. Chem. Lett. 4 (2013) 2423–2429.*

TABLE 8.2 Comparison of Solar Cells Based on Various Materials With Those Based on Perovskites

Characteristics	CdTe	CIGS	C-Si	Perovskite
Raw material cost	Low	Medium	Low	Low
Finished material cost	Low	High	High	Low
Fabrication cost	Medium	Medium	High	Low
Energy payback period	Medium	High	High	Low
Levelized cost of electricity	Medium	High	High	Low
Efficiency	Medium	Medium	High	High

CIGS, copper indium gallium selenide.

mass is. (6) The dispersed VBM and CBM result in the reduction of effective electron and hole mass from 0.32 to 0.36 m_o − 0.23 and 0.29 m_o, where m_o is free electron mass. (7) Such small effective mass of both electrons and holes is responsible for ambipolar conductivity in perovskites. (8) Finally, a large dielectric constant strongly screens the charged defects and impurities to increase the diffusion lengths. Thus, one can see that in all aspects, perovskite materials appear to be better materials for solar cell application. In addition, the direct optical transition of relevance in $CH_3NH_3PbI_3$ is from mixed-(Pb s,I p) states to Pb p states, and hence this intra-atomic Pb s state to Pb P state transition probability is high and hence shows strong absorption. The optical absorption of $CH_3NH_3PbI_3$ is one order of magnitude higher than that of GaAs in the visible range. Because visible light accounts for the major portion of the full solar spectrum, high absorption in its range is critical to achieve high efficiency cells.

In addition, these materials possess excellent transport properties such as crystallinity, diffusion length, and carrier mobility. Schottky defects such as PbI_2 and CH_3NH_3I vacancies were found not to generate defect states with energies within perovskite band gap; this is attributed to ionic bonding in perovskite. Defects such as Pb, I, and CH_3NH_3, vacancies associated with Frenkel defects were found to form shallow levels near band edges. Luckily the formation energies of deep-level defects are higher compared with shallow levels. Deep energy levels trap the charge carriers. Shallow defect levels are responsible for high diffusion lengths in perovskite solar cells. A high diffusion length is important because electrons and holes generated by absorption of light can travel enough distances to be extracted as current, instead of losing energy in the cell as heat. Organic–inorganic hybrid perovskites exhibit spontaneous electric polarization and the presence of ferroelectric domains will result in internal junctions that may aid electron–hole separation and transportation accounting for a sufficient lifetime for the excitons. The spontaneous dipole moment is due to the alignment of dipole moments of organic cations. A maximum current density of 28 mA/cm^2 is possible for a perovskite. The perovskite $CH_3NH_3PbI_3$ has a band gap 1.5 eV. Even considering 20% reflection for a transparent conducting film glass substrate, about 22 mA/cm^2 is the realistic photocurrent density from a 1.5-eV band gap material. Therefore 17% is a realistic efficiency because a photocurrent density of 22 mA/cm^2, photovoltage of 1.1 eV, and fill factor of 0.7 are achievable. In addition band tuning is a feasible process for these halide perovskites by combining all three cationic and anionic components. Generally, in perovskites the A cation does not have a major role in determining the band structure but acts to make charge compensation within the lattice. Nevertheless it can influence the optical properties by deforming the MX_6^- octahedron network owing to variation in its size. A larger or smaller cation can cause the whole lattice to expand or contract and lead to change in the B–X bond length, which is important in determining the band gap. This has been demonstrated by substituting A cations with cesium, methylammonium, and formamidinium cations [23].

The nature and positions of CBM and VBM have been studied by theoretical methods and it has been found that the band gap shifts toward the red region upon proceeding down the group (chlorine to iodine). This is due to the increase in the covalent character of halogen bonding with lead. Interestingly a perovskite structure that incorporates two halides allows for continuous tuning of band gap and optical absorption to cover almost the entire visible spectrum.

At this stage, we can compare the various solar cell materials and their efficiencies. One such attempt is made in Table 8.3 [33].

TABLE 8.3 Terrestrial Module Efficiencies

Classification	Efficiency (%)	Area (cm^2)	V_{OC} (V)	I_{SC} (A)	Fill Factor (%)
Si (crystalline)	22.6 ± 0.6	778	5.6	3.97	80.3
Si (multicrystalline)	18.5 ± 0.4	14,661	38.97	9.149	76.2
GaAs (thin film)	24.1 ± 1.0	858.5	10.89	2.255	84.2
CdTe (thin film)	17.5 ± 0.7	702.1	103.1	1.553	76.6
Copper indium gallium selenide (thin film)	15.7 ± 0.5	9703	28.24	7.254	72.6
Organic	8.7 ± 0.3	802	17.47	0.569	70.4

Measurements were made under the global AM 1.5 spectrum (1000 W/m^2) at a cell temperature of 298 K.
Data were extracted from M.A. Green, K. Emery, Y. Hishikawa, W. Warta E.D. Dunlop, Prog. Photovolt. Res. Appl. 23 (2015) 1–9.

Solar cell research is at its peak. This trend is revealed by the number of editorials and other commentaries in the *Journal of Physical Chemistry Letters* in Volume 6 alone, cited subsequently in the references to this chapter [34–37].

The claimed benefits of perovskite-based solar cells are their positive electron/hole transportation performance, low recombination rate and high photoluminescence, long diffusion length, and long carrier lifetime. However it is necessary to know the inner working of the hybrid perovskite solar cells especially what takes place under supra band gap irradiation, and how these features are related to the electrical features of these efficient devices. More effort is necessary to make perovskite-based solar cells viable alternatives in terms of their efficiency, stability, and scalability. It is anticipated that in the near future economically viable materials will be available for solar cell applications. In this connection the use of nanomaterials in the form of thin-film PVs has shown some success as a viable commercial technology. More than 20% power conversion efficiency can be routinely achieved with organic-inorganic lead halide perovskites. Many perovskite materials have already been identified and many more are going to be discovered in the near future. These devices may pave the way for the widespread deployment of thin-film PV. The efficiency of thin-film solar cells is continuously increasing while the price of these panels decreases so that the cost of PV electricity will soon reach that of current grid electricity. The last 4 decades saw semiconductor solar cells transform from single junction to multiple junction to thin film, thus leading to viable devices.

REFERENCES

[1] D. Poponi, Analysis of diffusion paths for photovoltaic technology based on experience curves, Sol. Energy 74 (2003) 331–340.
[2] S. Rehman, A. Bader Maher, A. Al-Moallem Said, Cost of solar energy generated using PV panels, Renewable Sustainable Energy Rev. 11 (2007) 1843–1857.
[3] A.Y. Al-Hasan, A.A. Ghoneim, A.H. Abdullah, Optimizing electrical load pattern in Kuwait using grid connected photovoltaic systems, Energy Convers. Manage. 45 (2004) 483–494.
[4] M. Ito, K. Kato, H. Sugihara, T. Kichimi, J. Kichimi, K. Kurokawa, A preliminary study on potential for very largescale photovoltaic power generation (VLSPV) system in the Gobi desert from economic and environmental viewpoints, Sol. Energy Mater. Sol. Cells 75 (2003) 507–517.
[5] X. Zhou, J. Yang, F. Wang, B. Xiao, Economic analysis of power generation from floating solar chimney power plant, Renewable Sustainable Energy Rev. 13 (2009) 736–749.
[6] T. Muneer, M. Asif, J. Kubie, Generation and transmission prospects for solar electricity: UK and global markets, Energy Convers. Manage. 44 (2003) 35–52.
[7] E. Cunow, B. Giesler, The megawatt solar roof at the new Munich Trade Fair Centre—an advanced and successful new concept for PV plants in the megawatt range, Sol. Energy Mater. Sol. Cells 67 (2001) 459–467.
[8] M.M.H. Bhuiyan, M. Ali Asgar, R.K. Mazumder, M. Hussain, Economic evaluation of a stand-alone residential photovoltaic power system in Bangladesh, Renewable Energy 21 (2000) 403–410.
[9] R. Alazraki, J. Haselip, Assessing the uptake of small-scale photovoltaic electricity production in Argentina: the PERMER project, J. Cleaner Prod. 15 (2007) 131–142.
[10] R.T. Kivaisi, Installation and use of a 3 kWp PV plant at Umbuji village in Zanzibar, Renewable Energy 19 (2000) 457–472.
[11] N.K. Bansal, G. Sandeep, Integration of photovoltaic technology in cafeteria building, at Indian Institute of Technology, New Delhi, Renewable Energy 19 (2000) 65–70.
[12] S. Ubertini, U. Desideri, Performance estimation and experimental measurements of a photovoltaic roof, Renewable Energy 28 (2003) 1833–1850.
[13] A. Slaoui, R.T. Collins, MRS Bull. 32 (2007) 211.
[14] S.C. Ameta, R. Ameta, Solar Energy Conversion and Storage, Photochemical Modes, CRC Press, 2015.
[15] A.S. Karmakar, J.P. Ruparelia, Institute of technology, Nirma University, in: Proceedings of the International Conference on Current Trends in Technology, 2011.
[16] F.-T. Kong, S.-Y. Dai, K.-J. Wang, Adv. OptoElectronics (2007). http://dx.doi.org/10.1155/2007/75384.
[17] A. Mishra, M.K.R. Fischer, P. Bauerle, Angew. Chem. Int. Ed. 48 (2009) 2474–2499.
[18] B. Viswanathan, M. Aulice Scibioh, Photoelectrochemistry – Principlesand Practices, Alpha Science International Limited, UK, 2014.
[19] B.E. Hardin, H.J. Snaith, M.D. McGehee, Nat. Photonics 6 (2012) 162–169.
[20] F. Di Giacomo, S. Razza, F. Matteocci, A. D'Epifanio, S. Licoccia, T.M. Brown, J. Power Sources 251 (2014) 152–156.
[21] Z. Yang, W.-H. Zhang, Chin. J. Catal. 35 (2014) 983–988.
[22] G. Peng, X. Xu, G. Xu, J. Nanomater. (2015). http://dx.doi.org/10.1155/2015/241853.
[23] P. Gao, Energy Environ. Sci. 7 (2014) 2448–2463.

[24] W.J. Yin, J. Phys. Chem. Lett. 5 (2014) 3625–3631.
[25] N.J. Jeon, Nature 517 (2015).
[26] A. Buin, Nano Lett. 14 (2014) 6281–6286.
[27] M. Gratzel, Nat. Mater. 13 (September 2014).
[28] P.P. Boix, Mater. Today 17 (2014).
[29] N.-G. Park, J. Phys. Chem. Lett. 4 (2013) 2423–2429.
[30] N.-G. Park, Mater. Today 8 (2014) 65–72.
[31] F. Hao, J. Am. Chem. Soc. 136 (2014) 8094–8099.
[32] www.pveducation.org.
[33] M.A. Green, K. Emery, Y. Hishikawa, W. Warta, E.D. Dunlop, Prog. Photovolt. Res. Appl. 23 (2015) 1–9.
[34] D.D. Egger, E. Edri, D. Cahen, G. Hodes, J. Phys. Chem. Lett. 6 (2015) 279–282.
[35] P.V. Kamat, J. Phys. Chem. Lett. 6 (2015) 4874–4875.
[36] K. Zhu, T. Miyasaka, J. Young Kim, I. Mora-Sero, J. Phys. Chem. Lett. 6 (2015) 2315–2317.
[37] K. Sivula, J. Phys. Chem. Lett. 6 (2015) 975–976.

Chapter 9

Hydrogen as an Energy Carrier

Hydrogen economy is a concept that was proposed by Professor John Bockris in 1970, although it was proposed earlier by J. B. S. Haldane. This concept is based on the principle that hydrogen can be employed as a potential fuel and is considered to be an energy carrier. Despite these favorable features, molecular hydrogen is not freely available and has to be generated from other sources in which hydrogen is combined with other elements such as hydrocarbons and water. A number of technologies are available for generating hydrogen from sources in which hydrogen is combined with other elements. There is a variety of ways in which hydrogen can be generated, such as the steam reforming of hydrocarbons and electrolysis of water [1]. A simplified listing of available technologies for producing hydrogen (the most abundant element in the universe but in a combined form) is given in Table 9.1.

Although it is desirable to consider of all these options in detail, we shall restrict the consideration to only a few of these available methods.

Various advantages are claimed for employing hydrogen as a future fuel. Advantages of hydrogen as a universal energy medium are that:

1. The combustion of hydrogen results in the formation of steam and liquid water. In this respect, the use of hydrogen is completely safe from an environmental standpoint.
2. It is nontoxic.
3. It is easily assimilated into the biosphere; its combustion products are recycled by plants in the form of carbohydrates.
4. It is possible to produce hydrogen from the most abundant chemical on earth: water. Hydrogen can be obtained electrolytically, photoelectrochemically, thermochemically, by direct thermal decomposition, or biochemically from water.
5. Hydrogen can be used as a feedstock for the chemical industry, enabling the production of an entire gamut of chemicals from hydrogen and conventional petrochemicals.
6. It is the most suitable fuel for use in fuel cells: the direct conversion of chemical energy into electricity without the heat, with enhanced efficiency.
7. Transmission of energy in the form of hydrogen is more economical than through high-voltage AC lines for large distances.

Energy Sources. http://dx.doi.org/10.1016/B978-0-444-56353-8.00009-5

TABLE 9.1 Possible and Current Technologies for Generating Hydrogen

1.	steam reforming of hydrocarbons
2.	partial oxidation of hydrocarbons
3.	thermal decomposition of hydrogen-containing compounds
4.	thermochemical cycles (iron–halogen or sulfur–iodine cycles)
5.	electrolysis of water
6.	electrochemical decomposition
7.	photolysis of water
8.	photochemical decomposition of water
9.	photo-electrochemical (PEC) decomposition of water
10.	biological decomposition
11.	biochemical or photo-biochemical processes

As is known, hydrogen is the most abundant element in the universe. Hydrogen is the simplest of atoms, composed of one proton and one electron. However, pure, diatomic hydrogen (H_2), the fuel of choice for so-called fuel cells, does not exist naturally. Because hydrogen combines easily with other elements, one is most likely to find it chemically bound in water, biomass, or fossil fuels. To obtain hydrogen in a useful form, it must be extracted from one of these sources. This process requires energy. Accordingly, the cleanliness and renewability of this energy is of critical importance. Whereas a hydrogen–oxygen fuel cell operates without producing emissions, producing hydrogen can result in significant greenhouse gases and other harmful by-products. Once obtained, hydrogen is a nearly ideal energy carrier [2].

DIRECT ELECTROLYSIS

Hydrogen production by water electrolysis involves passing an electric current through water to separate it into hydrogen (H_2) and oxygen (O_2). Hydrogen gas evolves from the negative cathode and oxygen gas collects at the positive anode. The reactions involved in the electrolysis of water are:

$$\text{Reduction electrode (cathode): } 2H_2O + 2e^- \rightarrow 2OH^- + H_2$$

$$\text{Oxidation electrode (anode): } 2OH^- \rightarrow H_2O + 1/2O_2 + 2e^-$$

$$\text{Complete cell reaction: } H_2O \rightarrow H_2 + 1/2O_2$$

The values of the cathode and anode half-cell potentials are known to be 0.401 V and −0.828 V, respectively, at 25°C at pH 14. If the activities of water and the gaseous species are considered to be in unity, the cathode and anode potentials required according to the Nernst equation will be:

$$E_C = -0.828 - 0.059 \log a_{OH}$$

$$E_a = 0.401 - 0.059 \log a_{OH}$$

Therefore the potential required to split water into H_2 and O_2, ie, $E_a - E_C$, is equal to 1.229 V. Although the theoretical potential is 1.23 V for water electrolysis, in practice the actual water decomposition will occur only above 1.7 V. The extra potential, which is essential for the water decomposition, is called overpotential. Overvoltage is evaluated mainly as a function of current and temperature. Overvoltages are composed of activation or charge transfer overvoltage; concentration or diffusion, or mass transfer overvoltage; and resistance or Ohmic overvoltage. In general, an aqueous solution of caustic potash or soda is used as the electrolyte for water electrolysis. Usually 25–28% of caustic potash or 15–20% caustic soda is used in commercial electrolyzers in view of the favorable conductivity values. The nature of anode and cathode is decided based on the hydrogen and oxygen overvoltages in the electrolytic medium in addition to their stability in the particular medium. The cathode and anode are separated by a diaphragm, which prevents the mixing of hydrogen and oxygen gases produced at the cathode and anode surfaces, respectively. The diaphragm should be stable in the electrolyte and minimizes the diffusion of gas molecules without affecting the conductivity of the medium.

EFFECT OF TEMPERATURE AND pH ON THE DECOMPOSITION POTENTIAL

The amount of electricity required to produce 1 mole of hydrogen by splitting 1 mole of water is 2 F, which is equal to 236.96 kJ of energy, whereas, heat generated by combustion of 1 mole of hydrogen is 285.58 kJ at 298K. The extra energy of 48.63 kJ must be absorbed from the surrounding of electrolytic cell if the water is electrolyzed with 1.229 V at 298K. Applying electrical energy of 285.58 kJ, ie, 1.481 V, to a water electrolyzer at 298K would generate hydrogen and oxygen isothermally. The values 1.229 and 1.481 V are called the reversible and thermoneutral voltages. The variations in reversible and thermoneutral voltage with temperature are shown in Fig. 9.1. Fig. 9.1 shows that when the temperature increases the reversible voltage decreases, whereas the thermoneutral voltage slightly increases with temperature. Fig. 9.1 demonstrates that in the region below the reversible voltage, hydrogen production is not possible. In the second region, the hydrogen is evolved with the absorption of heat from the surroundings. In the third region, the hydrogen is evolved with the liberation of heat, ie, the extra energy as potential above the thermoneutral potential is released as heat energy. In general, the commercial industrial electrolytic cells are operating between 60 and 80°C. The hydrogen and oxygen evolution potentials at various pHs are shown in Fig. 9.2. It can been seen from the figure that the net potential needed for the hydrogen and oxygen evolution at any given pH between 0 and 14 is 1.229 V at 25°C.

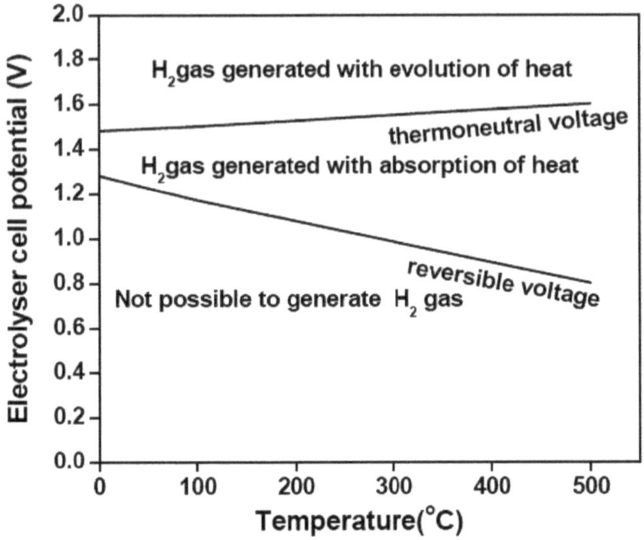

FIGURE 9.1 Variation in cell potential as a function of temperature.

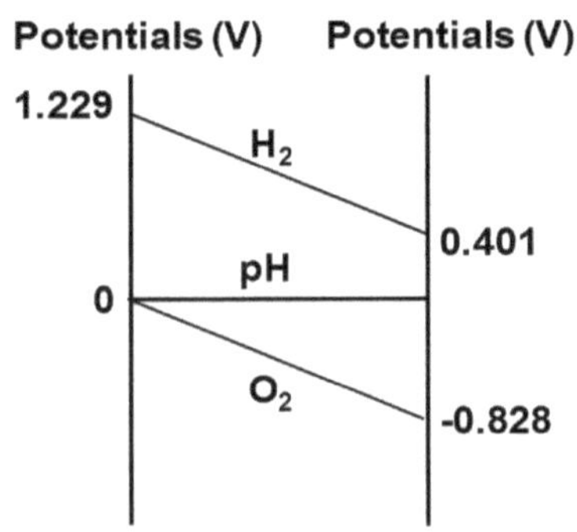

FIGURE 9.2 Hydrogen and oxygen electrode potential against pH of the electrolyte.

Because of the corrosive action on the electrode materials, especially at the anodes, acidic solutions are avoided for water electrolysis. Electrolysis produces extremely pure hydrogen, which is necessary for some types of fuel cells. However a significant amount of electricity is required to produce usable amount of hydrogen from electrolysis. In an ideal case, this could come from renewable sources such as wind, solar, and photo-catalysis. The hydrogen produced from electrolysis would in no way help reduce pollution in the atmosphere if the electricity needed for the reaction is obtained through fossil fuels.

STEAM-REFORMING (STEAM-METHANE REFORMATION)

Hydrogen can also be extracted or "reformed" from natural gas. A two-step process at temperatures reaching 1100°C in the presence of a catalyst makes four parts hydrogen from one part methane (CH_4) and two parts water. It is a relatively efficient and inexpensive process and can be made still more efficient by harvesting the waste heat (commonly referred to as cogeneration). This latter feature makes a steam-methane reaction particularly attractive for local use:

$$CH_4 + H_2O \xrightarrow[930^\circ C]{\text{Catalyst}} CO + 3H_2$$

$$CO + H_2O \xrightarrow[350^\circ C]{\text{Catalyst}} CO_2 + H_2$$

Although this process is well understood and can be implemented on a wide scale, it produces moderate emissions of carbon dioxide. Other innovative carbon-sequestration techniques are under development. Unlike renewable electrolysis, steam-methane reformation depends on the fluctuating price of natural gas. Nonetheless, steam-methane reformation is poised to be the near-term hydrogen production method of choice on the road toward completely renewable methods.

It would be desirable if one could also consider the reforming of renewable feedstocks instead of naphtha. Hydrogen is considered to be the energy carrier of the future and could have a major role in reducing environmental emissions. The state of matter for the fuel has been shifting from solid to liquid to gas, and the "decarbonization" trend that has been pursued implies that the transition to hydrogen energy seems inevitable. However, hydrogen has its own problems and there is a debate regarding whether hydrogen is the solution for energy-related environmental, security, and sustainability issues. One can examine the possibility, mainly through the production of hydrogen from ethanol (C_2H_5OH), a renewable source that may contribute to net-zero carbon dioxide (CO_2) emissions. In nature, hydrogen is always present in a bound form, in organic compounds and water. Hydrogen can be produced from different sources, eg, coal, natural gas, liquefied petroleum gas, propane, methane (CH_4), gasoline, light diesel, dry biomass, and biomass-derived liquid fuels (such as methanol, C_2H_5OH, and biodiesel), as well as from water. Among liquid hydrogen sources, C_2H_5OH is a good candidate for several reasons: (1) ethanol is renewable and is becoming increasingly available; (2) it is easy to transport and biodegradable, and is low in toxicity; (3) it could be easily decomposed in the presence of water to generate a hydrogen-rich mixture; and (4) it is free from catalyst poisons such as sulfur. In addition, the steam reforming of C_2H_5OH to produce hydrogen is thermodynamically feasible. Currently, almost 90% of hydrogen is produced by the reforming of natural gas or light oil fractions with steam at high temperatures. However, hydrogen production from natural gas is associated with emissions of greenhouse gases. Hydrogen production from ethanol steam reforming would be environmentally friendly and would use of renewable resources. Steam reforming of natural gas to produce hydrogen is the most energy-efficient technology currently available. Furthermore, the reforming process could extract hydrogen not only from hydrocarbon fuels but also from water. The reaction is fast, although hydrogen production is limited by thermodynamic balance. In addition, catalytic steam reforming of natural gas is by far the most cost-effective process when applied to large-scale production, compared with other technologies. Steam reforming is a cost-effective and efficient process. In addition to its being a bio-renewable resource, the steam reforming of ethanol is a promising choice in H_2-based energy systems. The overall steam-reforming reaction of ethanol can be represented as:

$$C_2H_5OH + 3H_2O \rightarrow 2CO_2 + 6H_2 \quad \left(\Delta H^0{}_{298} = +347.4\ \text{kJ/mol}\right)$$

Although this equation represents the overall steam reforming reaction, there can be other concurrent reactions such as dehydrogenation, dehydration, cocking reactions of methane, and the Boudouard reaction, and other water gas reactions and a variety of catalysts have been employed to optimize the steam-reforming reaction. However, different catalysts induce different pathways to produce H_2. Therefore, the choice of the catalyst has a vital role in the reforming process. Various catalysts have been used for H_2 production via the steam reforming of ethanol. These catalysts include oxides, metal-based (Ni, Cu, and Co) catalysts, metal mixture–based catalysts, and noble metal–based catalysts.

BIOMASS GASIFICATION

Hydrogen can be extracted from hydrogen-rich biomass sources such as wood chips and agricultural waste. When heated in a controlled atmosphere, biomass converts to synthesis gas, which consists primarily of carbon monoxide (CO), carbon dioxide (CO_2), and hydrogen (H_2). Gasification technology has been under intensive development over the past 2 decades. Large-scale demonstration facilities have been tested and commercial units are in operation. Fortunately, hurdles in biomass gasification have been economic rather than technical. Until recently, biomass gasification has been employed to produce low-value products such as electricity or heat, which rarely justify the capital and operating costs. However, the increasing demand for hydrogen promises to make biomass gasification economically viable in the near future.

HYDROGEN FROM COAL

Vast coal resources have often been viewed as a potential source against future energy needs. Unfortunately, coal mining pollutes and spoils the landscape, and burning coal produces many harmful emissions. Yet coal contains hydrogen, and techniques are being developed to sequester the remaining carbon. These processes generally involve coal gasification to produce hydrogen and electricity, followed by reinjection of CO_2 or mineralization via carbonates.

BIOCHEMICAL HYDROGEN PRODUCTION

Life requires metabolism, a complex web of redox chemistry. This requires energy, which can be obtained by breaking bonds [the multistep breakdown of glucose to generate adenosine triphosphate (ATP) and CO_2] or from electronic excitation. For example, plants, algae, cyanobacteria, and photosynthetic bacteria can use light energy to raise electrons into higher energy states. In case of plants, algae, and cyanobacteria, the source of excitable electrons is water. The excited electrons are stripped from water, which then splits into oxygen and protons.

Hydrogen is produced in microorganisms by enzymes capable of reducing free protons to molecular hydrogen. Examples of these enzymes include hydrogenases and the nitrogenases. The production of hydrogen by these enzymes is usually coupled to some other biochemical processes. The energy used by these enzymes is usually in multiple steps from an organism's central energy inputs and is provided in the form of electron carriers such as ferredoxin or reduced nicotinamide adenine dinucleotide phosphate and energy-yielding molecules such as ATP. Obtaining useful amounts of hydrogen from microorganisms will require the efficiency of hydrogenases to be increased and other obstacles to be overcome. One problem is that some hydrogenases and nitrogenases are inhibited by oxygen. Oxygen is produced by photo-system II during oxygenic photosynthesis.

In the summer of 2001, researchers manipulated the photosynthetic process of spinach plants to produce hydrogen. However, these biological means of hydrogen production are known only as laboratory experiments. Intense research persists to better understand ways to improve these hydrogen production methods. Quantum leaps in this field could be the equivalent of striking oil. Biological hydrogen production is the most challenging area of biotechnology with respect to environmental problems. The future of biological hydrogen production depends not only on research advances, ie, improvement in efficiency through genetically engineered microorganisms and/or the development of bioreactors but also on economic considerations (the cost of fossil fuels), social acceptance, and the development of hydrogen energy systems.

THERMOCHEMICAL DECOMPOSITION OF WATER

The decomposition of water into hydrogen and oxygen can be achieved when energy is supplied in the form of heat and work. The positive value of ΔG^0 decreases with an increase in temperature, but slowly because of the nearly constant enthalpy change as a function of temperature, and ΔG^0 becomes 0 around 4700K. This means that even the highest temperature available from a nuclear reactor, in the range of 1300K, is not sufficient to decompose water. Therefore, single-step thermal decomposition of water is difficult unless other methods such as electrolysis are resorted to. The two-step decomposition of water, in which a metal oxide, metal hydride, or hydrogen halide is involved, is, according to the equations:

$$H_2O + M \rightarrow MO + H_2$$

$$MO \rightarrow M + 0.5O_2$$

or

$$H_2O + M \rightarrow MH_2 + 0.5O_2$$

$$MH_2 \rightarrow M + H_2$$

or

$$H_2O + X_2 \rightarrow 2HX + 0.5O_2$$

$$2HX \rightarrow H_2 + X_2$$

However, even these two-step routes require temperatures on the order of 1273K or more. Water cannot be decomposed in one or two thermochemical steps when the available temperature is below 1273K. However this can be done in a multiple-step process in which each step is easy to accomplish with either a negative or a slightly positive ΔG for the reaction. For example, if the desired reaction is [Eq. (9.1)]:

$$H_2O \rightarrow H_2 + 0.5O_2 \quad (9.1)$$

it can be achieved in a sequence of steps as follows [Eqs. (9.2–9.5)]:

$$2H_2O_{(g)} + I_{2(g)} + SO_{2(g)} \rightarrow 2HI_{(g)} + H_2SO_{4(g)} \quad (9.2)$$

$$H_2SO_{4(g)} \rightarrow H_2O_{(g)} + SO_{2(g)} + 0.5O_{2(g)} \quad (9.3)$$

$$Ni_{(s)} + 2HI_{(g)} \rightarrow NiI_{2(s)} + H_{2(g)} \quad (9.4)$$

$$NiI_{2(s)} \rightarrow Ni_{2(s)} + I_{2(g)} \quad (9.5)$$

In this sequence the first reaction has a large positive ΔG (87.6 kJ/reaction), whereas all other reactions have negative ΔG values. Replacing step (1) by the following step will give a negative ΔG value [Eq. (9.6)]:

$$2H_2O_{(g)} + I_{2(g)} + SO_{2(g)} \rightarrow 2HI_{(aq)} + H_2SO_{4(aq)} \quad (9.6)$$

Carrying out the reaction in the four steps [Eqs. (9.6), (9.3), (9.4), and (9.5)] at 300, 510, 570, and 1070K, respectively, requires −74.3 kJ. Therefore any thermochemical cycle can be chosen by incorporating the following four reaction steps: water decomposition or hydrolysis, hydrogen generation, oxygen generation, and the regeneration of any intermediates formed. Another thermochemical cycle that is available for hydrogen generation is the Mark 15 process (iron–halogen system):

$$3FeCl_2 + 4H_2O \rightarrow Fe_3O_4 + 6HCl + H_2 \quad 723-1023K$$

$$Fe_3O_4 + 8HCl \rightarrow FeCl_2 + 2FeCl_3 + H_2O \quad 373-383K$$

$$2FeCl_3 \rightarrow 2FeCl_2 + Cl_2 \quad 573\ K$$

$$Cl_2 + H_2O \rightarrow 2HCl + 1/2O_2 \quad 1073K$$

There are other, similar proven cycles for hydrogen generation.

PHOTOCHEMICAL HYDROGEN PRODUCTION

Photochemical hydrogen production is similar to a thermochemical system in that it also employs a system of chemical reactants that carry out the splitting of water. However, the driving force is not thermal energy but light, generally solar light. In this sense, this system is similar to the photosynthetic system present in green plants. One can effectively use photochemical means to promote endergonic (energy-requiring) reactions. The sensitized oxidation of water by Ce^{4+} using irradiation of 254 nm light by the following reaction is known:

$$Ce^{4+} + 0.5H_2O \rightarrow Ce^{3+} + 0.25O_2 + H^+ \quad \Delta H = 3.8\ Kcal/mol$$

Ce^{3+} can be used with light of a lower wavelength to promote the hydrogen generation reaction:

$$Ce^{3+} + H_2O \rightarrow Ce^{4+} + 0.5H_2 + OH^-$$

The quantum efficiency of these processes is very low. Similarly $Ru(bpy)_3^{2+}$ (Fig. 9.3) and related complexes have low excited-state lifetimes and can serve as electron donors or electron acceptors. A typical reaction is:

$$[AR_2]^{2+}(ClO_4)_2^{\ 2-} \overset{h\nu}{\rightarrow} \underset{\text{excited state}}{*[AR_2]^{2+}(ClO_4)_2^{\ 2-}}$$

$$*[AR_2]^{2+}(ClO_4)_2^{\ 2-} + H_2O \rightarrow H_2 + 0.5O_2 + [AR_2]^{2+}(ClO_4)_2^{\ 2-}$$

where $R = C_{18}H_{35}$ and A = Ruthenium bipyridyl complex.

PHOTO-ELECTROCHEMICAL HYDROGEN PRODUCTION

In its simplest form, a PEC hydrogen production cell consists of a semiconductor electrode and a metal counter-electrode immersed in an aqueous electrolyte (Fig. 9.4). When light is incident on the semiconductor electrode, it absorbs part of the light and generates electricity. This electricity is then used for the electrolysis of water. Fujishima and Honda first demonstrated the electrolysis of water using solar energy in a PEC cell in 1972 [3].

As seen from the diagram, the cell consists of a semiconductor (TiO_2) photoanode, which is irradiated with UV–visible radiation. The counter-electrode is

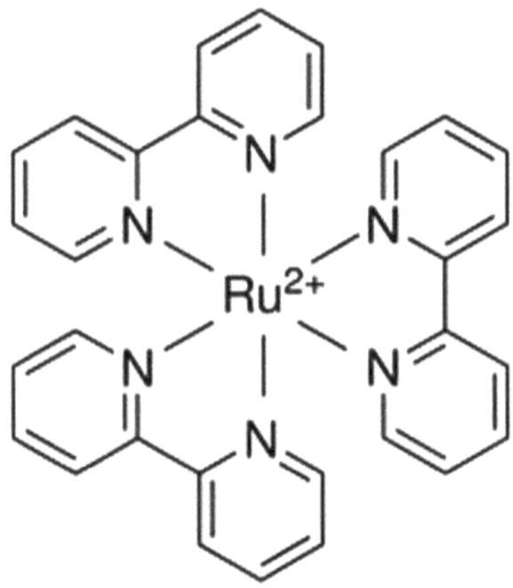

FIGURE 9.3 Ruthenium bipyridyl complex.

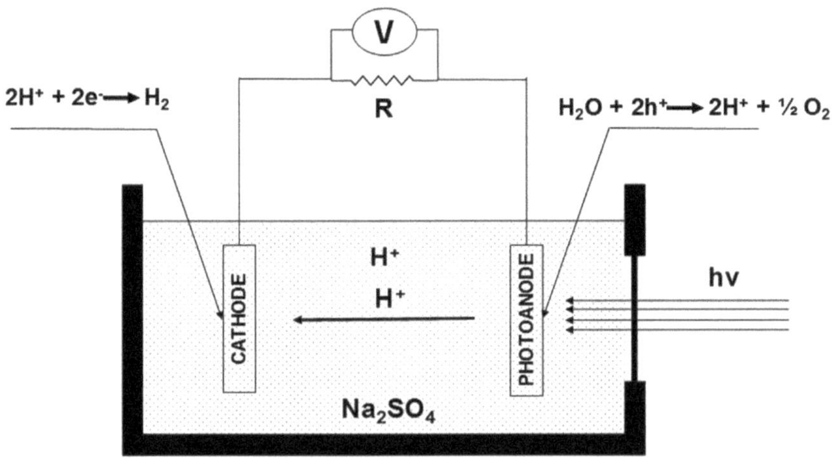

FIGURE 9.4 Schematic showing the structure of a simple PEC cell.

a metal. The processes that take place in the cell when light is incident on the semiconductor electrode are:

1. Photo generation of charge carriers (electron and hole pairs):

$$\text{Semiconductor} \xrightarrow{h\nu} 2e^- + 2h^+$$

2. Charge separation and migration of the holes to the interface between the semiconductor and the electrolyte and of electrons to the counter-electrode through the external circuit.

Now, holes are simply vacancies created in the valence band caused by the promotion of electrons from the valence band to the conduction band. However, in the study of electronic behavior of materials, "holes" are considered to be independent entities with their own mass.

3. The holes move to the interface and react with water, producing oxygen:

$$2h^+ + H_2O \rightarrow 0.5O_{2(gas)} + 2H^+_{(aq)}$$

4. The electrons travel in the external circuit and arrive at the interface between the counter-electrode (cathode) and electrolyte. There, they reduce the H^+ ions to H_2:

$$2e^- + 2H^+_{(aq)} \rightarrow H_{2(gas)}$$

The complete reaction is the absorption of photon and splitting of water into hydrogen and oxygen. The representation of the same process in band energy terms is shown in Fig. 9.5.

The cell depicted in Fig. 9.5 is a single photo-electrode type cell, with the anode being the active photo-electrode. The lower band is the valence band of the n-type semiconductor, whereas the upper band is the conduction band. The energy difference between the top of valence band and the bottom of

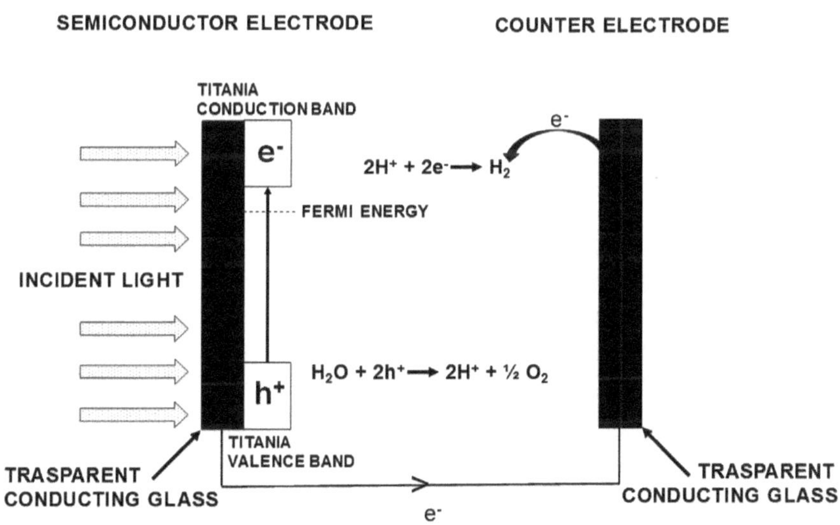

FIGURE 9.5 Operating principle of a photo-electrochemical cell.

conduction band is termed the band gap of the semiconductor, E_g. Other configurations of the PEC cell are possible:

1. The semiconducting material may be a p-type material. In this case, it will act as a photo-cathode, and reduction of H^+ ions to H_2 will take place at this electrode. The counter-electrode may be a metal in this case.
2. Both electrodes, the cathode and anode, are photoactive semiconducting materials. In this case, the n-type electrode will act as an anode and favors oxidation of water to oxygen, and reduction H^+ will take place at this electrode. The p-type electrode will act as a cathode, where H^+ ions will be reduced to H_2.

This method of hydrogen generation and also the photocatalytic hydrogen production have seen considerable development in the past 3–4 decades, ever since the possibility of PEC decomposition of water was demonstrated by Fujishima and Honda in 1972. A brief summary of these developments will be given at the end of this chapter.

PHOTOCATALYTIC HYDROGEN PRODUCTION

Essentially photo-catalyzed reactions have generated considerable interest after the photocatalytic splitting of water on TiO_2 electrodes was first demonstrated by Fujishima and Honda in 1972. Subsequently, various kinds of photo-catalysts have been employed for hydrogen production and the remediation of pollutants from water. A dispersed heterogeneous semiconductor surface provides a fixed environment that influences chemical reactivity. A simultaneous oxidation and reduction reaction occurs on the surface of the catalyst on photo-excitation. Other advantages are easy separation of the catalyst after the reaction by centrifugation, the availability of a large surface area, low cost, and stability.

In heterogeneous photocatalytic systems, absorption of light is an essential requirement for successful photo-catalysis. In addition, it should be stable at the reaction conditions employed and it should be chemically inert. Among available materials such as metals, semiconductors, and insulators, semiconductors have been used because the band gap of a semiconductor is optimal, band edge positions are suitable for the oxidation and reduction of water, and one can possibly use sunlight as an energy source to excite the electron from the valence band.

In addition to the favorable band gap and band positions, semiconductors are inexpensive, nontoxic, easily recoverable, and capable of retaining catalytic activity. Also, loading of metal on the semiconductor surface and coupling of two semiconductors can increase the efficiency of the semiconductor photo-catalyzed reaction. Although light absorption is essential, other parameters such as band gap, surface area, crystal phase, morphology, rate of interfacial charge transfer, carrier density, and stability are also essential for observed photocatalytic activity.

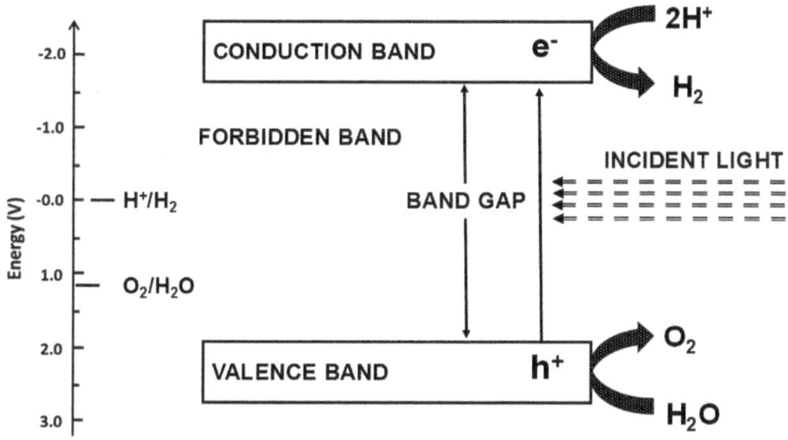

FIGURE 9.6 Excitation and redox reactions in a semiconductor.

Photo-catalysis involves the initial absorption of photons by a semiconductor to excite electrons from the valence band to the conduction band. This results in the formation of an electron–hole pair within the semiconductor. Excitation and redox processes taking place in semiconductor photo-catalyst are shown in Fig. 9.6.

For efficient photo-catalytic reaction, the electron–hole pair recombination must be suppressed. Either trapping the photo-generated electron or hole, or both, can lead to this. The electron in the conduction band moves to the surface and a reduction reaction takes place with adsorbed molecule or surface groups. Self-recombination with the hole in the valence band depresses the activity of the semiconductor.

The reduction and oxidation strength of the photo-excited electron and hole can be measured from the energy of the lower edge of the conduction band and the upper edge of the valence band. Depending on the relative positions of the top of valence band, the bottom of conduction band, and the redox potentials of the species, the oxidation and reduction processes are promoted (Fig. 9.7).

In general, the selection of a semiconductor for a particular reaction is based on the position of the valence and conduction band edges and the redox potential of the adsorbed species of interest, stability toward photo-corrosion, and the value of band gap. Band gaps and energy levels of various semiconductors are shown in Fig. 9.7. To reduce water, the potential of the bottom of the conduction band must be more negative than the hydrogen reduction potential; for oxidation reaction, the top of the valence band should be more positive than the oxidation potential of water. Energies of various semiconductors are shown in Fig. 9.7 with respect to a normal hydrogen electrode (NHE).

Because the energy of valence and conduction levels of TiO_2 is optimum to oxidize most organic species, its high oxidation ability of photo-generated holes ($E = 2.9$ V versus NHE at pH 0) makes it the best choice for a photo-

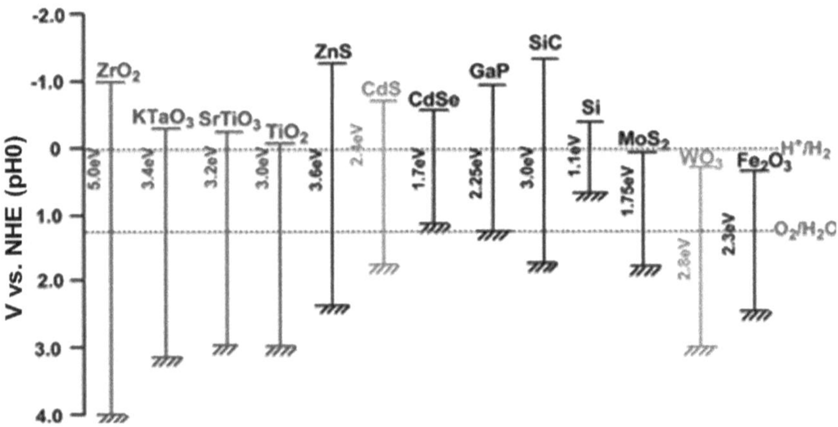

FIGURE 9.7 Energy levels of top and bottom of the valence and conduction bands of some typical semiconductors.

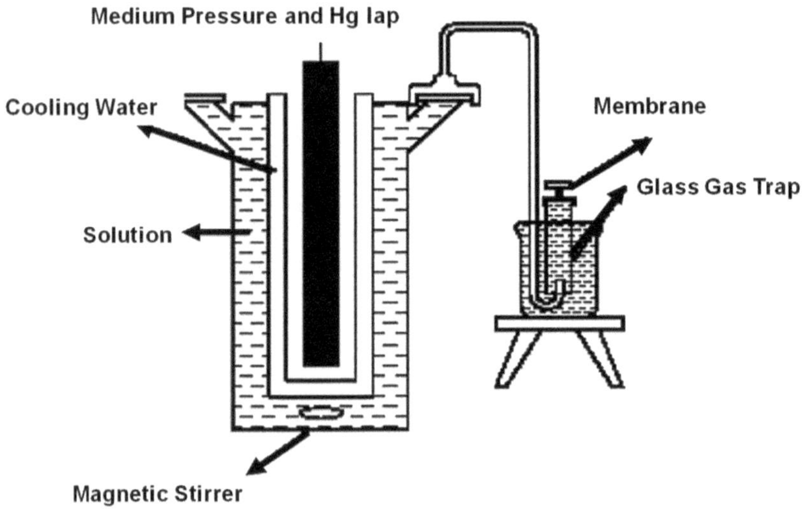

FIGURE 9.8 Typical photo-catalytic water-splitting setup. Other variations are possible.

catalyst. In addition TiO_2 is inert and resistant to photo-corrosion, which makes it a good photo-catalyst. Among three structural modifications of TiO_2 (brookite, rutile, and anatase), anatase is the form that is active. Although there are other semiconductors to fulfill these criteria, some experience "photo-corrosion" under the experimental conditions employed. Fig. 9.8 shows the typical photocatalytic water-splitting setup.

The major problem associated with the photo-catalytic splitting of water is the higher band gap of available semiconductor materials such as TiO_2. Because of the higher band gap, these materials require UV light irradiation to carry out the reaction whereas sunlight contains only ~5% UV radiation. The remaining part of the solar spectrum is composed mainly of visible and infrared radiation. Research is focusing on reducing the band gap of available materials by various methods and finding new photoactive materials with a lower band gap.

Although various methods are available, processes such as direct electrolysis, steam-methane reformation, biomass gasification, and hydrogen from coal and thermochemical decomposition require other forms of energy such as heat and/or electricity, which can be obtained from fossil fuels or other expensive methods such as nuclear energy. Also, some of the methods lead to the creation of greenhouse gases such as carbon dioxide. Methods such as photochemical hydrogen production have less quantum efficiency. Only processes such as PEC and photo-catalytic and biochemical hydrogen production have the potential to replace fossil fuels. For that, an effective semiconductor photo-catalyst, which has the desired band gap and absorbs light in the visible region, needs to be developed. Biochemical methods are highly sensitive to the environment and need to be optimized to work under normal atmospheric conditions.

Current research has shown more progress in this field, and it is hoped that we will see some methods in the future that will produce hydrogen with completely renewable sources with no emission of polluting gases.

DEVELOPMENTS IN PHOTO-ELECTROCHEMICAL AND PHOTOCATALYTIC DECOMPOSITION OF WATER

Despite consistent efforts since 1972, the search for a suitable semiconducting material that can decompose water by PEC (or even photo-catalytically) has not yet been identified. Nearly all known semiconductors, either elemental, compound, or some special configurations in the form of coupled or even sensitized semiconductors or doped systems, have been tried, but none of these systems has yielded the desired (industrially viable) level of decomposition of water. Even in the case of simple semiconductors such as titania, various forms, such as anatase, rutile, and brookite, have been tried, but with little success. Table 9.2 lists typical semiconductors employed for the PEC decomposition of water.

Although a variety of semiconducting materials have been employed for the PEC decomposition of water, success (the efficiency obtained is less than that obtained in solar cell efficiency) appears to be far away.

In addition, the relative position of the bottom of the conduction band and top of the valence band are two important energy parameters for the effective decomposition of water. In the case of a mixture of anatase and rutile, which is the case with commercial samples of TiO_2, a controversy appears to exist with respect to the relative positions of the top of the valence band and the bottom of the conduction band. This aspect has been an important issue; results reported are given in Fig. 9.9.

TABLE 9.2 Typical Semiconducting Materials Employed for Photo-Electrochemical Decomposition of Water

Group	Material	Chemical Formula	Band Gap eV	Type of Band Gap	Remarks
IV	Diamond	C	5.47	Indirect	Thermal conductivity
IV	Silicon	Si	1.1	Indirect	Electrical and mechanical property
IV	Germanium	Ge	0.67	Indirect	Radar devices
IV	Silicon carbide	SiC	2.3	Indirect	LED
III–V	Aluminum nitride	AlN	6.28	Direct	–
III–V	Aluminum phosphide	AlP	2.45	Indirect	–
III–V	Boron nitride	BN	6.36	Indirect	LED
III–V	Boron phosphide	BP	2.0	Indirect	
III–V	Gallium antimonide	GaSb	0.73	Direct	Infrared (IR) detector, LED
III–V	Gallium arsenide	GaAs	1.43	Direct	Electronic applications
III–V	Gallium nitride	GaN	3.44	Direct	Electronic applications
III–V	Gallium phosphide	GaP	2.26	Indirect	LED
III–V	Indium arsenide	InAS	0.36	Direct	Infrared detectors
III–V	Indium nitride	InN	0.7	Direct	Solar cells
III–V	Indium phosphide	InP	1.35	Direct	Optoelectronics
III–V	Aluminum gallium nitride	AlGaN	3.44	Direct	–

III–V	Aluminum gallium phosphide	AlGaP	2.26–2.45	Indirect	LED
VI	Selenium	Se	1.74	–	Rectifiers
II–VI	Cadmium selenide	CdSe	1.74	Direct	Optoelectronics and solar cells
II–VI	Cadmium sulfide	CdS	2.42	Direct	Solar cells
II–VI	Cadmium telluride	CdTe	1.49	–	Solar cells
II–VI	Zinc oxide	ZnO	3.37	Direct	Photo-catalyst
II–VI	Zinc sulfide	ZnS	3.54	Direct	Scintillator
II–VI	Zinc selenide	ZnSe	2.7	Direct	Lasers and LED
II–VI	Zinc telluride	ZnTe	2.25	Direct	Lasers and LED
II–VI	Cadmium zinc telluride	CdZnTe	1.4–2.2	Direct	Gamma ray detector
II–VI	Mercury cadmium telluride	HgCdTe	0–1.5	–	IR detectors
II–VI	Mercury zinc telluride	HgZnTe	$\leq$2.25	–	IR detectors
I–VII	Cuprous chloride	CuCl	3.4	Direct	–
I–VI	Copper sulfide	Cu_2S	1.2	Direct	Solar cells
IV–VI	Lead selenide	PbSe	0.27	Direct	IR detectors
IV–VI	Lead sulfide	PbS	0.37	–	IR detectors
IV–VI	Lead telluride	PbTe	0.32	–	Thermoelectric devices
IV–VI	Tin sulfide	SnS	1.0	Indirect	–
IV–VI	Tin sulfide	SnS_2	2.2	–	–

Continued

TABLE 9.2 Typical Semiconducting Materials Employed for Photo-Electrochemical Decomposition of Water—cont'd

Group	Material	Chemical Formula	Band Gap eV	Type of Band Gap	Remarks
IV–VI	Titanium oxide anatase	TiO_2	3.2	–	Photo-catalytic materials
IV–VI	Rutile	TiO_2	3.02	–	Photo-catalytic materials
IV–VI	Brookite	TiO_2	2.96	–	–
I–VI	Copper(I) oxide	Cu_2O	2.1		Rectifier diodes
I–VI	Copper oxide	CuO	1.2	p-Type	–
III–VI	Uranium dioxide	UO_2	1.3		Thermoelectric applications
IV–VI	Tin oxide	SnO_2	3.7		
II–IV–VI	Barium titanate	$BaTiO_3$	3.0	–	Ferro-electric applications
II–IV–VI	Strontium titanate	$SrTiO_3$	3.3	–	Ferro-electric applications
I–V–VI	Lithium niobate	$LiNbO_3$	4.0	–	Ferro-electric
III–I–VI	Lanthanum copper oxide	La_2CuO_4	2.0	–	Super conductor
VIII–VI	Iron disulfide	FeS_2	0.95	–	Solar cells
VIII–VI	Iron oxide	Fe_2O_3	2.1	–	–
V–VI	Niobium pentoxide	Nb_2O_5	3.4	–	–
V–VI	Tantalum pentoxide	Ta_2O_5	3.9–4.5	–	–
VIII–VI	Nickel oxide	NiO	1.34–2.92	–	–

IV–V–VI	Tin niobate	$SnNb_2O_6$	1.49		
VI–VI	Tungsten oxide	WO_3	2.8		
VI–VI	Tungsten sulfide	WS_2	1.80		
IV–VI	Zirconium oxide	ZrO_2	4.60		
V–VI	Bismuth sulfide	Bi_2S_3	1.5–1.74		
III–VII	Indium oxide	In_2O_3	3.6–3.9	Direct	
I–V–VI	Potassium tantalate	$KTaO_3$	3.42		
I–V–VI	Sodium tantalate	$NaTaO_3$	2.35		
I–V–VI	Sodium niobate	Na NbO_3	3.98		
V–VI–VI	Bismuth molybdate	Bi_2MoO_6	2.70		
VI–VI	Tungsten selenide	WSe_2	1.37		
IV–VI	Tin oxide	SnO_2	3.50		
IV–VI–VI	Lead tungstate	$PbWO_4$	4.7		
V–VI	Antimony oxide	Sb_2O_3	3.0		
V–VI	Bismuth oxide	Bi_2O_3	2.80		
VIII–VI	Cobalt oxide	CoO	2.60		
II–IV–VI	Magnesium titanate	$MgTiO_3$	3.70		

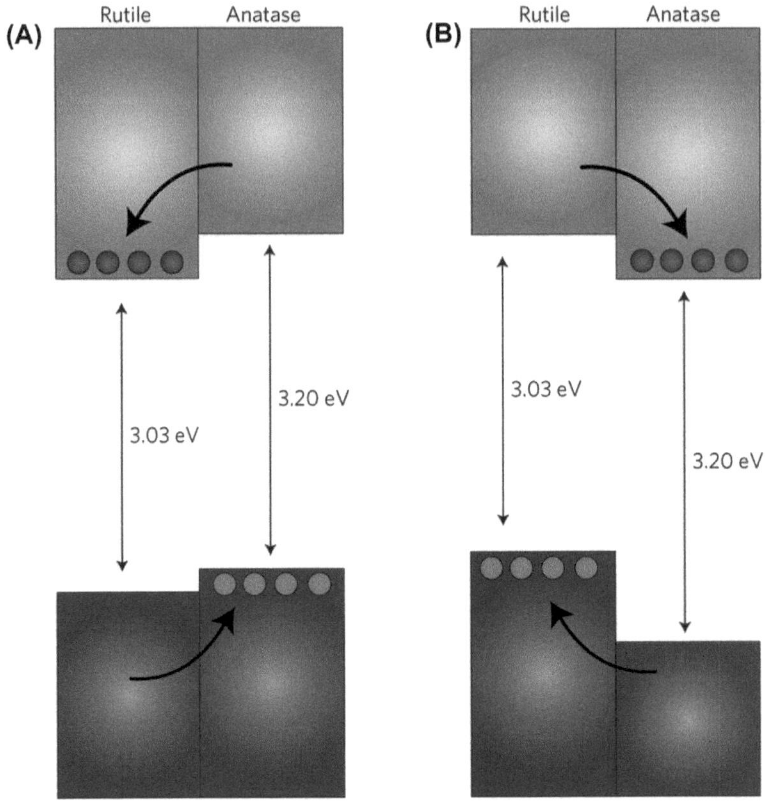

FIGURE 9.9 Two proposed valence and conduction band alignment mechanisms for the anatase–rutile interface. *Red arrows* indicate the flow of electrons (holes) in the conduction band (valence band). *Blue and orange dots* represent electrons and holes, respectively. *Reproduced from D.O. Scanlon, C.W. Dunnill, J. Buckeridge, S.A. Shevlin, A.J. Logsdail, S.M. Woodley, C. Richard, A. Catlow, M.J. Powell, R.G. Palgrave, I.P. Parkin, G.W. Watson, T.W. Keal, P. Sherwood, A. Walsh, A.A. Sokol, Nat. Mater. 12 (2013) 798–801.*

Although TiO_2 has been examined extensively for the PEC decomposition of water, this wide–band gap material restricts solar to hydrogen conversion efficiency. In an attempt to use the solar spectrum extensively, lower–band gap materials including tungsten oxide (WO_3), hematite (Fe_2O_3), and a variety of nonoxide semiconductors such as III–V elemental semiconductors such as GaAs materials have been tried, and the efficiency of conversion could be pushed to near 20%. The key challenge to promoting PEC cells for market exploitation is the materials science and engineering. The development of corrosion-resistant materials and their processing technologies are stumbling blocks together with requirements based on band gap values, band–edge alignments, surface activity stability in solution, and cost.

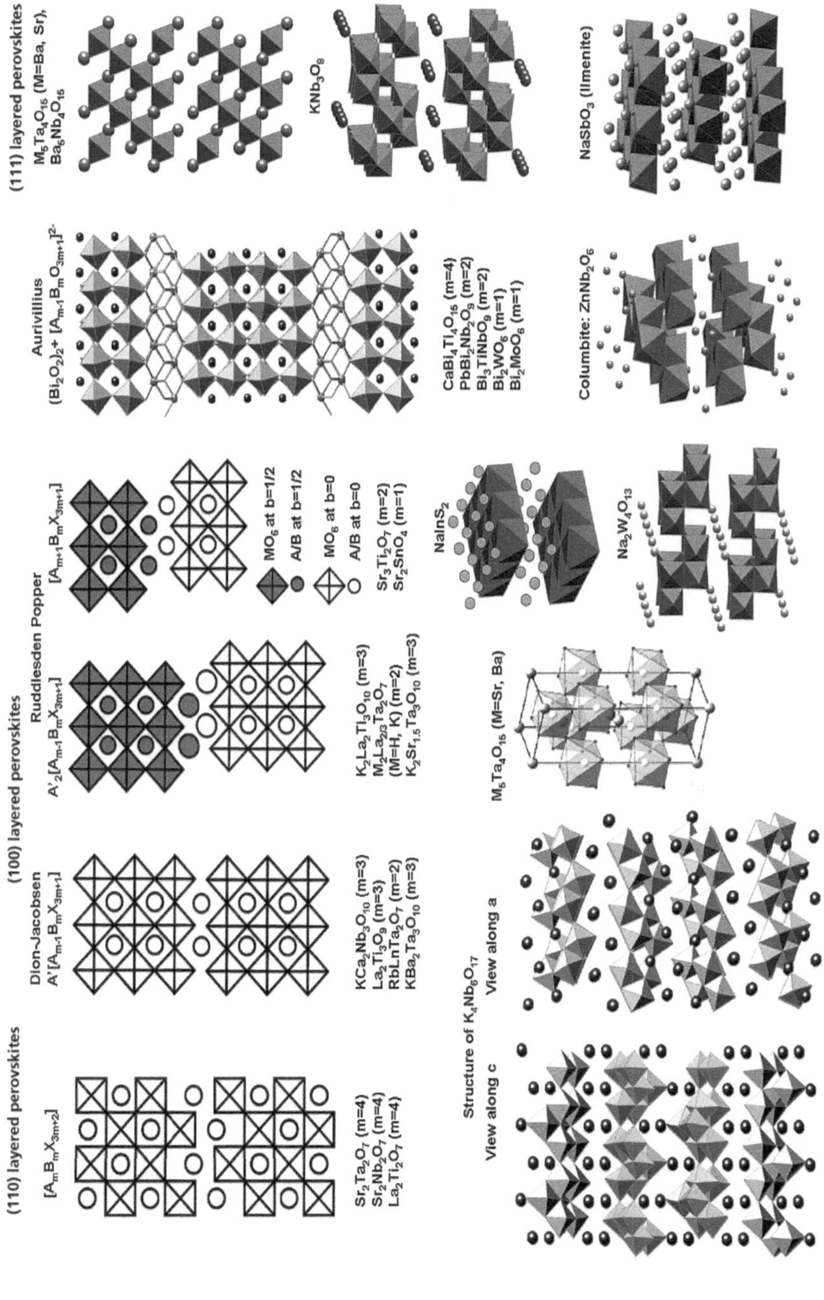

FIGURE 9.10 Typical structures of a set of PEC semiconductors. *Reproduced from F.E. Osterloh, Chem. Mat. 20 (1) (2008) 35–54.*

Various attempts have been made to increase the efficiency of the PEC decomposition of water. All of these attempts are directed to rationalize the methodologies for engineering the material of choice. Table 9.2 lists most of the element-based, oxide, and multiple semiconductors that have been experimented upon. All these attempts are based on the propositions that the electron energy level should be suitable for the reduction of H^+ ions (that is, the bottom of the conduction band must be more negative to the reduction potential of H^+ ions and the top of the valence band must be more positive to the oxygen evolution potential) [5–10]. However, the crystal structure of these materials has also been assembled in numerous publications and most oxide materials have some kind of layered structure that probably facilitates electron transfer reactions. Structures of a set of compounds employed as PEC materials are given in Fig. 9.10.

Various configurations employed for PEC applications can be summarized as shown in Fig. 9.11 [4,11]. The state of material (bulk state or nanostate) can suitably promote the decomposition of water to evolve hydrogen and oxygen in the nanostate while bulk material may not be able to evolve either hydrogen or oxygen since the conduction band is not negative to hydrogen evolution potential or oxygen evolution potential is not more positive to the position of the valence band top. A representation is shown in Fig. 9.12.

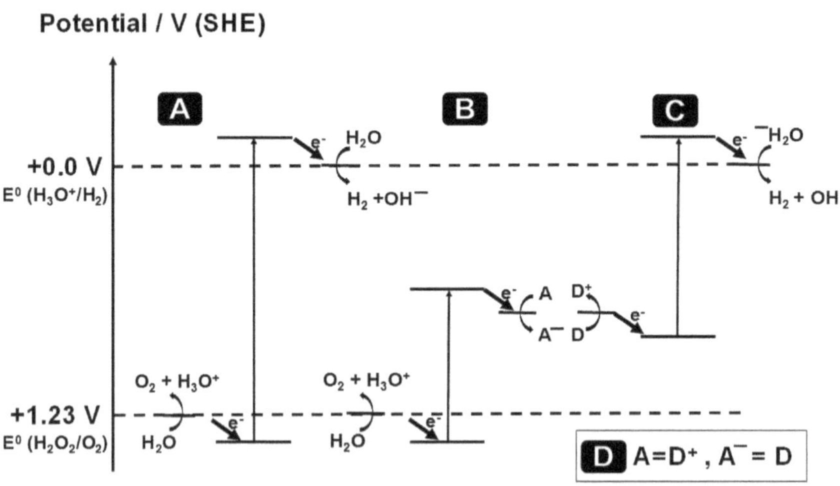

FIGURE 9.11 Potential energy diagrams for photochemical water splitting at pH 0: (A) single semiconductor system promoting both oxidation and reduction reactions; (B) with an electron acceptor promoting oxygen evolution; (C) with an electron donor and facilitating hydrogen evolution; (D) dual semiconductor system (z scheme) employing a redox shuttle promoting a redox substrate. *Reproduced from J. Phys. Chem. B 109 (2005) 16052.*

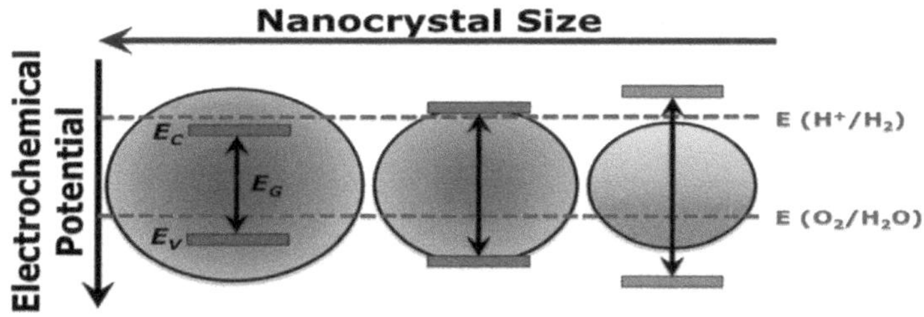

FIGURE 9.12 Energy positions of valence and conduction bands in bulk and nanostate of the material. In the nanostate the material is capable of evolving both oxygen and hydrogen whereas in the bulk state it can evolve either of them. *Reproduced from F.E. Osterloh, Chem. Soc. Rev. 42 (2013) 2294.*

REFERENCES

[1] National Hydrogen Association; United States Department of Energy. "The History of Hydrogen" (PDF) hydrogenassociation.org. National Hydrogen Association. p. 1.

[2] Daedalus or Science and the Future, A paper read to the Heretics, Cambridge, on February 4, 1923 – Transcript 1993.

[3] A. Fujishima, K. Honda, Nature 238 (37) (1972).

[4] F.E. Osterloh, Chem. Soc. Rev. (42) (2013) 2294.

[5] R. Narayanan, B. Viswanathan, Chemical and Electrochemical Energy Systems, University Press, 1998.

[6] www.rmi.org/sitepages/pid557.php.

[7] atom.ecn.purdue.edu/~vurade/PEC%20Generation%20of%20Hydrogen/Introduction%20to%20PEC%20hydrogen%20production.htm.

[8] www.fao.org/docrep/w7241e/w7241e00.htm#Contents.

[9] T. Ohta, Solar-Hydrogen Energy Systems, Pergamon Press, 1979. http://web.mit.edu/~pweigele/www/being/content/how/bio.html, 6.

[10] B. Viswanathan, M. Aulice Scibioh, Photoelectrochemistry: Principles and Practices, Alpha Science International Ltd, Oxford U K, 2014.

[11] F.E. Osterloh, Chem. Mater. 20 (2008) 35.

Chapter 10

Hydrogen Storage

Hydrogen is considered an alternatives to fossil fuels, as can be seen from the comparison of properties shown in Table 10.1. As an energy carrier, hydrogen has three components: production, storage, and distribution. Among these three components for the transformation to a hydrogen-based economy, production possibilities were briefly considered in the previous chapter. In this chapter, opportunities for the storage of hydrogen in its various forms will be considered. Hydrogen as an energy carrier can be stored in all three states of matter. A simple geometrical representation of storage possibilities is shown in Fig. 10.1.

The gas and liquid states of storage of hydrogen have considerable energy and weight penalty in addition to safety issues. However, these two forms of

TABLE 10.1 Comparison of Fuel Properties of Hydrogen With Those of Other Typical Fossil Fuels

Property	Unit	Hydrogen H_2	Methane CH_4	Gasoline—$(CH_2)_n$–
Lower heating value	kWh/kg	33.33	13.9	12.4
Self-ignition temperature	K	858	813	498–774
Flame temperature	K	2318	2148	2473
Ignition limits in air	Vol%	4–75	5.3–15	1.0–7.6
Minimal ignition energy	mW	0.02	0.29	0.24
Flame propagation in air	M/s	0.02	0.4	0.4
Explosion energy	Kg TNT/m^3	2.02	7.03	44.22
Diffusion coefficient in air	cm^2/s	0.61	0.16	0.05
Toxicity		No	No	High

Energy Sources. http://dx.doi.org/10.1016/B978-0-444-56353-8.00010-1

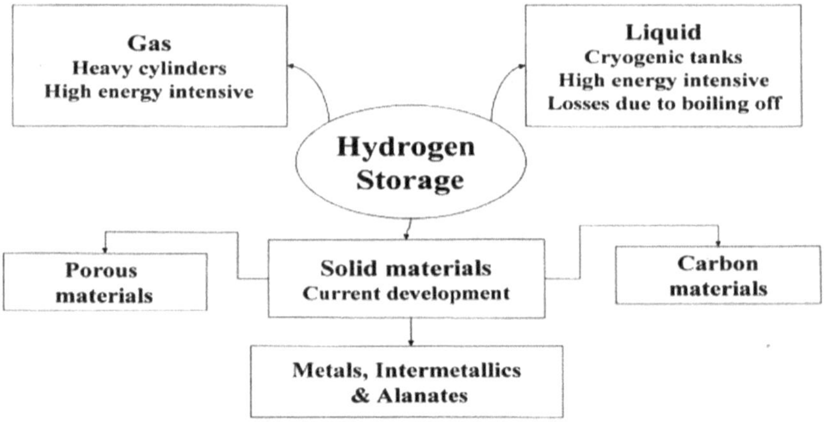

FIGURE 10.1 Possibility of storing hydrogen in gaseous, liquid, and solid states.

storage have been in practice for a long time. From various points of view, solid-state hydrogen storage appears to be advantageous and may become a viable option. Hence this aspect alone will be considered in this chapter.

For solid-state hydrogen storage, the following characteristics are expected for the system to be adopted:

1. Favorable thermodynamics: that, is the storage and release should take place without much heat requirement.
2. The kinetics of adsorption and desorption should be fast enough to be useful for applications on hand.
3. The extent of storage should be sufficiently large enough so that it can be adopted for mobile applications (volumetric and gravimetric density).
4. The material employed should withstand a sufficient cycle number for both adsorption and desorption.
5. The material of choice should have sufficient mechanical strength and durability.
6. The system chosen should be capable of a good heat transfer medium.

For stationary systems, the weight and volume of the system used for hydrogen storage are not key factors. However, for mobile applications such as fuel cell electric vehicles or hydrogen-fueled (internal combustion) cars, a hydrogen storage system has to be compact, safe, and affordable. In 1996, the International Energy Agency established the "Hydrogen Storage Task Force" to search for innovative hydrogen storage methods and materials. The US Department of Energy (DOE) hydrogen plan has set a standard for this discussion by providing a commercially significant benchmark for the amount of reversible hydrogen absorption. The benchmark requires a system weight efficiency (the ratio of stored hydrogen weight to system weight) of 6.5 wt% hydrogen and a volumetric density of 62 kg H_2/m^3. Because a vehicle powered by a fuel cell would require more than 3.1 kg of hydrogen for a 500 km range. Fig. 10.2 shows the various storage options that meet long-term needs. Current

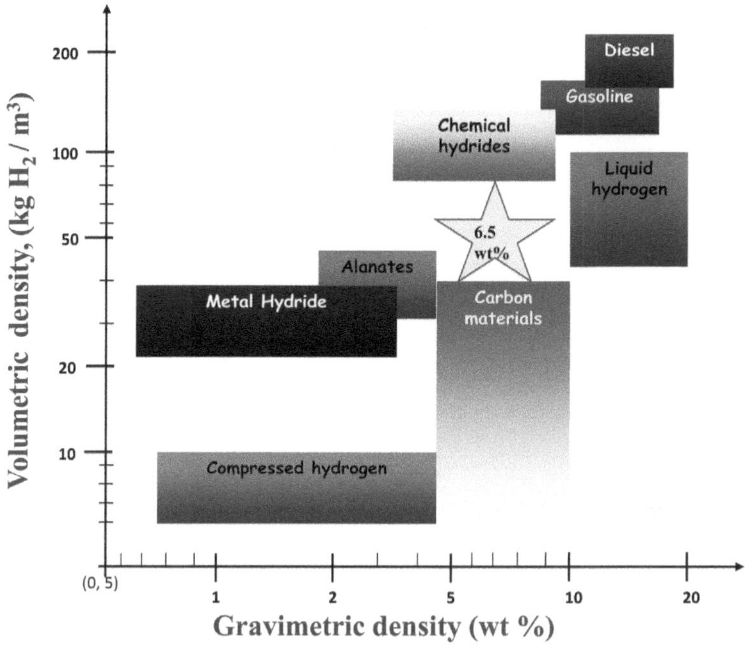

FIGURE 10.2 Comparison of volumetric and gravimetric density of hydrogen with various storage media. *Reproduced from S. Hynek, W.J. Fuller, J. Bentley, Int. J. Hydrogen Energy 22 (1997) 601–610.*

high-pressure hydrogen tanks and liquid hydrogen meet some although clearly not all near-term targets. The hydrogen storage materials for 2007 targets were based on systems such as solid-state (eg, metal hydrides) or liquid systems. The focus of the DOE was on materials-based technologies to meet 2010 targets and with potential eventually to meet 2015 targets, as given in Table 10.2, although these have not yet been met. The targets include a 20% penalty for the assumption that hydrogen storage systems (unlike conventional gasoline tanks) are not conformable and have limitations regarding how they may be packaged within the vehicle. The targets also assume a factor of 2.5–3 in terms of efficiency improvement in using a fuel cell power plant compared with a conventional gasoline internal combustion engine. If efficiency improvements are not as high as projected, this would clearly dictate even more challenging requirements for onboard hydrogen storage to achieve a comparable driving range. Currently, extensive efforts have been made to develop methods for hydrogen storage, including metal hydrides, porous solids, and nanostructured carbons. The development of a high-capacity lightweight material that could be used to store hydrogen reversibly under ambient conditions seems to be attainable. A viable onboard automotive hydrogen storage system must be compact, lightweight, low-cost, and safe. It must be capable of storing enough hydrogen to provide a reasonable traveling range and good dormancy (ability to retain hydrogen for long time without leakage.

TABLE 10.2 Department of Energy Targets for On-Board Hydrogen Storage

		DOE Technical Targets		
Storage Parameter	**Unit**	**2007**	**2010**	**2015**
Gravimetric energy density	MJ/kg	5.4	7.2	10.8
Gravimetric hydrogen density	wt%	4.5	6	9
Volumetric energy density	MJ/L	4.32	5.4	9.72
Volumetric hydrogen density	kg H_2/m^3	36	45	81
Cycle life	Cycles	500	1000	1500
Minimum/maximum delivery temperature	°C	−20/85	−30/85	−40/85
Minimum H_2 delivery pressure	atm	8	4	3
System fill time (5 kg H_2)	Minimum	10	3	2.5

GASEOUS HYDROGEN STORAGE

The traditional way to store hydrogen is in the gaseous form in pressure vessels. Because hydrogen behaves like an ideal gas at ambient temperatures and below, it satisfies the ideal gas law, $PV = n\mathrm{R}T$, where n is the amount of hydrogen in moles and R is the gas constant.

In a given volume and temperature, the energy density of storage is measured by increasing the pressure of the storage vessel. The only problem with this is the safety features and regulations that limit the allowable pressure. However, the gravimetric energy density of hydrogen storage largely depends on the material of the container, because light materials usually do not tolerate high pressures. Today modern storage tanks can be made with carbon fiber composite materials that are ultralight with permeation resistance and a tough impact-resistant reinforced shell. These are 10 times stronger than steel, which allows them to store hydrogen at pressures in excess of 350 bar, which can be used safely for over 100,000 refill cycles and to withstand surface temperatures of over 1073 K. However, the potential risk associated with very high-pressure systems is the procedure of compressing hydrogen. This raises important practical problems because there is currently no practical solution to refilling cylinders in a domestic situation to such high pressures rapidly and safely.

HYDROGEN STORAGE

Liquid hydrogen has been used as a fuel in space technology for several years. It is light and has fewer potential risks compared with compressed gas in terms of storage pressure. However hydrogen liquefies at 20.25 K. Thus storage vessels require sophisticated insulation techniques to minimize unavoidable heat transfer leading to hydrogen loss via boil-off. Research in this area is focused on finding improved insulation and cooling methods. Hydrogen boil-off and the high cost of sophisticated insulating techniques required for maintaining very low temperatures coupled with the initial energy cost of creating liquid hydrogen make this method of hydrogen storage impractical for application in hydrogen fuel cell–powered vehicles.

SOLID HYDROGEN STORAGE

Among the various options for hydrogen storage, storage by the solid matrices alone appear to be appropriate. This method offers more safety and comfort for transport applications.

Metal Hydrides

Hydrogen can be combined with many metals to form hydrides that will release hydrogen upon heating. Metal hydrides are composed of metal atoms that constitute a host lattice and hydrogen atoms trapped in interstitial sites. Because absorption of hydrogen increases the size of lattices, the metal is usually ground into powder to prevent the decrepitation of metal particles. There are two possible ways of hydriding a metal: direct dissociative chemisorption and electrochemical splitting of water. These reactions are, respectively,

$$M + (x/2)H_2 \rightarrow MH_x \quad \text{and}$$

$$M + (x/2)H_2O + (x/2)e^- \rightarrow MH_x + (x/2)OH^-$$

where M represents the metal. In electrochemical splitting there has to be a catalyst such as palladium.

A schematic of hydrogen chemisorption is shown in Fig. 10.3. As shown in the figure, molecular hydrogen reaches a shallow potential minimum near the surface and the atomic hydrogen has a deeper minimum almost at the surface. In the metal lattice hydrogen has periodic potential minimum at the interstitial site. As hydrogen molecule approaches the metal surface, weak van der Waals forces begin to act upon the molecule, drawing it closer. The molecule reaches the potential minimum. Large forces will be required to force it into molecular form. However, the dissociation energy of the hydrogen molecule is exceeded by the chemisorption energy. Thus the hydrogen molecule dissociates and individual hydrogen atoms are attracted to the surface by chemisorptive forces and they reach the potential well. Even thermal energy at an ambient temperature is enough to increase the vibrational amplitude of hydrogen atoms, which thus can reach and enter the metallic surface.

Metal and hydrogen usually form two different kinds of hydrides: α-phase and β-phase. In the α-phase only some hydrogen is dissolved and in the β-phase

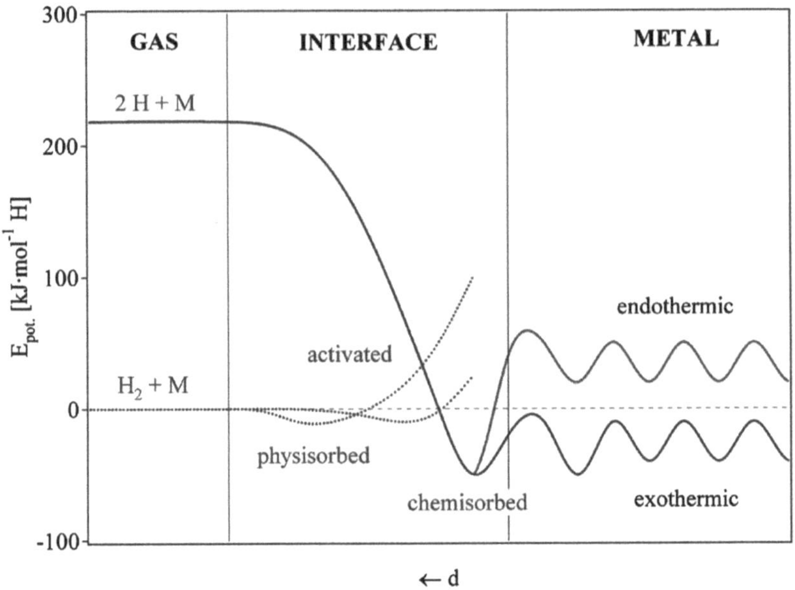

FIGURE 10.3 Schematic of hydrogen chemisorption on metal. *Reproduced from L. Schlapback (Ed.), Hydrogen in Intermetallic Compounds, Springer Verlag, Germany, 1988.*

a stoichiometric phase is formed. For example, in the case of Mg_2Ni for the two hydrides the composition is $Mg_2NiH_{0.3}$ and Mg_2NiH_4. Initially the α-phase is formed; later, repeated cycling leads to the formation of the β-phase:

$$Mg_2NiH_{0.3} + 3.7H \rightarrow Mg_2NiH_4$$

A common characterization method of a metal hydride system is to follow the pressure–concentration–temperature traces in the form of *P*-C-T isotherms. A theoretical pressure–concentration isotherm with α- and β-phases is shown in Fig. 10.4. The concentration, ie, the hydrogen capacity, is usually defined as hydrogen atoms per metal species (H/M). To characterize the metal hydride, it is convenient to use the maximum capacity $(H/M)_{max}$. The reversible capacity $\Delta(H/M)$ defined as the plateau width is also a useful tool when considering the hydrogen storage capacity of metal hydrides.

Thermodynamic reaction equilibrium is defined by the equilibrium constant K, $RT \ln K = \Delta H - T\Delta S$, where ΔH is the reaction enthalpy, and ΔS is the reaction entropy. For a solid–gas reaction, the equilibrium constant reduces to the pressure of the gas. Thus the Van't Hoff equation is obtained as $\ln P = [(\Delta H/RT) - (\Delta S/R)]$. The plot of ln *P* versus 1/*T* gives the Van't Hoff plot. The reaction enthalpy can be derived from the gradient of the plot and the plot indicates the suitability of the *P-T* behavior of a hydride for practical applications. The theoretical Van't Hoff plot usually describes the real properties of metal hydrides (Fig. 10.5).

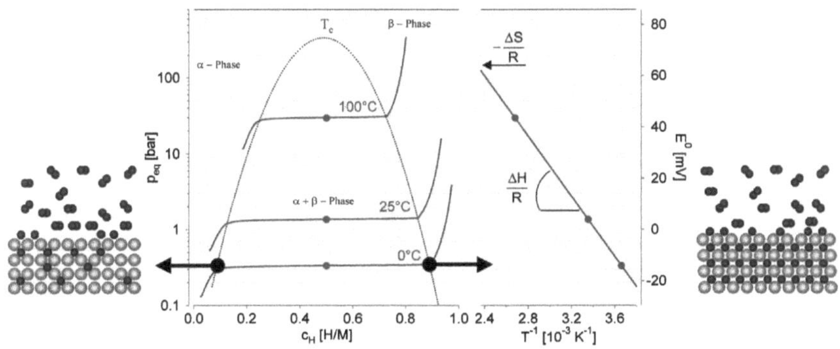

FIGURE 10.4 Pressure–composition isotherms for hydrogen absorption in a typical metal hydride. *Reproduced from L. Schlapback, A. Zuttel, Nature 414 (2001) 353–358.*

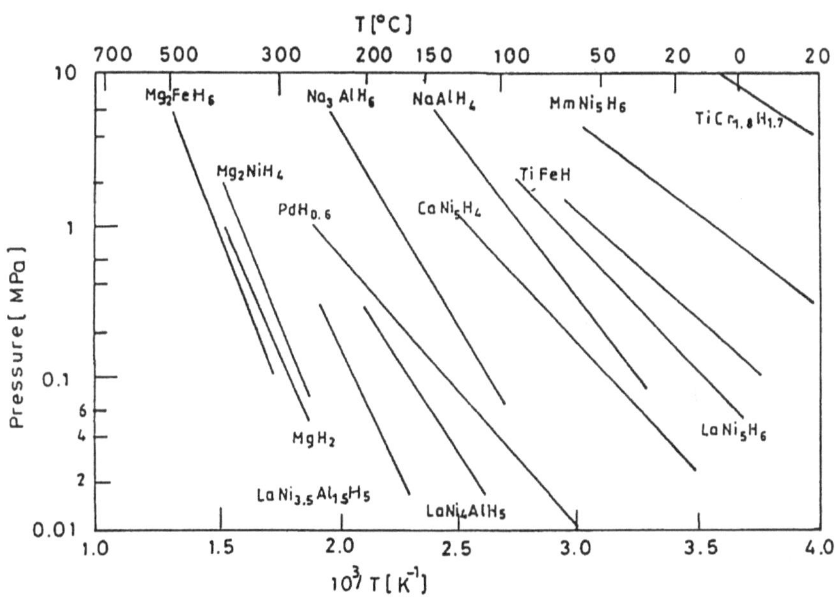

FIGURE 10.5 Van't Hoff plot of selected hydrides. The stabilization of the hydrides of $LaNi_5$ by partial substitution of Ni by Al in $LaNi_5$ is shown, as well as the substitution of La by misch metal. *Reproduced from A. Zuttel, Naturwissenshaften 91 (2004) 157–172.*

In Fig. 10.4 the solid solution (α-phase), the hydride phase (β-phase), and the region of the coexistence of the two phases are shown. The region of coexistence is characterized by the flat plateau and ends at the critical temperature, T_C. The construction of the Van't Hoff plot is shown on the right-hand side. The slope of the line is equal to the enthalpy of formation divided by the gas constant, and the intercept is equal to the entropy of formation divided by the gas constant. The reaction enthalpy of hydride formation is an important quantity. It is usually negative, so the reaction is exothermic, and thus hydride formation releases energy. Therefore dehydrogenation needs energy input. Because most applications are at ambient temperature, or at least in the range 273−373 K, the reaction enthalpy should be small so that the hydride could take heat from the surroundings while releasing hydrogen. In some fuel cell systems, the hydride can take heat directly from the fuel cell. The reaction enthalpy also directly affects the stability of a hydride as the pressure varies exponentially. The essential requirements that should be satisfied by metal hydrides proposed for hydrogen storage application at a commercial level are:

- high content of hydrogen storage
- Facile reversibility of formation and decomposition reactions. The hydride should be decomposable at moderate temperatures that can be provided from locally available heat sources such as solar, automobile exhaust, and waste heat sources
- absorption-desorption kinetics that should be compatible with the charge−discharge requirements of the system
- equilibrium dissociation pressure of the hydride at peak desorption rate that should be compatible with the safety requirements of the hydride containment system. The hydride itself should have a high safety factor
- sufficient chemical and dimensional stability of hydride to permit its being unchanged over a long number of charge−discharge cycles
- minimal hysteresis in adsorption−desorption isotherms
- reasonable resistance of hydride to deactivation by low concentrations such as O_2, H_2O, CO_2, and CO
- affordable total cost of hydride (raw materials, processing, and production) for intended application. Long-term availability of raw materials (ie, metal resources) must be ensured. The cost of the hydride system (which includes its containment) per unit of reversibly stored hydrogen should be as low as possible.
- moderate cost of storage vessel, ancillary equipment, and fabrication/installation
- low operating and maintenance costs and purchased energy requirements (that is, energy other than waste energy and energy extracted from ambient air) per storage cycle

A careful combination of technical and economic considerations will determine the suitability of a hydride product for a given hydrogen storage or hydrogen containment application. The hydrogen storage capacity of some metals and intermetallics are given in Table 10.3.

Metal hydrides are effective in storing large amounts of hydrogen in a safe and compact way. All reversible hydrides working around ambient temperature and atmospheric pressure consist of transition metals; therefore the gravimetric hydrogen density is limited to less than 3 atm%.

TABLE 10.3 Hydrogen Storage Capacity of Metallic and Intermetallic Systems

Material	P_{des} (atm)	Temperature (K)	H-atoms/cm^3 ($\times 10^{22}$)	wt% of Hydrogen
MgH_2	~10	552	6.5	7.6
Mg_2NiH_4	~10	528	5.9	3.6
$FeTiH_2$	4.1	265	6.0	1.89
$LaNi_5H_6$	1.8	285	5.5	1.37

It remains a challenge to explore the properties of lightweight metal hydrides. This is formed by palladium-coated nanostructured magnesium films, which are reported to be able to adsorb approximately 5 wt% at 373 K under a hydrogen atmosphere of 1 bar and completely desorb below 373 K under vacuum. It has also been found that in some metal hydride systems such as magnesium hydride, the absorption could be drastically improved by ball milling magnesium with nanometer-size catalysts of nickel or iron under a 10 bar argon atmosphere. To date the highest recorded hydrogen capacity for a metal hydride applicable for fuel cell electric vehicles is 2.0 wt%. This was recorded for a chromium–titanium–vanadium alloy at 313 K at ambient pressure. Thus with the promising results observed to date, metal hydrides clearly have potential as hydrogen storage systems with the limits of this potential yet to be fully defined.

Complex Metal Hydrides

Nontransition metal hydrides, also known as complex metal hydrides, have been shown to have a high theoretical hydrogen storage capacity, as indicated in Table 10.4.

Group I, II, and III metals, eg, Na, Mg, B, and Al, result in a large variety of metal hydrogen compounds. They are especially interesting because of their lightweight and high hydrogen storage capacity. The hydrides of boron tetra hydroborates $M(BH_4)_x$ and the tetra hydro aluminates (alanates) are interesting storage materials. Alanates can decompose to give hydrogen through a series of reactions:

$$3NaAlH_4 \rightarrow Na_3AlH_6 + 3H_2 \quad \Delta H = 37\ \text{kJ/mol}\ (3.7\ \text{wt\%}\ H_2\ \text{released})$$

$$Na_3AlH_6 \rightarrow NaH + Al + 3/2H_2 \quad \Delta H = 47\ \text{kJ/mol}\ (1.9\ \text{wt\%}\ H_2\ \text{released})$$

These two reactions occur at two different temperatures, and the second step of decomposition requires temperatures above 373 K. The total

TABLE 10.4 Common Complex Hydrides for Hydrogen Storage Applications

Hydride	H_2 Content (wt%)[a]
$LiAlH_4$	10.5
$NaAlH_4$	7.5
$KAlH_4$	5.7
$Be(AlH_4)_2$	11.3
$Mg(AlH_4)_2$	9.3
$Ca(AlH_4)_2$	7.7
$Ti(AlH_4)_4$	9.3
$LiBH_4$	18.0
$NaBH_4$	10.4
$Al(BH_4)_3$	17.0

[a]*Maximum theoretical hydrogen storage capacity.*

hydrogen capacity is 5.6 wt%. Use of alanates as a reversible hydrogen storage system has several advantages over other conventional metal hydride systems. The low molecular weight of the alanates allows for a greater amount of hydrogen storage per unit weight of storage material. These potential benefits of using alanates as hydrogen storage systems were overshadowed by the fact that the synthesis of alanates is a difficult process and the kinetics of the reaction used to store the elemental hydrogen is slow. However, it was shown that by doping sodium alanate ($NaAlH_4$) with titanium catalyst, hydrogen binding of up to 4 wt% at 473 K could be made reversible with an enhanced kinetic release rate. Both ball milling and doping with transition metals such as vanadium result in lowering of the decomposition reaction temperature. Another report stated that milling in the presence of carbon enhanced the reversible hydrogen storage capacities of sodium alanates. It was reported that storage capacities of 2.5−3 wt% have been achieved at 353−413 K and 4.5−5 wt% at 423−453 K. However, the required temperatures and reaction rates for operation of sodium alanates and variants are inadequate for application and the exact mechanism of the catalyst is still unknown. In addition, the hydrogen release of the systems is not yet optimized. It is because the full storage capacity is lost after the first cycle. This leaves researchers who work with sodium alanates with many questions to answer.

Zeolites

Zeolites are microporous compounds with an effective pore size of about 0.3−1.0 nm. At ambient temperatures, the pore entrances are not sufficient to allow hydrogen molecules into the structure. However, at elevated temperatures and pressures, hydrogen molecules are forced to penetrate the cavities. Subsequent cooling to room temperature and release of excess pressure cause the hydrogen molecules to become encapsulated (or trapped) within the pores. When the zeolite is reheated the hydrogen molecules are released. Zeolites have structures based on TO_4 tetrahedra, where T is a silicon or aluminum atom. Depending on the structure, the Si/Al ratio, and substituting atoms such as Na, K, and Pd, zeolites are named A, X, Y, or mordenite. Examples of the pore structure of zeolites are given in Fig. 10.6.

The hydrogen storage capacity of zeolites is poor. At temperatures in the range of 473−573 K and pressures of about 100−600 bar, about 0.1−0.8 wt% of hydrogen is stored. By modifying the zeolite, hydrogen storage capacity is increased. One modification is the increase in ionic radius from Na^+ to K^+ but a decrease for larger Rb^+ and Cs^+ ions. The observed trend has been explained by the decrease in effective critical pore size as the ionic radius increases, which implies the increased ability for encapsulations while the available intracrystalline void volume per gram of zeolite decreased. The maximum storage capacity of 9.2 cm^3 (STP)/g (ie, 0.08 wt%) (STP stands for standard temperature and pressure) obtained for sodalite if loaded at 573 K and 100 bar suggests that the hydrogen encapsulation capacity is generally higher for zeolites with a reasonable portion of small cavities in their structure. The stability of zeolites for cycling hydrogen storage has not been studied thoroughly.

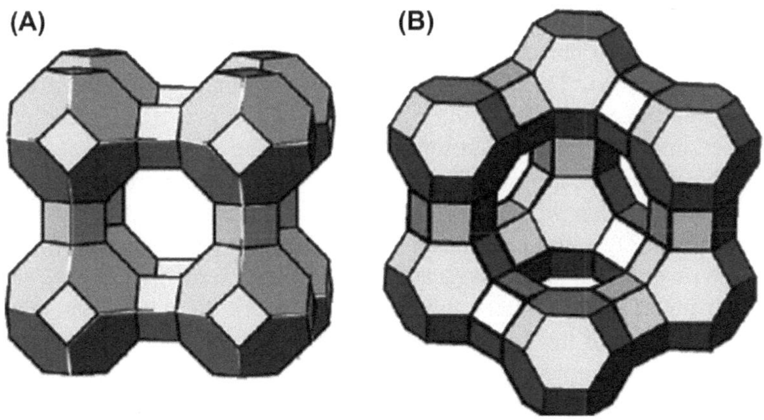

FIGURE 10.6 Framework structures of zeolites: (A) zeolite A, and (B) zeolite X and Y. The corners on each framework represent Si or Al. These are linked by oxygen bridges represented by the lines of the frameworks.

Glass Spheres

Glass spheres are small hollow glass microballoons whose diameter varies from about 25 to 500 mm and whose wall thickness is about 1 mm. The microspheres are filled by heating in high-pressure hydrogen to temperatures sufficient for rapid diffusion of hydrogen into the microspheres. Upon cooling the low diffusivity of hydrogen at ambient temperature causes the gas to be retained in the microspheres. Hydrogen is then released when needed by reheating the microspheres. Because of the inherently poor thermal conductivity of inorganic glasses, a problem further exacerbated by the morphology and size of hollow microspheres, poor hydrogen release rates have limited further development and implementation of this hydrogen storage method. A possible solution to the poor hydrogen release rates has been realized with the discovery of photo-induced outgassing in which a high-intensity infrared light is used to elicit hydrogen release in selectively doped glasses. This process results in faster response times for the release of hydrogen in glasses than can be obtained by normal heating and may provide a path to superior performance for hydrogen storage in glass microspheres. However, high pressure and high temperature are stumbling blocks to the use of glass spheres as a hydrogen storage medium for mobile applications. The storage capacity of spheres is about 5–6 wt% at 200–490 bar.

Metal Organic Frameworks

Studies have shown a promising new class of hydrogen storage compounds of porous metal coordination structures. The material is composed of a metal organic framework (MOF) with a cubic three-dimensional, extended porous structure. The open channels in MOFs are perfectly ordered, allowing the effective access of hydrogen to the interior space. The internal surface of MOFs can be modified to change the channel curvature, thus enhancing hydrogen sorbent interactions. The “one-pot” synthesis is simple, cost-effective, and reproducible. More important, the structures of these materials, including the metal building unit, pore dimension, shape, size, and volume, can be systematically tuned for the purpose of modifying and improving hydrogen uptake and adsorption/desorption properties. In particular, cubic frameworks consisting of tetrahedral $[Zn_4O]^{6-}$ clusters linked by organic units were discovered to adsorb hydrogen reversibly up to 1.6 wt% at 77 K and 1 atm. Experiments have shown that both metal oxide clusters and organic units are the adsorption sites for hydrogen. Therefore the hydrogen storage capacities of MOFs can be significantly modified by replacement of Zn^{2+} with similar metal ions such as Mg^{2+}, Ni^{2+}, or Cu^{2+} and/or a variation of organic linking units. A hydrogen storage capacity of 3.8 wt% in metal-benzenedicarboxylate has been reported at 77 K and 1.6 MPa. However no significant hydrogen uptake on MOFs has been reported at room temperature. Understanding the relationship between these structural factors and hydrogen uptake in a more quantitative way would be helpful for the development of MOFs as efficient hydrogen storage materials. As a potential hydrogen storage material, MOFs still have to be developed and the results have yet to be independently verified.

Chemical Storage

Chemical hydrogen storage may offer options with high-energy densities and potential ease of use, particularly if systems involve liquids that may be easily dispensed using infrastructure similar to today's gasoline refueling stations. Most of these reactions are irreversible. Therefore the spent storage material would have to be regenerated off-board the vehicle because they cannot be reconstituted simply by applying an overpressure of hydrogen gas at a modest temperature and pressure. A number of chemical systems using both exothermic and endothermic hydrogen release are currently under investigation.

Chemical compounds containing hydrogen can also be considered a kind of hydrogen storage. These include methanol, ammonia, and methyl cyclohexane. Under STP conditions all of these compounds are in liquid form and thus the infrastructure for gasoline could be used for transportation and storage of these compounds. There is a clear advantage compared with gaseous hydrogen, which demands leak-proof, preferably seamless piping and vessels. The hydrogen storage capacity of these chemical compounds is good: 8.9 wt% for methanol, 15.1 wt% for ammonia, and 13.2 wt% for methylcyclohexane. These figures do not include the containers in which the liquids are stored. Because the containers can be made of lightweight composites or even plastic in some cases, the effect of the container is negligible especially with larger systems.

A number of processes have been tried to release hydrogen from ammonia borane in the solid state and in solution. Catalysts, including a range of acids and transition metal complexes, have been demonstrated and are being optimized to enhance the amount of hydrogen released as well as the overall kinetics for hydrogen release. However, efficient and cost-effective regeneration of the spent fuel resulting from the dehydrogenation of ammonia-borane is critical to the successful application of ammonia-borane as an on-board hydrogen storage material.

The chemical storage of hydrogen also has some disadvantages. The storage method is irreversible, the compounds cannot be charged reproducibly with hydrogen. The compounds must be produced in a centralized plant and the reaction products have to be recycled somehow. This is difficult especially with ammonia, which produces highly environmentally unfavorable nitrogen oxides. Other compounds produce carbon oxides, which are also unfavorable.

Carbon Materials

Why Use Carbon Materials for Hydrogen Storage?

Carbon materials are attractive candidates for hydrogen storage because of their combined adsorption ability, high specific surface, pore microstructure, and low mass density. Hydrogen sorption by carbon materials has assumed considerable importance especially in the context of these properties.

The Coordination Number is Variable/Expandable

The carbon cluster that can be formed can be varied with the cluster size starting from 2 to 2000 carbon atoms. The three well-known forms of carbon are diamond, graphite, and fullerenes. In diamond each carbon has four bonds to its neighbors and forms a three-dimensional lattice. Graphite is built of two-dimensional hexagonal sheets of carbon atoms in which the carbon—carbon distance in the plane is 1.42 Å and the distance between the sheets is 3.35 Å.

They Promote New Morphologies

Various morphologies are promoted by carbon alone, which form sheets in the case of graphite, whereas a hard crystalline solid is seen in diamond. Carbon materials with different morphologies such as cone, onion, belt, and sphere are known.

Covalent Character Retention

Although a variety of morphologies are possible, the covalence of the carbon atoms is maintained and retained in the structure. The strong covalent in-plane bonding and weak van der Waals interplane bonding result in anisotropic physical properties that are useful for applications in lubrication and other processes requiring "slippage" between layers. The in-plane carbon—carbon bonds are shorter than those of diamond but the interlayer distance is large.

Variable Hybridization Is Possible

The existence of variable hybridization states is possible in carbon from sp-sp^3. Variable valency states of fractional values are also possible. The classical example is the difference between sp^3 and sp^2 bonding properties seen in diamond and graphite.

For diamond the three-dimensional, fourfold coordinated sp^3 structure is rigid and almost isotropic in its properties. In contrast, the sp^2 bonding in graphite is planar and threefold coordinated in the planes with weak bonding between planes. Besides the usual hybridization (sp^3, sp^2, and sp), structures involving more than a single type of hybridization (mixed forms) and intermediate hybridization of the type sp^n (with $3 > n > 1$, $n \neq 2$) are included.

TABLE 10.5 Classification Scheme for Carbon Allotropes, Molecular Crystals, and Derived Forms

sp^3	sp^2	sp
Diamond	Graphite	Carbyne
Cubic	Hexagonal	α-Carbyne
Hexagonal	Rhombohedral	β-Carbyne
		Chaoite
$sp^3 + sp^2 + sp$	sp^n (with $3 > n > 1$, $n \neq 2$) Intermediate forms of carbon	
Mixed forms of carbon	$3 > n > 2$ Fullerene C_X $X = 60, 70, 84...$ (when $X = \infty$ $n = 2$) Carbon onions Carbon nanotubes	$2 > n > 1$ Cyclo(N)carbons $N = 18, 24, 30...$ (when $N = \infty$, $n = 1$)
Amorphous carbon		
Glassy carbon		
Carbon black		
Adamantine carbon		

The former cover mixed short-range order carbon species with more or less randomly distributed C atoms, whereas the lattice describe a structure in which curvature introduces strains responsible for the mixture of a different hybridization. For example, the structure of fullerene C_{60} facilitates attribution of intermediate hybridization of sp^2 and sp^3 to the carbon atoms. These details are given in Table 10.5.

Geometrical Possibilities/Size Considerations

Various geometric forms can be obtained, such as platelets, sheets, and disks, flowers, cones and ball shapes.

Metastable State

Carbon can form metastable compounds in which the most stable form is graphite. However, the hardest substance with a crystalline nature (diamond) is a metastable state.

Similarity to Biological Architecture: "Haeckelites"

Biological structures are formed by carbon materials such as the helical structure of DNA, and an equal number of hexagons and heptagons can lead to the formation of different morphologies.

Boron- and Nitrogen-Doped Graphite Arrangements Promise Important Applications

Substitution in graphite is possible although it is a more stable form of carbon. Subsequently, the substitution can be possible in all metastable forms of carbon. The electronic properties of the materials can be drastically changed by the presence of impurities. When an impurity is added to a system, a slight modification is observed in the energy bands. From the electronic point of view, two types of impurities can be present: donor and acceptor. In the former, also called n-type doping, the impurity provides an extra electron that can only be accommodated in the conduction band (for example, N in carbon materials). The p-type doping impurity provides one electron less to the system and therefore leaves one state unoccupied in the valence band (B in carbon). Because a carbon atom and the average C–C distance is only 1.41 Å in graphite, it is believed that the only likely substitution dopant in graphite is boron. However, a few studies have already been reported regarding N-doped graphite. Because diamond has a larger C–C distance equal to 1.54 Å between neighboring atoms, both boron and nitrogen can enter the diamond lattice as a substitution. Fullerenes and carbon nanotubes (CNTs), which are metastable, can undergo substitution.

Usable Capacity Ratio

The performance of a carbon sorbent to meet the requirement of storage property can be judged by its usable capacity ratio (UCR). It is a measure of the effectiveness of sorption compared with gas compression at the same pressure. The UCR is defined as the mass available fuel in an adsorbent-loaded vessel divided by the mass of available fuel in a vessel without adsorbent (compressed gas only). The available fuel in this case hydrogen is the mass of hydrogen in the vessel at the storage or working pressure minus the mass of hydrogen in the vessel at the discharge pressure. Various forms of carbon materials such as activated carbon, fullerenes, and carbon nanomaterials (fibers, plates, and tubes) have shown UCR greater than unity. Extensive research has been carried out to find carbon materials whose UCR is greater than unity.

Activated Carbon

Activated carbon is a synthetic carbon containing small graphite crystallites and amorphous carbon. The pore diameters are usually less than 1 nm and possess a specific surface area up to 3000 m^2/g. Advanced carbon is prepared from carbon-rich organic precursors by a thermal method (dry distillation) to form carbonized organic precursors, which can be activated to increase the pore volume either thermally or chemically. Experimental results on activated carbon show a linear dependence of the excess hydrogen absorption capacity on the specific surface area of the activated carbons. In the early 1980s the first work was started to investigate the potential of hydrogen storage in the capillary passages of activated carbon at low temperatures. The gravimetric measurements resulted in less than 2 wt% adsorption at 77 K or room temperature with a specific surface area of 2600 m^2/g.

Fullerene

Fullerene is a spheroidal or polyhedral-shaped carbon molecule. It can be described as a sphere and is composed of 20 triangular equilateral faces with 12 apices, each at the junction of 5 triangles. Fullerene was discovered in 1984 by Harry Kroto and Richard Smalley at Rice University. The fullerene molecule closely resembles a soccer ball. It has 20 hexagons and 12 pentagons. Structural studies of C_{60} have shown it to possess a face-centered cubic lattice, as shown in Fig. 10.7. The lattice parameter is $a = 14.17 \pm 0.001$ Å, with a van der Waals space of 2.9 Å. Because of the high symmetry of the molecule, the ^{13}C-nuclear magnetic resonance spectrum and the infrared absorption spectrum are particularly simple. Doping the C_{60} with a metal changes the conductivity and yields a phase that is superconducting at temperatures less than 18 K. The functionalization of C_{60} is relatively well known, such as reduction, oxidation, alkylation, and metal complex formation. Characteristics of fullerene are so close to those of graphite that the molecules are hollow, which leads them to form inclusion derivatives. Heating C_{60} at high pressures and temperatures leads to polymerization of the molecule by forming intermolecular bonds, as evidenced by a decrease in their distance from 2.9 Å in pristine C_{60} to 1.64−1.68 Å. The resulting three-dimensional phase is harder than diamond.

Fullerenes have been investigated as a potential hydrogen storage material based on their ability to react with hydrogen via the hydrogenation of carbon.

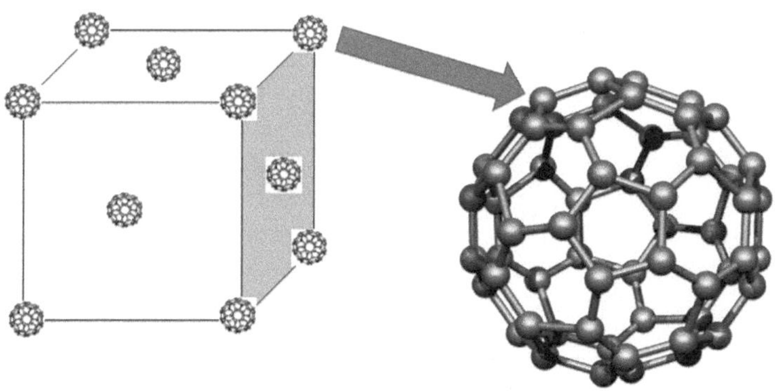

FIGURE 10.7 Face centred cubic lattice unit cell of the Buckminster fullerene crystal. Each lattice point has a C_{60} molecule.

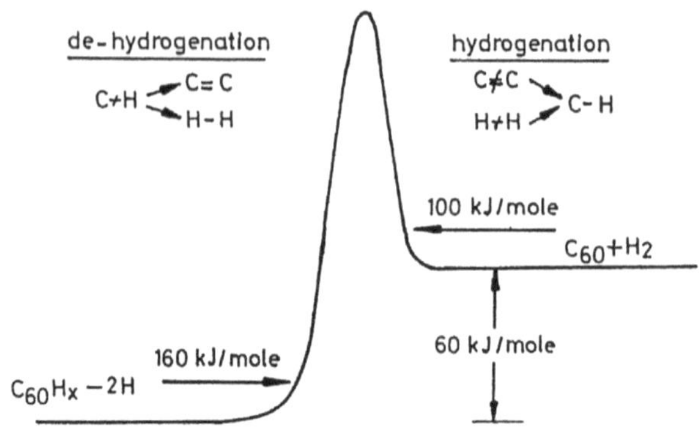

FIGURE 10.8 Activation energy profile of fullerene to form hydrides.

According to theory a maximum of 60 hydrogen atoms can be attached both to the inside (endohedrally) and outside (exohedrally) of the fullerene spherical surface and a stable $C_{60}H_{60}$ isomer can be formed, which accounts for a storage capacity of 7.7 wt% hydrogen. The storage capacity is greater than the required amount for an economically viable 6.5 wt% storage medium. Although the storage capacity of fullerenes is high, the hydrogenation of fullerene currently requires high pressures and temperatures for the reaction to take place by overcoming the reaction barrier. As shown in Fig. 10.8 hydrogenation of fullerene involves the formation of C–H bonds as a result of the breakage of C + C double bonds and dissociation of hydrogen molecule to form hydrogen atoms. From experimental results, the activation energy for the hydrogenation was estimated to be 1.0 eV/H_2, and for dehydrogenation to form hydrogen molecule the activation energy is higher, about 1.6 eV/H_2. To overcome this potential barrier, high temperatures (>673 K) and pressures (>60 MPa) are required.

There have been a number of attempts to prepare hydrofullerenes. Initially experiments were carried out by direct gas phase hydrogenation of fullerenes under elevated temperatures and high pressures (673–723 K, 60–80 MPa) which resulted in the formation of compounds with hydrogen content up to 6.1 wt%. To reduce the high energy required for the activation of hydrogen molecules, various approaches have been adopted such as selecting and optimizing the metal catalyst for the reaction to take place. However, the addition of metal catalysts resulted in a decrease in the temperature and pressure. Nevertheless, the storage capacity is not significant because of the high polarizability of the metal toward hydrogen compared with carbon. Alternatively the reaction was carried out in the liquid phase, which involves the use of solvent molecules; by this process the experimental conditions required are a temperature in the range 453–523 K and a pressure of 12 MPa. The essential results of various methods of hydrogen stored are listed in Table 10.6.

TABLE 10.6 Hydrogen Storage Capacity of Fullerene Determined by Various Methods

Method	Chemical Reaction	Conditions Pressure (p) Temperature (T)
Direct noncatalytic hydrogenation	$C_{60} + H_2 \rightarrow C_{60}H_{2-18}$ (2.4 wt%)	$pH_2 = 50-85$ MPa $T = 573-623$ K
Reaction of gaseous hydrogen with $C_{60}Pd_{4.9}$	$C_{60}Pd_{4.9} + H_2 \rightarrow C_{60}H_{2-26}$ (3.48 wt%)	$pH_2 = 2.0$ MPa $T = 473-623$ K
Catalytic hydrogenation in toluene solution in presence of Ru/C	$C_{60} + H_2 \rightarrow C_{60}H_{36-48}$ (6.3 wt%)	$pH_2 = 2-12$ MPa $T = 363-553$ K
Radical hydrogenation with promoter C_2H_5I	$C_{60} + H_2 \rightarrow C_{60}H_{36}$ (4.8 wt%)	$pH_2 = 6.9$ MPa $T = 723$ K
Reaction with lithium in ammonia in presence of *t*-BuOH	$C_{60} + H_2 \rightarrow C_{60}H_{18-36}$ (4.8 wt%)	$T = 78$ K
Reduction in toluene solution through hydroborating or hydrozirconating	$C_{60} + H_2 \rightarrow C_{60} + H_{2-4}$ (0.6 wt%)	$T = 278$ K
Hydrogen transfer on fullerene from dihydroanthracene	$C_{60} + H_2 \rightarrow C_{60}H_{18-36}$ (4.8 wt%)	$T = 623$ K
Fullerene hydrogenation in Zn-concentrated HCl-toluene system	$C_{60} + H_2 \rightarrow C_{60}H_{18-36}$ (4.8 wt%)	$T = 293$ K

Carbon Nanofibers

Graphite nanofibers (GNF) consist of graphite platelets stacked together in various orientations to the fiber axis with an interlayer spacing similar to bulk graphite. They are produced by the decomposition of mixtures of ethylene, hydrogen, and carbon monoxide on selective metal and alloy catalysts with three distinct structures such as tubular (90 degree), platelet (0 degree), and herringbone (45 degree), in which the angle indicates the direction of the fiber axis relative to the vector normal to the graphene sheets (Fig. 10.9). The spacing between each layer is the same as in conventional carbon: ~3.4 Å with a diameter of 5–500 nm. The length of these GNFs can vary between 5 and 100 μm.

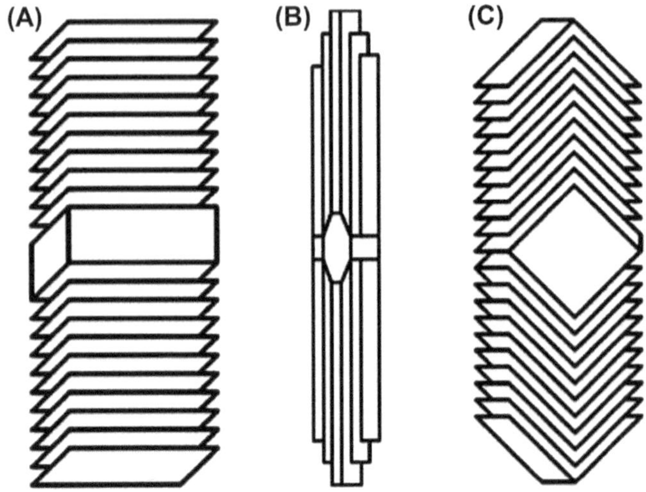

FIGURE 10.9 Graphitic nanofibers: (A) Platelet, (B) ribbon, and (C) herringbone structures.

Carbon Nanotubes

In 1991, Iijima first described a new form of carbon called CNTs. CNTs are formed by rolled graphite sheets with an inner diameter from 0.7 nm to several nanometers and a length of 10–100 μm, as shown in Fig. 10.10.

CNTs are usually described as closed on both sides by a hemisphere: that is, half a fullerene. Tubes formed by only a single graphite layer are called single-walled nanotubes (SWNT) tubes consisting of multiple concentric graphite layers are called multiwalled nanotubes (MWNT). The interlayer distance in MWNTs is closer to the interlayer distance in graphite, which is equal to half the unit cell parameter. The diameter of SWNTs varies from 0.671 to 3 nm, whereas MWNTs have typical diameters of 30–50 nm. The helicity of the nanotubes is usually described by the Hamada vector, which indicates how the graphene sheet is rolled up along a lattice vector with components. The values of integers n and m identify the general geometry of SWNT. Tubes with $n = m$ are named "armchair" and tubes with either $n = 0$

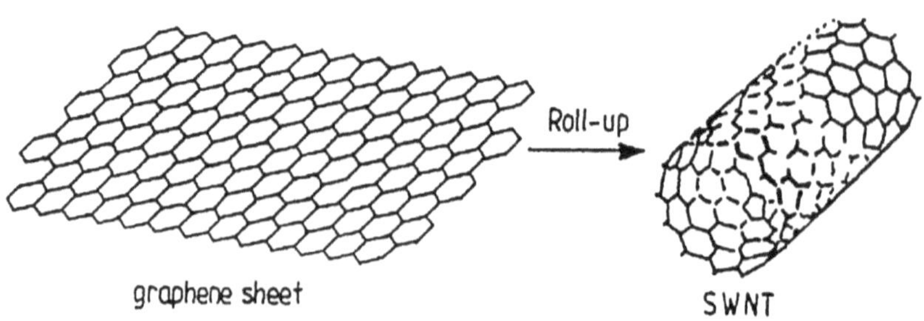

FIGURE 10.10 Graphene sheet rolled to form a cylindrical single-walled nanotube.

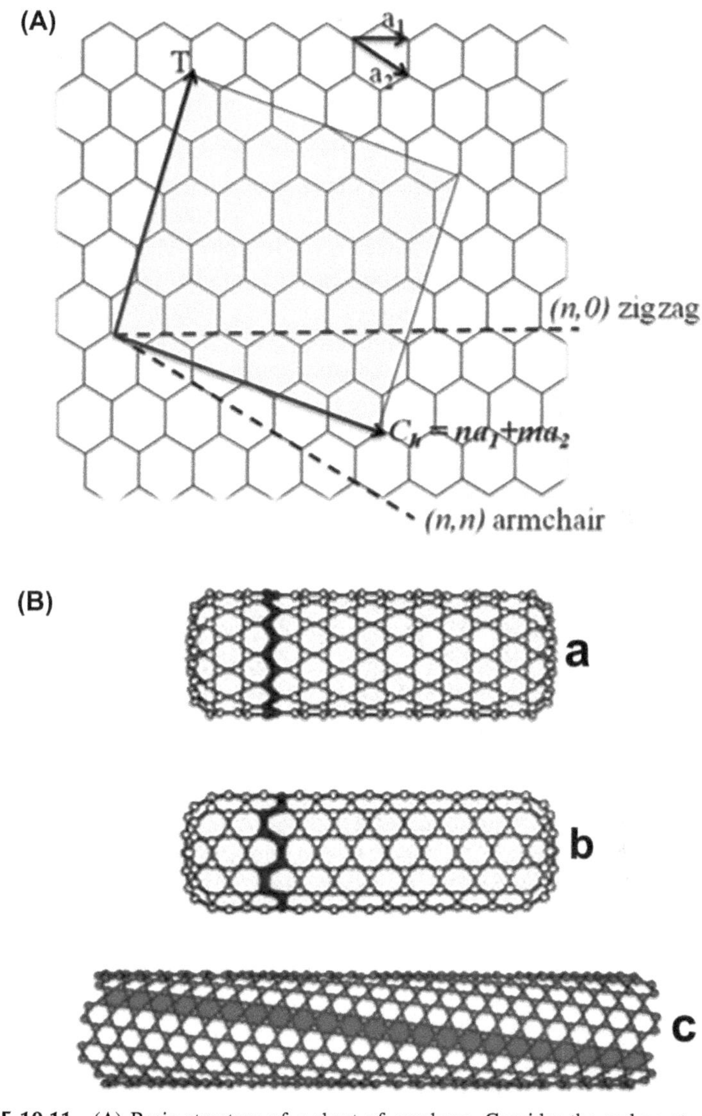

FIGURE 10.11 (A) Basic structure of a sheet of graphene. Consider the carbon atoms to be placed on each point on the edge of the hexagon. (B) Three different classifications of nanotubes: (a) shows a (9,0) zigzag tube, (b) shows a (5,5) armchair tube, and (c) shows a (6,4) chiral tube.

or $m = 0$ are named "zigzag." All others have chiral symmetry. A schematic representation is given in Fig. 10.11. One can view an SWNT as a rolled-up sheet or strip of sp^2-bonded graphene. The atoms are located using a pair of integers (n,m) and the lattice vector $C = na_1 + ma_2$, as shown in Fig. 10.11A. A tube can be classified using the pairs of integers by viewing the rolling up of the sheet as the placement of the atom at (*o*,*o*) on the atom at (*n*,*m*). Hence different diameter tubes and helical arrangements of hexagons can arise by changing (*n*,*m*), as shown in Fig. 10.11B.

Synthesis of Carbon Nanotubes

Carbon nanotubes were first noticed at the ends of the graphite electrodes used in an electric arc discharge employed in fullerene synthesis, and the structures were examined by transmission electron microscopy. The decomposition of hydrocarbon gases by transition metal catalyst particles (Fe, Co, and Ni) has been used to produce carbon nanofibers that are similar in dimensions to the nanotubes but far from structural perfection. SWNTs were first made by the electric arc through the introduction of catalyst species (Fe and Co) into carbon plasma. Several metal catalysts have been tried, but a good yield of CNTs was obtained with Co, Ni, and bimetallic systems such as Co-Ni, Co-Pt, and Ni-Y as catalysts. Another effective way to produce SWNTs is by using laser evaporation. It was shown that nanotubes could be obtained with good uniformity in size and structure. Depending on the temperature of the oven in which the nanotubes were grown, the diameter varied between 1 and 5 nm diameter; the higher the temperature (range between 1073 and 1473 K), the larger the nanotube diameter. Other attractive strategies for the synthesis of nanotubes are through catalytic chemical vapor deposition (CVD) or using well-defined porous inorganic membranes as templates in which a disordered form of carbon is deposited by CVD and further graphitized at higher temperatures to yield nanotubes. The advantage of such template-based methods is that the size of the particles and the pores that determine the size of the nanotubes can be controlled before depositing the carbon. The length of the nanotubes formed can be controlled by adjusting the amount of carbon vapor feedstock and the thickness of the membranes. By removing the templates after the nanotube growth, freestanding arrays of highly graphitized nanotubes can be formed.

Hydrogen Storage in Carbon Nanotubes

Among nanostructures, tubular materials are especially interesting because their morphology is assisted by an intrinsic multifunctionality that arises from four different contact regions: a tube opening, outer surface, inner surface, and interstitial region. These properties render nanotubes promising candidates for the realization of highly functional, effective and resource-saving nano-devices such as sensors, capacitors, or storage and release systems. Surprisingly, despite their relatively small surface area and pore volume, CNTs and carbon nanofibers may have a high hydrogen storage capacity (Table 10.7).

TABLE 10.7 Summary of Reported Hydrogen Storage Capacities in Carbon Nanostructures

Adsorbent	Hydrogen Storage (wt%)	Conditions of Temperature (K) and Pressure (MPa)
SWNT (low purity)	5–10	273/0.04
SWNT (high purity)	3.5–4.5	298/0.04
SWNT (high purity)	8	80/8
SWNT (50% purity)	4	300/12
SWNT	2	80/10
SWNTs	~0.1	30–520/0.1
Li-doped MWNT	20	200–400/0.1
K-doped MWNT	14	300/0.1
MWNT	5	300/10
GNF (tubular)	11.26	298/11.35
CNF (herringbone)	67.55	298/11.35
Nanostructured graphite	7.4	~300/1.0
SWNT-Ti-6Al-4V	1.47	300/0.08
SWNT-$TiAl_{0.1}V_{0.04}$	~7	300/0.067

CNF, carbon nanotube; *GNF*, graphite nanotube; *MWNT*, multiwalled nanotube; *SWNT*, single-walled nanotube.

Interest in carbon nanomaterials for hydrogen storage applications escalated following claims of experimental results in 1997 that CNT might be capable of 5–10 wt% hydrogen capacity. Storage of hydrogen in carbon materials took a dramatic turn in 1998–1999 when exotic values of hydrogen absorption by carbon nanostructures up to 10–12 hydrogen molecules per carbon atom was reported, although in a subsequent communication this figure was reduced to four hydrogen molecules per carbon atom. This is nearly twice (0.4 gH_2/g C) that of the most hydrogenated carbon compound, namely CH_4, in which the hydrogen to carbon weight ratio is 0.25 gH_2/g C. Subsequently there have been various reports in the literature to substantiate these observations and none could achieve the expected 6.5 wt% or even repeat the original adsorption capacity reported. Inelastic neutron scattering experiments have shown that pure carbon surface cannot activate hydrogen. By incorporating metals and alloys to a purified SWNT, it has been shown that hydrogen adsorption can take place in two different sites and this adsorbed hydrogen desorbs at two different temperature ranges. There are various theoretical studies with a view to finding answers to the questions:

- How do structural characteristics influence the physical/chemical process?
- Where does the absorption occur? In inner hollow cavities, in another pore space (inner tube space), or in only one of them?
- In the absorption of hydrogen onto CNTs, what chemical or physical interaction occurs between hydrogen and carbon?
- What is the mechanism by which the absorption process takes place?
- What is the maximum absorption capacity?

The interaction may be based on either van der Waals attractive forces (physisorption) or on the overlap of the highest occupied molecular orbitals of carbon with the occupied electronic wave function of hydrogen electron, overcoming the activation energy barrier for hydrogen dissociation (chemisorption). Physisorption of hydrogen limits the hydrogen to a carbon ratio of less than one hydrogen atom per two carbon atoms (ie, 4.2 mass %) whereas in chemisorption the ratio of two hydrogen atoms per one carbon atom is realized, similar to the case of polyethylene. Physisorbed hydrogen has a binding energy normally on the order of 0.1 eV whereas chemisorbed hydrogen has C−H covalent bonding with a binding energy of more than 2−3 eV.

Several reports both experimental and theoretical attribute the uptake of hydrogen atom to physisorption of hydrogen on carbon. The mechanism of hydrogen storage on carbon nanomaterials that explains the higher storage capacity than the expected value reported uptake by several groups remains unclear. Although many studies are reported in the literature, until reliable and repeatable results are obtained the question of the mechanism of hydrogen storage in carbon materials cannot be answered conclusively. The current situation regarding the measurement of hydrogen sorption in carbon materials is shown in Fig. 10.12. It is seen that any value from 0 to 67 wt% has been reported with no consistent reproducible result. It is hoped that the scene will soon change and that reproducible, reliable storage capacity will be achieved soon.

In the literature, the concept of the interaction of atomic hydrogen is not clearly dealt with, and if it becomes a facile process, one can achieve higher storage capacities. To achieve atomic hydrogen adsorption an activator is essential to dissociate the incoming hydrogen molecule and the dissociated hydrogen atoms can move on to the carbon surface (spillover). Therefore it is clear that carbon material cannot activate hydrogen in a catalytic style in its pure form. Metal or metal oxide−doped carbon materials offering higher sorption capacities than their pure counterparts may fit into this concept. Although metal and metal oxide−doped CNTs offer higher sorption capacities (~5.6 wt%), the desorption process occurs in two steps mentioned earlier, leading to lower usability. This is because M−H bonds are stronger. To increase the user capacity ratio of carbon nanomaterials, one has to find a

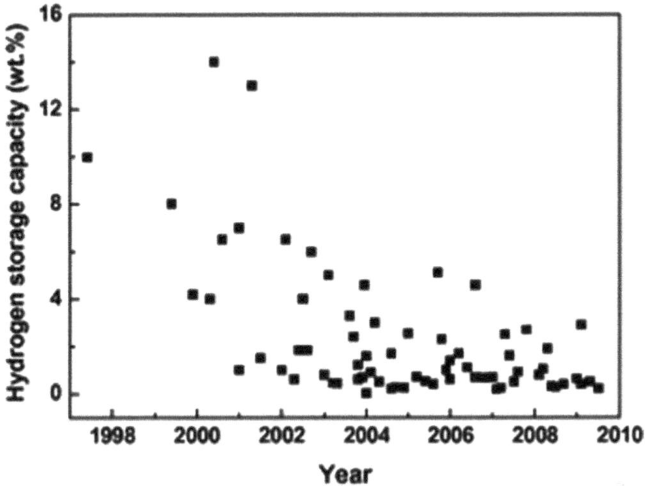

FIGURE 10.12 Hydrogen storage capacity in weight percent in carbon materials as a function of the year of the report of the study. The scatter and uncertainty are clearly seen. *Reproduced from C. Liu, Y. Chen, C.Z. Wu, S.T. Xu, H.M. Chen, Carbon 48 (2010) 452–455.*

suitable activator that could split the incoming hydrogen gas molecules and transfer them to carbon sites. A rationale is evolved for the research of a suitable activator, and essential criteria for such an activator are that:

- they must possess equipotential sites with that of carbon
- the sites themselves should be easily hydridable
- they should act as catalytic sites rather than mere reactive sites that lead to a stoichiometric reaction

What Are the Alternatives?

Hydrogen adsorption on carbon materials, especially nanotubes, is possible only when it contains some sites in which hydrogen molecules can be activated. Taking the cue from nature, it is postulated that the presence of heteroatoms in CNTs may be appropriate sites for the activation of hydrogen molecule. In nature the high percentage of hydrogen stored mainly in the form of hydrocarbon formation is facilitated by the presence of a heteroatom containing compounds such as pyridine, thiophene, and pyrrole, because the heteroatom has an important role in hydrogen storage in carbon. Where the heteroatom has the high polarizability that can alter the Fermi level of the carbon network, one can tune the redox potential by substituting the appropriate heteroatom such as nitrogen, boron, phosphorus, and sulfur with the required atomic weight percentage. The hydrogenation behavior of carbon is well established in which potentially heteroatom substituted CNTs hydride by the electronegative difference between them. The alteration in electronegativity makes the molecule

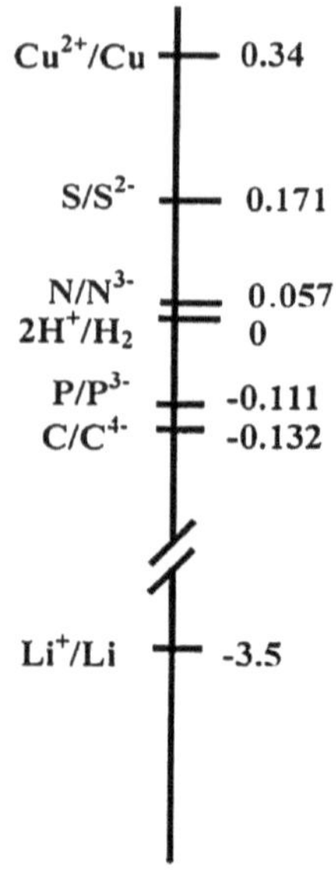

FIGURE 10.13 Standard redox potential (V) values of some selected species.

to be polarized and behaves differently from unsubstituted CNTs. The hydriding property of the heteroatom is explainable by the redox potential (Fig. 10.13) and by the Ellingham diagram such as the formation of N−H and B−H bonds (Fig. 10.14). By the heteroatom substitution one can exploit the manipulation of electronic and structural perturbation resulting in the carbon lattice for hydrogen storage applications.

In Fig. 10.15 the difference between the nano state and the bulk state for hydrogen storage is shown. The essential differences are that: (1) The

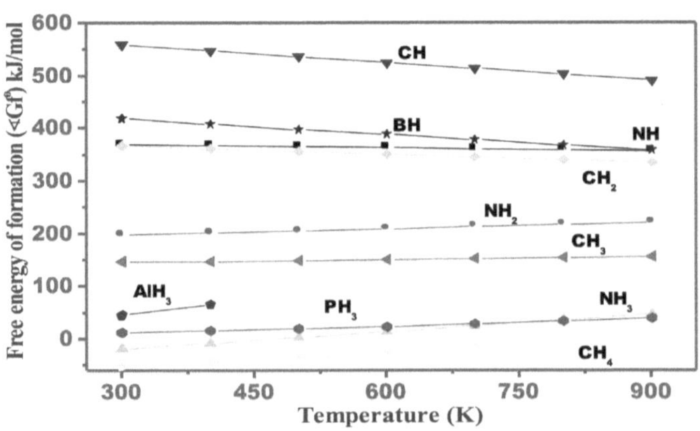

FIGURE 10.14 Ellingham diagram for various heteroatom (hydriding property).

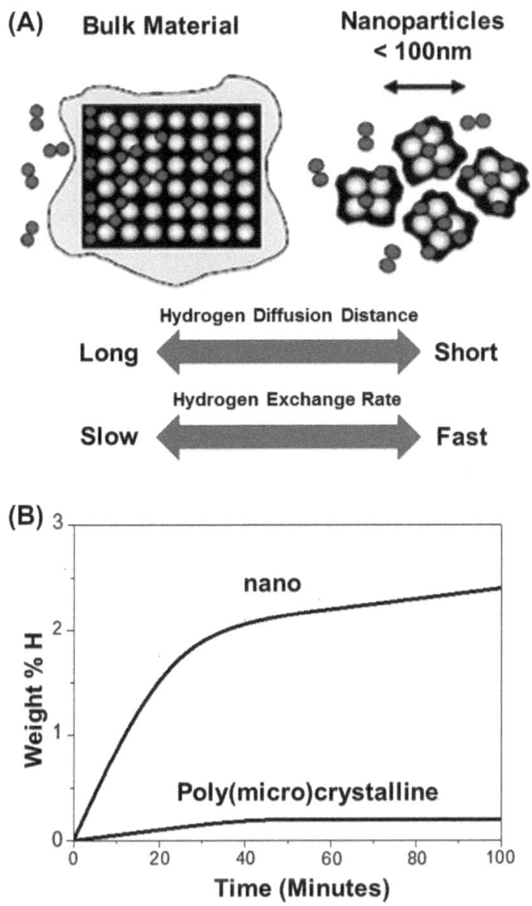

FIGURE 10.15 The difference between nano and bulk material with respect to hydrogen storage. (A) Pictorial difference in hydrogen storage between bulk and nano-material; (B) rate of hydrogen uptake between bulk and nano-material.

hydrogen diffusion length is long in bulk material whereas it is short for the nano state of the same material. (2) The hydrogen diffusion rate is higher in the nano state compared with the bulk material. (3) The heat formation for the bulk hydride is higher than that in the nano state. This situation is shown diagrammatically in Fig. 10.16.

EPILOGUE

Enormous efforts to achieve the desired levels of storage of hydrogen in solid-state materials have not yielded expected results. Possibly this has given rise to some frustration regarding the possibility of achieving the required levels of

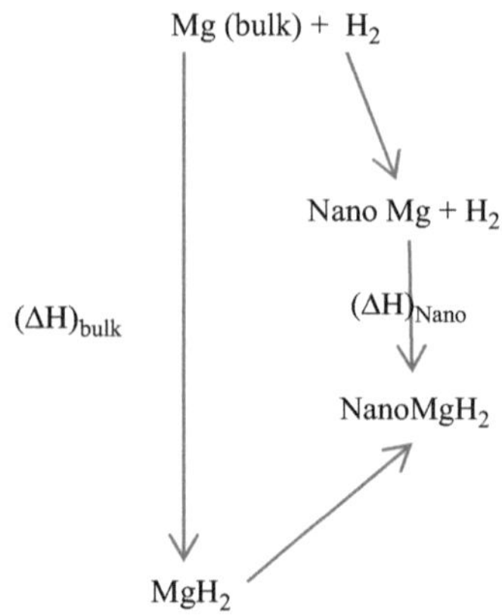

FIGURE 10.16 Schematic representation of a comparison of the heat of formation of metal hydride in bulk and nano states.

storage. Nature has shown that in carbon materials hydrogen storage can vary from low values to 25 wt% in methane. However this level of storage involves covalent bonding and hence requires sufficient energy to release the hydrogen back. To achieve the desired levels of hydrogen storage capacity in carbon materials or in any other solid-state material, one has to resort to a configuration and morphology of materials in such a way that the volumes contained by the atoms constitute the solid. This means that one has to make materials in which each constituting atom should be capable of generating more than one void volume in the solid, and this may be possible only if architecting solids at will becomes possible. Among the solid-state materials, lightweight metals and carbon appear to be the appropriate choice if both the thermodynamics and kinetics of hydrogen absorption and desorption can be coupled. To date, this seems to be a far cry from reality.

Chapter 11

Photo-Catalytic Routes for Fuel Production

The field of heterogeneous photo-catalysis has received considerable attention in recent decades regarding energy conversion and environmental protection. Photo-catalysis can be considered the acceleration of a reaction in the presence of a catalyst and photons. A significant application of photo-catalysis in energy conversion is water splitting, although other processes such as CO_2 reduction and other energy conversion processes for possible fuel production and the reduction of dinitrogen to yield ammonia are included. Another area of great interest is pollutant removal from water and the air. The area has turned to be multidisciplinary such that it includes semiconductor physics, surface sciences, photo-physics, physical chemistry, materials synthesis, the characterization of materials, and chemical engineering.

In 1972, Fujishima and Honda demonstrated the splitting of water on a TiO_2 surface, which can be considered the starting point for fuel production by photo-catalysis, although previous to this photocatalytic transformation on ZnO semiconductor surfaces was reported. More recently, investigations into photo-catalysis have focused on the removal of concentrations of organic and inorganic species from aqueous or gas phase with a view to treat water for drinking and also for environmental applications (Fig. 11.1).

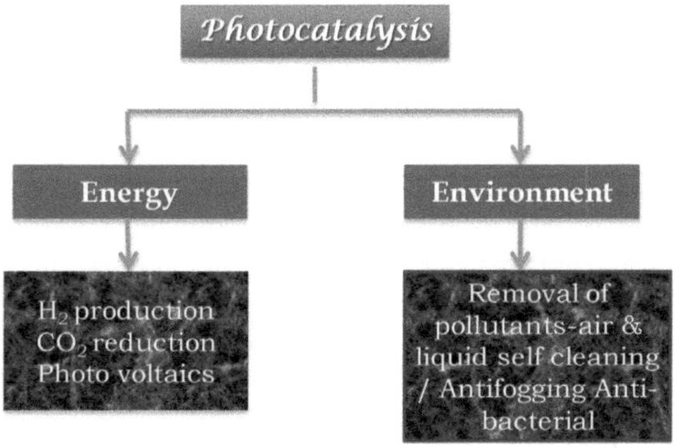

FIGURE 11.1 Major applications of photo-catalysis.

Energy Sources. http://dx.doi.org/10.1016/B978-0-444-56353-8.00011-3

Essentially, illumination on a semiconductor surface generates electrons in the conduction band by excitation and holes in valence band that can be considered to be reducing and oxidizing species generated by photo-excitation. Thus the bottom of the conduction band and top of the valence band refer to the reduction potential and oxidation potential, respectively, of the semiconductor in general and thus can carry out both reactions on the same surface. To facilitate electron transfer at the conduction band and the hole at the valence band, suitable transfer species such as Pt and IrO_2 can be included.

In pollutant removal, although the generated charges (electron and hole generated by photo-excitation) can directly react with adsorbed pollutants, reactions with water predominate because water is in excess when pollutants from water are considered. Oxidation of water or OH^- by the hole produced by photo-excitation generates a hydroxyl radical [with an oxidation potential of 2.8 V with respect to the normal hydrogen electrode (NHE)], a powerful oxidant. In addition to holes and electrons, other chemical species that can be used for the disinfection of water include ozone (2.07 V), H_2O_2 (1.78 V), HOCl (1.49 V), and chlorine (1.36 V). Hydroxy radicals are most powerful oxidizing species generated on semiconductor surfaces. In addition, conduction band electrons by photo-excitation can reduce the adsorbed oxygen to oxygen radicals, thus preventing recombination with the holes and generating oxygen radicals that can participate in the degradation of pollutants in solution.

A material to be considered as a photo-catalyst should satisfy certain properties such as photo-activity and the adsorption of the pollutants (substrates), biological and chemical inertness, stability toward photo-corrosion, stability under illumination, light harvesting ability, low cost, and nontoxicity.

MECHANISM OF SEMICONDUCTOR PHOTO-CATALYSIS

A semiconductor material harnesses sunlight or artificial light and the energy is used to break down different substances such as organic or inorganic substrates and pollutants, dyes, pesticides, crude oil, and chlorine-containing organisms, as shown in the mechanism displayed in Fig. 11.2. Owing to the generation of OH radicals or the oxygen molecule radical, this photo-process can mineralize pollutants, as shown in Fig. 11.2. Photo-catalysis can be used to reduce air pollution or for self-cleaning surfaces, and also for water purification. To avoid free nanoparticle catalysts, semiconductor nanoparticles are commonly immobilized on a suitable substrate or employed in the form of thin films. The absorption of photons of energy equal to or greater than the band gap of the semiconductor results in the production electron–hole (e^-/h^+) pairs, as in Eq. 11.2,

$$\mathrm{SC} + h\nu \rightarrow e^-\,[\mathrm{cb}]\,(\mathrm{SC}) + h^+\,[\mathrm{vb}]\,(\mathrm{SC})$$

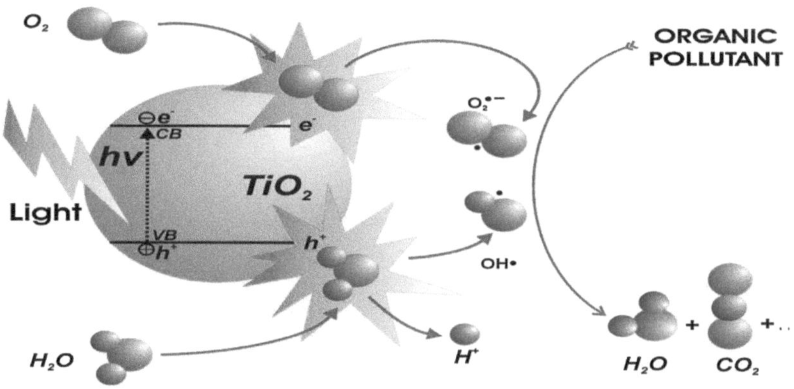

FIGURE 11.2 Schematic of semiconductor excitation by illumination of radiation of energy equal to or greater than the band gap leading to the creation of conduction band electrons and holes in the valence band.

where cb and vb stand for the conduction band and valence band. This charge separation can lead to recombination releasing the absorbed light energy as heat with no attendant chemical reaction taking place. Instead, the charge carriers can participate in the redox reactions with adsorbed chemical species as the valence band holes are strongly oxidizing while the conduction band electrons are strongly reducing. On the semiconductor surfaces, the excitons (that is, the electron−hole species) can promote redox reactions with water, hydroxide ion, organic substrates, or oxygen, thus leading to mineralization of the pollutant (Fig. 11.3).

Various modifications can be effected in semiconductors and the possibilities are illustrated using the example of titania in Table 11.1.

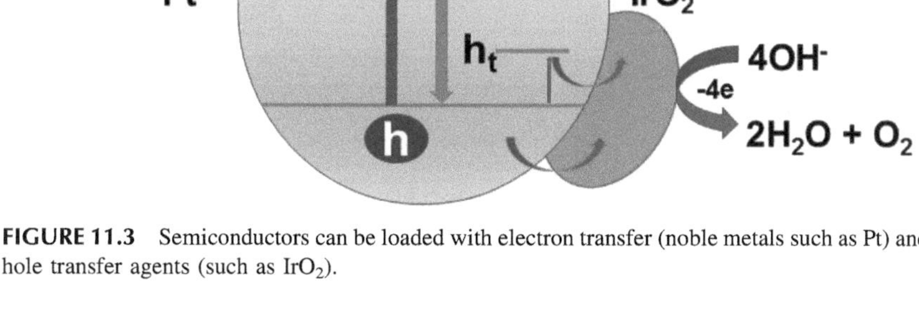

FIGURE 11.3 Semiconductors can be loaded with electron transfer (noble metals such as Pt) and hole transfer agents (such as IrO_2).

TABLE 11.1 Implications of Different Modifications in Titania

Modifications	Implications and Mode of Action
Doping of metals and metal ions	• Act as electron traps and facilitate charge carrier separation
	• Introduce impurity states and induce visible light absorption
	• Absorb visible light via surface plasmon resonance
Doping of anions	• Narrow band gap owing to mixing of p-states of dopants (N, S) with O 2p states in the valance band of TiO_2.
	• Introduce impurity states above the valence band of titania
	• Both states induce visible light absorption
Coupling with semiconductors	• A narrow—band gap semiconductor, with appropriate energy levels, absorbs visible light and transfers excited electrons into the conduction band of titania. UV light source not needed
	• Besides visible light activity, effective separation of charge carriers is achieved
Sensitization with light harvesting components and dyes	• Light-absorbing components can absorb visible light and inject photo-excited electrons into conduction band of titania
	• Besides visible light activity, effective separation of charge carriers is achieved

APPLICATIONS OF PHOTO-CATALYSIS FOR POLLUTANT REMOVAL

Heterogeneous photo-catalysis has been claimed to be low-cost technology for the treatment of pollutants in air and water. This technology is a standalone for water treatment purposes. Because it uses sunlight or artificial light, it is inexpensive and environmentally friendly and can be deployed in any part of the world.

The similar features of photo-catalysis and photosynthesis are shown in Fig. 11.4.

PROCESSES FOR CO_2 CONVERSION

CO_2 is the most oxidized form of carbon; therefore the only chemical transformation at normal energies that would be possible is to reduce it.

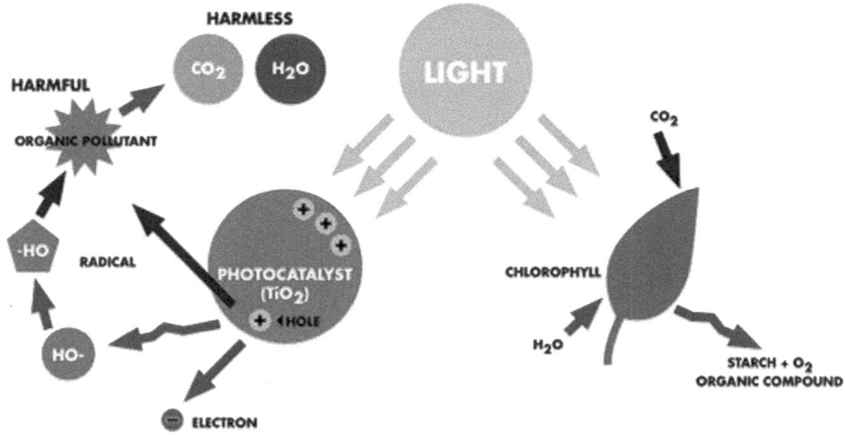

FIGURE 11.4 Parallels between photo-catalysis and photosynthesis.

A wide range of CO_2 conversion and reduction techniques are under investigation:

1. Chemical reduction by metals that occurs at relatively at high temperatures (Eqs. (11.1–11.3)):

$$2Mg + CO_2 \rightarrow 2MgO + C \tag{11.1}$$

$$Sn + 2CO_2 \rightarrow SnO_2 + 2CO \tag{11.2}$$

$$2Na + CO_2 \rightarrow Na_2C_2O_4 \tag{11.3}$$

2. Thermochemical conversion (Eq. (11.4)):

$$CO_2 \xrightarrow{Ce^{4+},\ T > 900^{\circ}C} CO + 1/2O_2 \tag{11.4}$$

3. Radiochemical method (Eq. (11.5)):

$$CO_2 \xrightarrow{\gamma-\text{Radiation}} HCOOH,\ HCHO \tag{11.5}$$

4. Photochemical conversion (Eq. (11.6)):

$$CO_2 \xrightarrow{h\nu} CO,\ HCHO,\ HCOOH \tag{11.6}$$

5. Biochemical conversion (Eq. (11.7)):

$$CO_2 + 4H_2 \xrightarrow{\text{Bacteria}} CH_4 + 2H_2O \tag{11.7}$$

6. Electrochemical conversion (Eq. (11.8)):

$$CO_2 + xe^- + xH^+ \xrightarrow{eV} CO,\ HCOOH,\ (COOH)_2 \tag{11.8}$$

7. Bio-photochemical conversion (the "bio" part of the energy consists of catalysis and the information content of an enzyme (Eq. (11.9)):

$$CO_2 + \text{oxogluric acid} \xrightarrow{h\nu} \text{isocitric acid} \tag{11.9}$$

8. Electro-photochemical conversion (Eq. (11.10)):

$$CO_2 + 2e^- + 2H^+ \xrightarrow{h\nu,\ eV} CO + 2H_2O \tag{11.10}$$

The conventional catalytic reduction of CO_2 to chemicals (formic acid, methanol, and methane) with external hydrogen is feasible but hydrogen from renewable resources has to be used to render the process viable and sustainable.

Among these processes, photochemical conversions (direct photochemical, photo-electrochemical, and photo-biochemical) use abundantly available solar energy, qualify as sustainable processes and hence are being pursued with tremendous interest. A brief account of photo-catalysis and photosynthesis, which form the basis for all types of photo-catalytic conversions in general, and CO_2 reduction in particular, is considered.

Historically, photo-catalysis has had a pivotal role in the evolution of life on earth. Photosynthesis, the process so vital for plant life, is essentially a photo-catalytic process involving the conversion of CO_2 and water to carbohydrates, with the release of oxygen into the atmosphere. Chlorophyll acts as a collector of solar energy and the conversion takes place with the help of a series of enzyme-catalyzed process steps. This phenomenon of photosynthesis marked the beginning of the process of evolution of various forms of life on earth that survive on oxygen. Both photo-catalysis and photosynthesis use the sustainable source of energy, sunlight, that is available abundantly at almost zero cost (Fig. 11.4).

The photo-catalytic reduction (PCR) of CO_2 by water to yield hydrocarbons, a process commonly referred to as "artificial photosynthesis," dates to nearly 9 decades ago, in 1921–1922. The process has since attained tremendous importance on a global level because it is considered a means of controlling current atmospheric CO_2 levels (the greenhouse gas effect) by converting CO_2 into fuel and chemicals and because of the role of CO_2 as a source of energy.

CO_2 PHOTO-REDUCTION WITH WATER: PROCESS FEATURES

CO_2 reduction is a thermodynamically uphill process, as illustrated by its standard free energy of formation ($\Delta G^\circ = -394.36$ kJ/mol). Economical CO_2 fixation is possible only if renewable energy, such as solar energy is used as the energy source. Equally difficult is the reduction and splitting of water using solar radiation to yield hydrogen, which requires a similar combination of activation steps. There are two different processes for CO_2 reduction:

- direct photo-reduction of CO_2 using water as a reductant
- photo-catalytic splitting of water to generate hydrogen and the further reaction of this hydrogen with CO_2, forming C_1–C_2 fuels

The most ideal and desirable process would be the simultaneous reduction of CO_2 and water to yield hydrocarbons, which essentially works out to artificial photosynthesis. The photo-catalytic process occurs via the direct absorption of photons with energy greater than or equal to the band gap of the photo-catalyst, resulting in the generation of electron–hole pairs. The initial excitation and electron energy transfer to the adsorbed reactants on the photo-catalyst make chemical reactions in the photo-catalytic process possible.

The use of solar energy via chemical storage can be achieved by photo-catalytic or photo-electrochemical activation of light-sensitive catalytic surfaces. When comparing the two systems, the photo-catalytic system is simpler and easier to construct and operate (Fig. 11.5). Primary steps involved in the photo-catalytic reduction of CO_2 to hydrocarbons include:

- photo-generation of charge carriers by excitation,
- bulk diffusion and interfacial transfer of charge carriers to surface species, and
- reduction/oxidation and further conversion of surface species to different hydrocarbons.

Photo-generated electrons bring out a reduction of CO_2 to hydrocarbons, and holes facilitate the oxidation of water. Whereas the elementary process steps and mechanism of photo-catalytic water splitting have been established

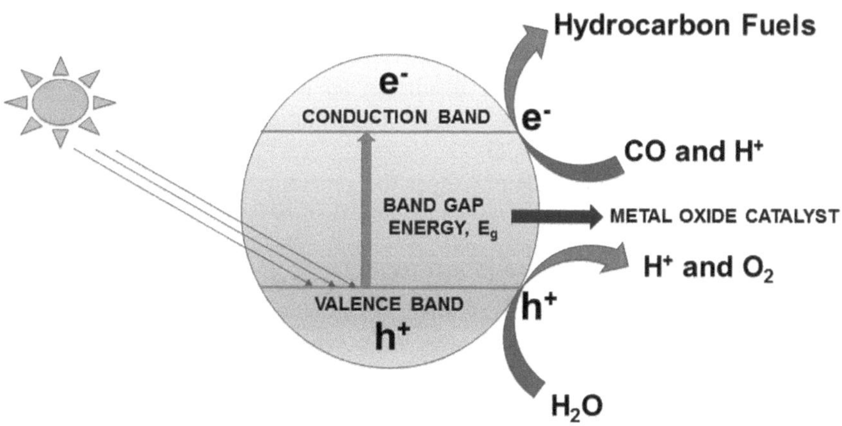

FIGURE 11.5 Transformations during photo-catalytic reduction CO_2 with water.

to an acceptable level, there is little clarity regarding the reaction pathways that lead to the conversion of CO_2 to hydrocarbons.

Thermodynamic Features

Hydrogen formation from water involves a free energy change (ΔG_0) of 237 kJ/mol and an enthalpy change (ΔH_0) of 285 kJ/mol; corresponding values for CO formation from CO_2 are 257 and 283 kJ/mol at 25°C (1 atm). Hence, the minimum energy required for water and CO_2 splitting processes are 1.229 and 1.33 eV, respectively (per photon). In theory, the band gap of a photo-catalyst used for splitting and reduction of water/CO_2 should be at least 1.33 eV. Eqs. (11.11)–(11.18) show one, two, four, six, and eight electron reduction potentials (versus NHE) for CO_2 reduction and H_2O oxidation at pH 7 and 298 K, assuming unit activities for all gaseous and aqueous species.

For reaction E^0_{redox} V versus NHE at pH 7:

$$2H^+ + 2e^- \rightarrow H_2 \quad -0.41 \tag{11.11}$$

$$H_2O \rightarrow 1/2O_2 + 2H^+ + 2e^- \quad -0.82 \tag{11.12}$$

$$CO_2 + e^- \rightarrow CO_2^- \quad -1.9 \tag{11.13}$$

$$CO_2 + H^+ + 2e^- \rightarrow HCO_2^- \quad -0.49 \tag{11.14}$$

$$CO_2 + 2e^- + 2H^+ \rightarrow CO + H_2O \quad -0.53 \tag{11.15}$$

$$CO_2 + 4H^+ + 4e^- \rightarrow HCHO + H_2O \quad -0.48 \tag{11.16}$$

$$CO_2 + 6H^+ + 6e^- \rightarrow CH_3OH + H_2O \quad -0.38 \tag{11.17}$$

$$CO_2 + 8H^+ + 4e^- \rightarrow CH_4 + 2H_2O \quad -0.24 \tag{11.18}$$

From these equations it is clear that CO_2 photo-reduction is not a single-step reaction. Upon transfer of one electron, the structure changes from a linear to bent configuration that results in an irreversible reduction. Also, single electron transfer to CO_2 is highly endergonic because of the negative adiabatic electron affinity of CO_2.

The initial step in the PCR of CO_2 is the generation of electron–hole pairs upon absorption of photons of energy greater than or equal to the band gap of the photo-catalyst. Subsequently, photoelectrons are used to reduce CO_2, and holes to oxidize water. However, the electron–hole recombination process, which is two to three orders of magnitude faster, competes with desirable redox processes. Therefore, the electron–hole recombination process has to be minimized by some means to increase efficiency and improve the rates of CO_2 photo-reduction. Besides recombination, many other factors such as the incident light intensity, fraction of the incident light absorbed by the photo-catalyst, type of catalyst and its particle size, specific surface area, and structural and photo-physical properties are known to influence the kinetics of CO_2 photo-reduction.

Choice of Catalysts: Guiding Principles

It is essential that photo-generated electrons possess requisite energy, as indicated by the thermodynamic criteria, to facilitate the reduction of CO_2. The chemical potentials of the photo-generated electrons and holes depend highly on the band edge positions of the semiconductor. This means that the conduction band bottom energy level of the photo-catalyst has to be more negative with respect to the energy for CO_2 reduction, and the valence band top has to be more positive with respect to the energy for the oxidation of water. Thus, the choice of the catalyst for CO_2 photo-reduction with water is determined by the relationship between energy levels of conduction and valence bands vis-à-vis the energies for CO_2 reduction and water oxidation. Pioneering studies on photo-electro-catalytic reduction of CO_2 on various semiconducting oxides has brought out basic criteria for the catalytic activity. This guiding principle has been illustrated effectively in the relationship between the energy levels of conduction and valence bands and those of the redox couples as depicted in Fig. 11.6.

Semiconductors such as SiC, GaP, CdS, ZnO, and TiO_2, which satisfy these criteria and are active for the photo-reduction of CO_2, and oxides such as WO_3

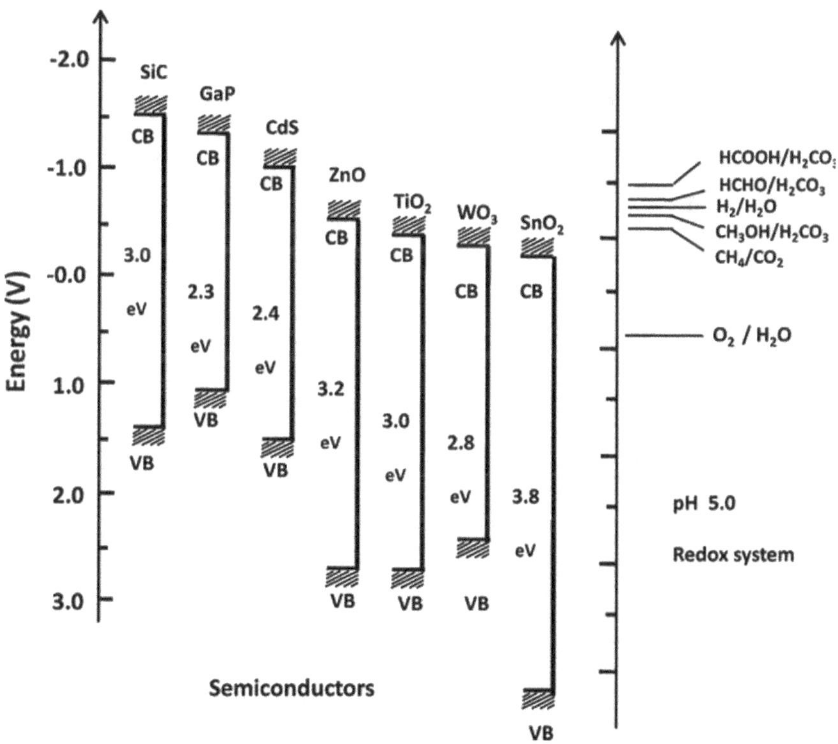

FIGURE 11.6 Conduction band and valence band potentials for photo-catalysts relative to the energy levels of redox couples in water.

and SnO_2, whose conduction band levels are below that for CO_2 reduction, have failed to perform. Energy levels of conduction and valence band, compared with those for redox couple and the lifetime of charge carriers, prolonged by reduced recombination, are two fundamental criteria to be considered for the selection of an efficient photo-catalyst for CO_2 photo-reduction.

CATALYSTS FOR PHOTO-REDUCTION OF CO_2 WITH WATER

A remarkable feature of the PCR of CO_2 is that it could emerge as a carbon-neutral process, in which carbon emissions are recycled back as useful fuels or chemicals. The process involves two major steps: the photo-catalytic splitting of water to yield hydrogen, which in turn helps in the photo-reduction of CO_2 to different hydrocarbon products in the second step. The design of effective catalysts for such a complex process, involving multielectron transfer steps, holds the key for the viability of the process.

Although many semiconductors have smaller band gaps and absorb in the visible range (eg, CdS and Fe_2O_3 with band gap values of 2.4 and 2.3 eV, respectively), only a few are catalytically active because the energy levels of either the conduction or valence bands are unsuitable for CO_2 reduction and/or water oxidation. This limitation, together with the poor photo-corrosion stability of many semiconductors, significantly limits the number of potential photo-catalytic materials for CO_2 photo-reduction.

Large–band gap semiconductors are the most suitable photo-catalysts for CO_2 reduction because they provide sufficient negative and positive redox potentials in conduction and valence bands, respectively. The disadvantage of using wide–band gap semiconductors is the requirement of high energy input.

Need for Modifications in Catalysts

The photo-catalytic activity of a typical semiconductor such as titania is initiated by the absorption of light energy corresponding to or higher than the band gap energy, which results in the generation of electrons and holes. Upon migration to the semiconductor surface, the electron–hole pair interacts with the adsorbed reactants to facilitate the reduction and oxidation process, respectively. In the absence of such an energy transfer, the pairs recombine with the loss of energy. The electron–hole recombination process, which is two to three orders of magnitude faster, competes with the desirable redox processes. Therefore the recombination process needs to be minimized by suitable modifications to increase efficiency. Hence modifications in wide–band gap semiconductors (band gap engineering) aim to deal with two major issues: extending the light absorption range beyond the UV region and arresting the recombination of charge carriers.

By suitable modification of the band gap, ie, reducing the band gap by creating additional/impurity energy levels, the light absorption range or wavelength could be increased. In the case of titania, the light absorption range is increased to cover part of the visible region. Arresting the recombination rate increases the lifetime of photo-electrons and holes, leading to a corresponding increase in photo-catalytic activity.

Major Types of Modifications in Catalysts

A number of strategies have been devised to bring about modifications in wide—band gap semiconductors, broadly classified as:

- doping with metal cations and anions,
- coupling with other semiconducting oxides,
- sensitizing with light-harvesting compounds/dyes or other light-absorbing molecules, and
- inducing plasmon resonance by specific metals, especially Au and Ag.

Main features of these modifications and the implications of the properties of semiconductors, especially for titania, are summarized in Table 11.1 and illustrated in Fig. 11.7.

INFLUENCE OF EXPERIMENTAL PARAMETERS

Photo-catalytic activities depend highly on experimental conditions such as the amount of catalyst, light intensity, lighting area, reactor volume, type of reactor, temperature, and pressure. Intrinsic properties of semiconductor

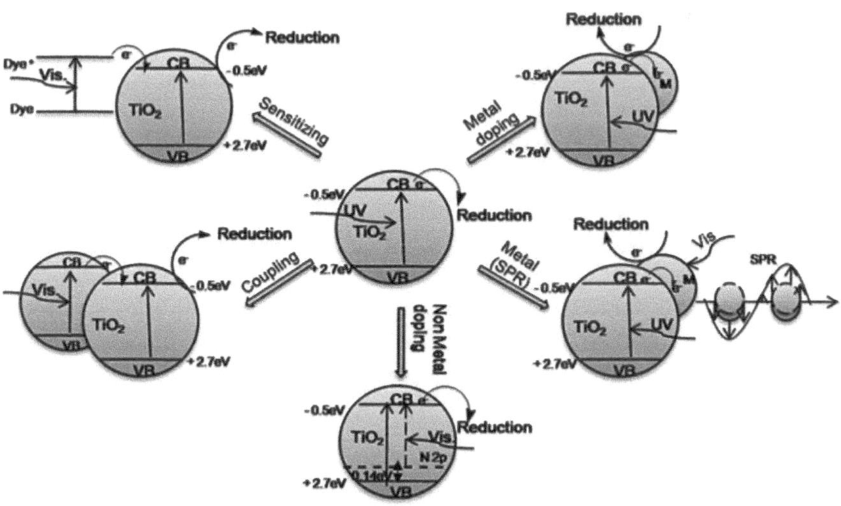

FIGURE 11.7 Implications of modifiers on the electronic structure of titanium dioxide.

photocatalysts, such as the particle/crystallite size, phase composition, surface area, surface hydroxyls, lattice defects and type and level of dopants (metals and nonmetals), morphology (powder, nanoparticles, nanotubes, nanorod, and hollow nanospheres), structural features, and photo-physical properties, influence their performance. Photo-catalytic activity and the phase transition behavior of TiO_2 are significantly influenced by the preparative conditions and methods. This explains the wide variations in activity and product patterns observed for the same photo-catalyst. For example, a standard/reference catalyst Degussa P-25 (TiO_2) has a wide difference in efficiency in results, as shown in Table 11.2. Different experimental conditions adopted could be another major factor that affects yield data, which shows variations in total yield as well as product patterns. Hence, rationalization of activity and product patterns in terms of photo-physical properties has been a difficult task.

Effect of Wavelength, Band Gap, and Light Intensity

Semiconductors absorb light radiation with the threshold wavelength that provides sufficient photon energy to overcome the band gap between the valence and conduction bands. This threshold wavelength, which is required to promote electrons to the excited state, corresponds to the minimal photon energy and depends on the band gap energy; eg, for TiO_2 anatase with a band gap energy of 3.2 eV, it is about 387.5 nm. Irradiation using light with a shorter wavelength (254 nm) is significantly more effective for CO_2 reduction using TiO_2 than that with a longer wavelength (350 nm). The wavelength (λ) of the light used influences the yield of products in studies on N (4% w/w) and Ni (6% w/w) co-doped titania for CO_2 reduction with water. For this catalyst formulation, methanol yields of 482, 253.5, and 120.5 (all in μmol/g-cat) with radiation wavelengths of 254, 365, and 400–780 nm, respectively, have been observed and recorded, which clearly brings out the influence of the wavelength of the light used. Electrons in excited states are produced via electronic transitions, the probability of which depends on the light intensity. The CO_2 reduction rate increases linearly with the light intensity, and at midrange light intensities, the photo-catalytic reaction rate depends on the square root of the light intensity. At high light intensities the rate is independent of the light intensity.

Variations in the wavelength of radiation used can also affect the rate of formation of products on $BiVO_4$ catalysts (Table 11.3). Monoclinic $BiVO_4$ is more active than the tetragonal form. The selective formation of ethanol is observed using 300 W UV radiation with and without a UV cutoff filter. However, with a 36-W lamp, both types of alcohol formed are nearly the same.

Influence of Reaction Pressure

CO_2 photo-reduction with water containing dispersed TiO_2 increased with an increase in CO_2 pressure. An increase in the solubility of CO_2 in the liquid

TABLE 11.2 CO_2 Photo-Reduction With Water on P-25 TiO_2

Sr. Number	Products[a]	CH_4	CH_3OH	Reactor Type	Reactants	Light Source	T (K)	References
1	CH_4, C_2H_6	0.9	Tr	Quartz cell	CO_2 in H_2O	500 W Hg > 310 nm	278	[1]
	CH_3OH							
2	H_2, CO, CH_4	2.0	Tr	Quartz cell	CO_2 in H_2O	1000 W Hg < 700 nm	343	[2]
3	H_2, CH_4, C_nH_m	4.74		Miniaturized reactor	CO_2 in H_2O	200 W Hg/ Xe < 900 nm	278	[3]
		5.68					373	
		6.42					473	
4	CH_4, HCOOH	0.43		Stainless steel (SS) vessel	CO_2/IPA (iso propyl alcohol)	4200 W, Xe		[4]
5	CH_4, CH_3OH	0.35		Quartz cell	CO_2: H_2O 1:5	75 W Hg 254 nm	323	[5]
6	CH_3OH	–	6.37	Inner irradiator cell	CO_2 in NaOH	8 W Hg 250 nm	323	[6]
7	CH_4, CH_3OH		93.75	Inner irradiator cell	CO_2 in $NaHCO_3$	15 W Hg 365 nm		[7]

Continued

TABLE 11.2 CO_2 Photo-Reduction With Water on P-25 TiO_2—cont'd

Sr. Number	Products[a]	CH_4	CH_3OH	Reactor Type	Reactants	Light Source	T (K)	References
8	CH_3-COOH		1.8	Quartz window	Liquid CO_2	990 W Xenon Arc lamp		[8]
9	CH_4	0.33	–	Quartz cell	CO_2: H_2O 1:5	100 W Hg > 250 nm	323	[9]
10	CH_4	0.001		Top illumination	CO_2: H_2O 1:5	4.8 W 253.7 nm	311	[10]
11	CH_4	14.6		Glass plate	CO_2: H_2O 1:5	15 W 365 nm	296	[11]
12	CH_4	3.51		Annular reactor	CO_2 in $NaHCO_3$/IPA	450 W Hg	293–298	[12]
13	CH_3OH		430	Inner irradiator cell	CO_2 in NaOH	400 W halide 300–600 nm	315	[13]

[a]*Products are in micromoles per hour per gram of catalyst; tr - traces.*
Data collected from literature.

TABLE 11.3 Rates of Methanol and Ethanol Formation With Radiations of Different Wavelengths

$BiVO_4$	Production Rate[a] (μmol/h)	300-W Xe Arc Lamp With UV Cutoff Filter	300-W Xe Arc Lamp Without UV Cutoff Filter	36-W Fluorescent Lamp
Monoclinic	Ethanol	21.6	406.6	2.3
	Methanol	0	0	1.8
Tetragonal	Ethanol	1.1	4.9	0.6
	Methanol	0	0	0.6

[a]*The rate was determined on the basis of the average production rate after 80 min of irradiation.*
Reproduced from Y. Liu, B. Huang, Y. Dai, X. Zhang, X. Qin, M. Jiang, M. Whangbo, Catal. Commun, 11 (2009) 210.

phase (water as well as aqueous 0.2 N NaOH solution) at elevated pressures and the consequent increase in the concentration of CO_2 in the solution phase are responsible for the observed increase in the CO_2.

CO_2 photo-reduction using TiO_2 powder in a liquid CO_2 medium was studied. CO_2 has limited solubility in water. Also, the reduction of CO_2 is competitive with hydrogen formation from water. To overcome this disadvantage, a liquid CO_2 system has been explored. The main reduction product was exclusively formic acid. The photo-catalytic reduction of CO_2 using TiO_2 pellets was also studied. The pellet increased the contact areas and adsorption capacity. Their yield was significant compared with the thin-film coating technique.

Influence of Titania Particle Size

The effect of TiO_2 particle size on the photo-catalytic reduction of CO_2 was also studied. As the particle size decreased, higher yields of methanol and methane over the TiO_2 nanoparticles under the illumination of light were obtained. The optimum particle size corresponding to the highest yields of both products was 14 nm. For particles with crystallite sizes less than 14 nm, conversions dropped owing to a combination of electronic (increase in recombination rates) and optical properties. The observed optimum particle size was a result of the competing effects of the specific surface area, charge—carrier dynamics, and light absorption efficiency.

Influence of Type of Photo-Reactors and Reaction Media

The design of a photo-reactor is an important aspect in the CO_2 photo-reduction process with water, which could significantly influence the selectivity of products and the efficiency of the photo-catalyst and process. An ideal photo-reactor must have uniform light distribution throughout the entire system to achieve optimal results. The two parameters that determine the type of photo-reactors used in CO_2 photo-reduction are the phases involved, ie, multiphase, gas–solid, liquid–solid, gas–liquid–solid, and the mode of operation, ie, batch, semibatch, or the continuous flow of reactants. The photo-catalyst can be used in either suspended or immobilized forms in different reactor configurations.

The influence of reactor geometry on the photo-reduction of CO_2 using two annular batch reactors was studied. The dependence of product yields on the reactor diameter and the volume of the liquid phase confirmed the fact that the requirement of perfect mixing is difficult with the annular configuration of the reactor. The highest yields of photo-catalytic reduction were achieved in a configuration in which the lamp just touches the surface of the liquid in the reactor and the configuration of the reactor was not annular. Advantages and disadvantages of different types of photo-reactors for application in CO_2 photo-reduction are given in Table 11.4.

An optical fiber reactor was used for the photo-reduction of CO_2 with H_2O using TiO_2, Cu/TiO_2, Ag/TiO_2, Cu–Fe/TiO_2–SiO_2, and dye-sensitized Cu–Fe/P25 catalysts coated on optical fibers. Compared with a traditional packed-bed reactor, an optical fiber reactor provides medium to transmit light uniformly throughout a reactor, and hence higher conversion is achieved. In addition, a higher processing capacity is possible because the photo-catalyst can be dispersed on optical fibers with a large surface area in a given reactor volume (Table 11.5).

When CO_2 present in the atmosphere dissolves in water, it is mostly in the form of carbonate. Many authors have studied the photo-catalytic reduction of carbonate to form various chemicals. The PCR of carbonate in an aqueous solution by the UV/TiO_2 process has been studied. The photo-catalytic reduction of carbonate proceeded faster in acidic solutions than in alkaline solutions. The main products of the PCR of carbonate by UV/TiO_2 reduction were methanol and methane. A Langmuir–Hinshelwood (L-H)-type kinetic equation was developed to model the photo-catalytic reduction of carbonate.

The effect of adding carbonate salt to the photo-catalytic decomposition of liquid water over a Pt–TiO_2 catalyst was studied. The addition of carbonate salts to Pt-loaded suspensions led to the highly efficient stoichiometric photo-catalytic decomposition of liquid water into H_2 and O_2. Neither the pH nor cation directly contributed to water splitting. The presence of a high concentration of carbonate ions is essential for the catalytic photo-decomposition of water. The carbonate ion affected both the Pt particles and the TiO_2 surface. The Pt was covered with some titanium hydroxide compounds; therefore, the rate of the back reaction (H_2O formation from H_2 and O_2) on the Pt was effectively suppressed in the presence of carbonate ions. On the other hand, the TiO_2 surface was readily covered with several types of carbonate species. These carbonate species aided the desorption of O_2 from the TiO_2 surface.

TABLE 11.4 Advantages and Disadvantages of Photo-Reactor Systems

Reactor Design	Advantages	Disadvantages
Fluidized and slurry reactor (multiphase)	**1.** Temperature gradients inside beds can be reduced through vigorous movements caused by solid passing through fluids **2.** Heat and mass transfer rates increase considerably owing to agitated movement of solid particles **3.** High catalyst loading	**1.** Filters (liquid phase) and scrubbers (gas) are needed **2.** Flooding tends to reduce effectiveness of catalyst **3.** Difficulty of separating catalyst from reaction mixture **4.** Low light use efficiency owing to absorption and scattering of light by reaction medium **5.** Restricted processing capacities owing to mass transport limitations
Fixed-bed reactor	**1.** High surface area **2.** Fast reaction time **3.** Conversion rate per unit mass of catalyst is high owing to flow regime close to plug flow **4.** Low operating costs owing to low pressure drop **Variants of fixed-bed designs**	**1.** Temperature gradient between gas and solid surface is common
Monolith reactor	**1.** High surface-to-volume ratio and low pressure drop with high flow rate can be achieved **2.** Configuration can easily be modified	**1.** Low light efficiency owing to opacity of channels of monolith
Optical fiber reactor	**1.** High surface area and light use efficiency **2.** Efficient processing capacities of catalyst	**1.** Maximum use of reactor volume is not achieved **2.** Heat buildup of fibers can lead to rapid catalyst

Adapted From O. Ola, M.M. Maroto-Valer, J. Photochem. Photobiol. C: Photochem. Rev 24 (2015) 16.

TABLE 11.5 Production Rate of Methane and Ethylene Over TiO_2-Based Catalysts Under Artificial Light: Influence of Type of Photo-Reactor on Reaction Rate

	Production Rate[a] [μmol/(g·cat)/h]	
	Ethylene	Methane
Cu (0.5 wt%)-Fe (0.5 wt%)/TiO_2/glass plate	0.049	0.060
Cu (0.5 wt%)-Fe (0.5 wt%)/TiO_2/optical fiber	0.575	0.914
N3-dye-Cu (0.5 wt%)-Fe (0.5 wt%)/TiO_2/glass plate	0.033	0.148
N3-dye-Cu (0.5 wt%)-Fe (0.5 wt%)/TiO_2/optical fiber	0.562	0.847

[a] *Production rates were determined on the basis of the average value after 4 h of irradiation. The artificial light was in the wavelength range of 320–500 nm with an intensity of 225 mW/cm².*
Adapted from J.C.S. Wu, T.H. Wu, T. Chu, H. Huang, D. Tsai, Topic Catal. 47 (2008) 131; J.C.S. Wu, H.M. Lin, C-H. Lai, Appl. Catal. A Gen. 296 (2005) 194; J.C.S. Wu, H.M. Lin, Int. J. Photoenergy 7 (2005) 115; J.C.S. Wu, Catal. Surv. Asia 13 (2009) 30.

In the place of pure water as the medium for the dispersion of the catalysts, in some cases aqueous solution of $KHCO_3$ or NaOH has been used. Different reaction media such as NaOH, NaOH + Na_2SO_3, NH_4OH, and NH_4OH + Na_2SO_3 have been used to understand the influence of the reaction medium on the PCR of CO_2. NaOH was a better reaction medium than NH_4OH for the yields of both the gas phase (CH_4, CO) and liquid phase (CH_3OH) products. Some authors tried with organic reaction media such as methanol, propan-2-ol, which itself becomes oxidized during the reaction and hence is unfavorable for the potential separation of products from CO_2-reduced products. The use of NaOH as a reaction medium is also beneficial because OH^- ions act as hole scavengers, thereby retarding the electron–hole recombination; simultaneously, the alkaline solution increases the solubility of CO_2 vis-à-vis pure water.

Photo-Reduction With Other Reductants

Some researchers have attempted to replace water with other reductants. This provides a high reaction yield and high selectivity to desired products by changing the mechanism. Experiments have been conducted with CdS in various solvents including water, methanol, ethanol, and 1-propanol with dielectric constants of 80, 33, 24.3, and 20.1, respectively. The results indicated that if low-dielectric constant solvents or low-polarity solvents are used, CO_2^- anion radicals can be strongly adsorbed on the surface through the carbon atom of another CO_2^- anion radical pre-sorbed on surface Cd sites, because these radicals are not completely solvated by low-polarity solvents.

CO is produced as the major reduction product of CO_2. If a high-dielectric constant solvent is used (eg, water), the CO_2^{-} anion radicals can be greatly stabilized by the solvent, resulting in weak interactions with the photo-catalyst surface. Similar effects of solvents have been observed on titania embedded in SiO_2 matrices. Subsequently, the carbon atom of the radical tends to react with a proton to produce formic acid. PCR of CO_2 using TiO_2 suspension in aqueous solutions containing 2-propanol as a hole scavenger leads to the preferential formation of methane.

CLASSIFICATION OF CATALYST SYSTEMS

For the sake of clarity and a meaningful comparative study, the catalyst systems studied so far have been grouped thus:

1. catalysts based on pure titania
2. titania catalysts modified with metals or metal oxides
3. titania catalysts modified with anions
4. titania catalysts dispersed onto mesoporous materials
5. titania catalysts modified with sensitizers
6. metal oxide catalysts: mono, binary, and ternary oxides

Details regarding the application of these catalysts toward CO_2 photo-reduction with water are compiled in Tables 11.6–11.13.

KINETICS AND MECHANISM OF CO_2 PHOTO-REDUCTION WITH WATER

The CO_2 reduction pathway on semiconductor surfaces is at an infant stage. However, based on detailed experiments, researchers have put forward plausible pathways for CO_2 reduction on TiO_2 surfaces.

Three different reaction pathways, depending on the binding mode of the CO_2^{-} radical anion on the semiconductor surface, have been proposed, ie, (1) formaldehyde, (2) carbene, and (3) glyoxal (Fig. 11.8).

In the case of the formaldehyde pathway, the binding mode of CO_2 through an oxygen atom on a titanium ion or a carbon atom through a surface oxygen atom of TiO_2 favors the formation of a COOH radical, which further combines with H to form formic acid. As shown in Fig. 11.8(A), final eight e−reduction product CH_4 was obtained from the consecutive reduction of formic acid through HCHO and CH_3OH as intermediate products. However the formation of surface formate species is proposed as the initial step according to Scheme 11.1.

$$CO_2 + 2e^- + H^+ \rightarrow HCOO^- \qquad (11.20)$$

A mechanistic study of the reaction with methanol as an electron donor on Ti silicalite was studied using in situ Fourier transform infrared spectroscopy

TABLE 11.6 Studies on Pure Titania-Based Catalysts

References	Catalyst	Type of Reactor	Reactant	Temperature (K)	Light Source and Intensity	Product Formation (μmol/g·cat)
[18]	TiO_2	SS autoclave batch Mode	98 kPa – 2.5 MPa CO_2 in NaOH	293	340 nm	HCOOH, CH_3OH, CH_3CHO, C_2H_5OH
[19]	TiO_2 liquid CO_2 medium	SS vessel batch mode	6/5 MPa CO_2	293	Xe (990 W)	HCOOH: 0.08
[4]	TiO_2 Degussa p 25	SS vessel batch mode	2.8 MPa CO_2 in 1 M IPA	293	Xe (4.2 kW)	CH_4: 1.3
[20]	TiO_2 in CO_2 Sc	SS vessel	9 MPa CO_2	308	Xe 900 W	HCOOH
[21]	TiO_2	Rayonet photo-reactor	CO_2 in IPA	308	350 nm	CH_4: 0.3
[22]	TiO_2 pellets	Cylindrical quartz tube	$CO_2 + H_2O$	308	253.7 nm UV	CH_4: 200 ppm
[10]	TiO_2	Top Illuminated cell	$CO_2 + H_2O$	311	253.7 nm UV	CH_4: 0.001
						HCHO

[12]	Mixed- phase TiO_2	Annular glass	CO_2 in aqueous $NaHCO_3$ + IPA	293–298	UV–visible 450W Hg	CH_4, P25-5; A+R-50; A-20
[23]	TiO_2 (14 nm)	Stirred batch annular	CO_2 in NaOH	308	254 nm Hg 8 W	CH_3OH: 0.16
[24]	P-25, RM-TiO_2 UV 100	Glass reactor with quartz window	Saturated CO_2 in NaOH	298	Hg/300 –700 nm 250 W	CH_4, CH_3OH, C_2H_5OH
[25]	TiO_2 (rod)	SS with quartz window	(CO_2 + H_2O) vapor	RT	300 W Xe lamp	CH_4: 5.7/h
[26]	1D TiO_2	Cylindrical with quartz window	(CO_2 + H_2O) vapor	RT	400 W Xe lamp	CH_4: 136/h
[27]	TiO_2 ultrathin nanosheets	Glass reactor	(CO_2 + H_2O) vapor	RT	300 W UV Xe lamp	CH_4: 5.8 ppm/h
[28]	Meso TiO_2	Airtight circulation system	(CO_2 + H_2O) vapor	RT	300 W Xe lamp	CH_4: 1.75
						CO: 2.75

ppm, parts per million; *RT*, room temperature.

TABLE 11.7 Studies on Metal-Doped Titania Catalysts

References	Catalyst	Type of Reactor Used	Reactants	Temperature (K)	Light Source and Intensity	Product Formation (μmol/g·cat)
[29]	RuO_2 doped with TiO_2	Borosilicate glass reactor	$CO_2 + H_2O$	333	75 W Hg	HCOOH: 1.47
						HCHO: 0.13
						CH_3OH: 0.22
						Q.E 0.04%
[30]	Pd/TiO_2	Pyrex glass cuvette	CO_2 in 0.05 $MNaHCO_3$ $Na_2C_2O_4$	299	450 W Xe/369 nm cut off filter	$HCOO^-$: −4.5
[31]	Cu/TiO_2	SS vessel with quartz window	28 kff/cm^2 CO_2 at pH 5.45	308	450 W UV Xe	CH_4: 21.8 C_2H_4: 26.3 C_2H_6: 2.7
[32]	Cu(II) 5 wt %/JRC-TiO_2	Flat-bottomed quartz cell		275	75 W Hg > 290 nm	CH_4: 1.5 CH_3OH: 6.0
[33]	Cu_2O/TiO_2	Slurry phase quartz reactor	135.6 kPa CO_2 in 0.2 N NaOH	323	8 W Hg/254 nm	CH_3OH: 118.5
[6]	Cu/TiO_2	Cylindrical quartz reactor	CO_2 in 1 M NaOH	308	Hg/365 nm	CH_3OH: 1000
[34]	3 wt% CuO/TiO_2	Horizontal SS vessel	CO_2 in 1M $KHCO_3$	373	10 W UV 1 W/cm^3	CH_3OH: 2655
[15]	1.2 wt% Cu(I)/TiO_2	Continuous optical fiber	1.29 bar CO_2 0.026 bar H_2O	348	365 nm	CH_3OH: 0.45

[14]	TiO_2 Cu or Ag/ TiO_2	Optical fiber	p CO_2: 1.19 bar pH_2O: 0.03 bar	348	UV, 365 nm 10 W/cm^2	CH_3OH: TiO_2: 2.0
						Cu/TiO_2: 3.8
						Ag/TiO_2: 4.3
[35]	Cu-Fe/TiO_2	Continuous circular Pyrex glass	$CO_2 + H_2O$	348	250–450 nm; 225 mW/cm^2	CH_4: 0.9
						C_2H_4: 0.6
						Q.E: 0.024%
[36]	Cu and Pt on N-doped TiO_2	SS chambers with quartz window	$CO_2 + H_2O$	317	250–450 nm 225 mW/cm^2	CH_4: 75
						Alkane: 25
						H_2: 155 ppm/cm^2/h
[37]	Pt/TiO_2 nanotube	Fixed-bed quartz reactor	CO_2 in water vapor	308	300 W Hg 365 nm	CH_4: 3.9
[38]	7% Ag-doped TiO_2	Annular reactor	110 kPa CO_2 in 0.2 M NaOH	308	8 W Hg/254 nm	CH_3OH: 1.9
						H_2: 120
						CO: 0.9
[39]	2 wt% Cu/TiO_2	SS316 high-temperature chamber		298	100 W Hg/365 nm	In situ infrared spectroscopic study
[40]	N-Ni/TiO_2		CO_2 (200 mL/min)	348	365 nm 15 W	CH_3OH 482
[41]	Ag-doped TiO_2	Batch-stirred annular reactor	CO_2 in 0.2 N NaOH	308	8 W Hg/254 and 365 nm 8 W Hg filled with Ar (400 nm)	254-nm light more effective than 365-nm 400-nm ineffective

TABLE 11.8 Studies on Anion-Doped Titania Catalysts

Number	Photocatalyst	Type of Reactor Used	Conditions		Light Source and Intensity	Product Formation (μmol/g·cat)	References
			Reactant	Temperature (K)			
1.	S-doped TiO_2	Batch-type reactor	$CO_2 + H_2O$	Sunlight	8 mW/cm^2 UV lamp	CH_3OH, C_2H_5OH	[42]
2.	N-doped TiO_2 nanotubes	Quartz reactor	CO_2 in 0.1 N NaOH	308	500W W-Halogen lamp	HCOOH 12475.8 ± 266.7, CH_3OH: 1132.6 ± 45.4, HCHO: 921.6 ± 24.1	[43]
3.	N-doped mesoporous TiO_2 with cocatalyst Pt, Ag, Au	Quartz tubular reactor	CO_2 + gaseous H_2O	333	350 W Xe lamp	CH_4: 5.7 in 2 h	[44]
4.	I-doped TiO_2	Stainless-steel with quartz window	$(CO_2 + H_2O)$ vapors	RT	450 W Xe lamp		[45]
5.	N/100	Semibatch	$CO_2 + H_2O$	RT	–	CH_3OH: 23/h	[46]
6.	g-C_3N_4–N–TiO_2 (CT–70)	Gas closed-circulation system	$(CO_2 + H_2O)$ vapors	303	300-W Xe arc lamp	CO: 14.73	[47]
7.	N–TiO_2/spirulina	Slurry reactor	$CO_2 + H_2O$	RT	13-W lamp	H_2: 144.99*	[48]
						CH_4: 0.48	
						C_2H_4: 0.12	
						C_2H_6: 0.17	

RT, room temperature. *mmol/g.

TABLE 11.9 Studies on Titania Supported on Different Materials

			Conditions				
Number	Catalyst	Type of Reactor Used	Reactant	Temperature (K)	Light Source and Intensity	Product Formation (μmol/g·cat)	References
1.	TiO_2-loaded Nafion films	Cylindrical high-pressure optical cell with quartz window	2000-psia CO_2		990 W Xe	HCOOH: 190, CH_3OH: 280, CH_3COOH: 30	[49]
2.	Multiwalled carbon nanotube supported TiO_2	SS reactor	CO_2 and H_2O vapor	308	UV 365 nm	CH_4: 58.7, HCOOH: 93.4, C_2H_5OH: 149.4	[11]
3.	CdSe/Pt/TiO_2 heterostructure	Custom-built phot- catalysis cell	3-Torr H_2O vapor 0.3-Torr CO_2	308	Visible light 420 nm	CH_4, CH_3OH, CO, H_2 (CO_2 conversion rate 0.8 ppm/g/h)	[50]
4.	Graphene-TiO_2 nano Composite				UV 365 nm	CH_4: 8.5	[51]
5.	CdS–(Bi_2S_3), TNT	Continuous-flow reactor	CO_2 in NaOH, Na_2SO_3	308	500 W Xe	CH_3OH TNT: 102.5, TNTs CdS: 159.5, TNTs Bi $_2S_3$: 224.6	[52]

Continued

TABLE 11.9 Studies on Titania Supported on Different Materials—cont'd

			Conditions				
Number	Catalyst	Type of Reactor Used	Reactant	Temperature (K)	Light Source and Intensity	Product Formation (μmol/g·cat)	References
6.	Nafion-coated Pd–TiO_2	Quartz reactor	CO_2 in Na_2CO_3	RT	300-W Xe lamp	CH_4: 7 C_2H_6: 0.7	[53]
7.	Mesoporous TiO_2/graphitic carbon microspheres	Stainless-steel reactor	$CO_2 + H_2O$ vapor	303	300-W Xe lamp	CH_4: 10.3	[54]
8.	CNTs@Ni-doped TiO_2	Quartz reactor	$CO_2 + H_2O$ vapor	RT	Visible light λ 400 nm	CH_4: 0.145/h	[55]
9.	Graphene-$Ti_{0.9}O_2$ hollow spheres	Cylindrical stainless-steel quartz window	$CO_2 + H_2O$ vapor	RT	450-W Xe lamp	CO, CH_4	[56]
10.	Graphene-TiO_2 nanosheets	Glass reactor	$CO_2 + H_2O$ vapor	RT	300-W Xe lamp	CH_4: 8 C_2H_6: 16.8	[57]

RT, room temperature.

TABLE 11.10 Studies on Titania Dispersed on Porous Materials

			Conditions				
Number	Catalyst	Type of Reactor Used	Reactant	Temperature (K)	Light Source and Intensity	Product Formation (μmol/g·cat)	References
1.	Q-TiO_2 embedded in SiO_2	Quartz cell	CO_2 in IPA	308	500 W Hg	CO, formate	[58]
2.	Pt on (a) Ti-ion exchanged (b) Ti in MCM-48 framework	Flat-bottom quartz cell	734 Pa CO_2	328	75 W Hg/$\lambda >$ 280 nm	CH_4: 12, CH_3OH: 2.	[59]
3.	Pt/TiO_2 in zeolites	Quartz cell	CO_2 in H_2O	328	Hg/$\lambda >$ 280 nm	Ti in O_h sites- CH_4, T_d sites- CH_3OH	[60]
4.	Ti-silicalite molecular sieve		CO_2, H_2O vapor		UV laser 266 nm	CO, HCHO, CH_3CHO	[61]
5.	Fluorination of Ti-FSM-1	Quartz cell with flat bottom	CO_2 in H_2O	323	100 W Hg/$\lambda >$ 250 nm	Fluorination increases selectivity toward CH_3OH	[5]
6.	TiO_2 in porous SiO_2 thin film	-	CO_2 in H_2O	323	100 W Hg/$\lambda >$ 250 nm	CH_4: 70, CH_3OH: 10	[62]
7.	Ti–MCM- 41	Infrared vacuum cell, 3.4 mL	0.1 MPaCO_2	298	266 nm	CO, O_2, H_2O confirmed as an e-donor.	[63]

Continued

TABLE 11.10 Studies on Titania Dispersed on Porous Materials—cont'd

			Conditions				
Number	Catalyst	Type of Reactor Used	Reactant	Temperature (K)	Light Source and Intensity	Product Formation (μmol/g·cat)	References
8.	Ru/TiO_2, TiO_2/SiO_2	Inner irradiation type (batch mode)	CO_2 in IPA	308	1000 W Hg/ 365 nm.	CH_3OH: 250, CH_4: 75	[64]
9.	Cu, Fe substitution in TiO_2-SiO_2	Packed-bed, circulated reactor 216 fibers	$CO_2 + H_2O$ vapor	348	150 W Hg/ 320–500 nm, 500–800 nm	Cu–Fe/TiO_2 favors C_2H_4 formation, Cu– Fe/TiO_2-SiO_2 favored CH_4	[65]
10.	0.5% Cu/Ti O_2-SiO_2	SS wall and quartz window. (continuous flow)	CO_2 in H_2O vapor	308	Xe arc source, $\lambda < 650$ nm	CO: 60, CH_4: 10	[66]
11.	Cu_2O/SiC	XPA-II photochemical reactor	CO_2 in NaOH and Na_2SO_3 solutions	308	500 W Xe/ 200–700 nm	CH_3OH: 191	[67]
12.	Ti-SBA-15	Combinatorial photo reactors	CO_2 in H_2O		120 W Hg/ 280–650	CH_4, C_2H_4, C_2H_6	[68]
13.	Cu/TiO_2-SiO_2	SS wall with quartz window (continuous flow)	$(CO_2 + H_2O)$ vapor	RT	Xe arc lamp	CO: 10/h CH_4: 60/h	[69]

14.	2 wt% Cu-TiO_2-SBA 15	Inner irradiation type (continuous mode)	CO_2 in $NaOH + H_2O$	315	500-W halide lamp	CH_4: 627/h	[70]
15.	Kaolinite/TiO_2	Stirred-batch annular reactor	NaOH	RT	8-W Hg lamp	CH_4: 4.5	[71]
						CO: 2.5	
16.	1, 3 wt% Ce –TiO_2/SBA 15	Stainless-steel with quartz window	$CO_2 + H_2O$	RT	450-W Xe lamp	CO: 1	[72]
17.	Ti-KIT-6/Si-Ti = 100	Pyrex glass reactor	$CO_2 + H_2O$	RT	300-W UV lamp	CH_4: 4.4/h	[73]
						H_2: 2.2/h	
18.	TiO_2 (20%)/KIT6	Pyrex glass reactor	$(CO_2 + H_2O)$ vapor	RT	300-W lamp	CH_4: 44.56 CH_3OH: 1.09 CO: 120.54 H_2: 44.56	[74]
19.	Mesoporous g-C_3 N_4 flakes	Closed gas circulation system	CO_2 in NaOH	RT	300-W Xe lamp, visible light λ 420 nm	CH_3OH, C_2H_5OH	[75]
20.	Si/TiO_2 nanosphere	Stainless-steel cell with quartz window	$(CO_2 + H_2O)$ vapor	RT at 50 psi	355-nm laser lamp	CH_3OH: 197 μM/100 mL	[76]
21.	Cu/TiO_2 monoliths	Pyrex glass reactor	$(CO_2 + H_2O)$ vapor	RT	200-W Hg/ 500 W halogen lamp	H_2: 12.5, CH_3OH: 4, C_2H_5OH: 3, CH_3CHO: 3	[77]

Continued

TABLE 11.10 Studies on Titania Dispersed on Porous Materials—cont'd

Number	Catalyst	Type of Reactor Used	Conditions		Light Source and Intensity	Product Formation (μmol/g·cat)	References
			Reactant	Temperature (K)			
22.	TiO_2/SiO_2	Quartz reactor	$(CO_2 + H_2O)$ vapor	298	300-W Xe lamp	CO: 3/h	[78]
23.	Cu/GO	Cylindrical reactor	$(CO_2 + H_2O)$ vapor	303	300-W halogen lamp	CH_3OH: 3/h, CH_3CHO: 4/h	[79]
24.	Ag-MWCNT@ TiO_2	Quartz reactor	$(CO_2 + H_2O)$ vapor	333	15-W energy-saving light	CH_4: 6.34, C_2H_6: 0.68	[80]
25.	Au-In/TiO_2	Cylindrical stainless-steel with quartz window	$CO_2 + H_2$	RT	200-W Hg lamp	CO: 8982/h	[81]
26.	Pt-RGO/TiO_2	Quartz reactor	$(CO_2 + H_2O)$ vapor	333	15-W energy-saving light	CH_4: 1.7	[82]
27.	Cu–TiO_2/ZSM-5	–	CO_2 in $NaHCO_3$ solution	298	254-nm UV lamp	CH_3OH: 50.05	[83]
28.	Ca/Ti_xSi MCM-41	Quartz reactor	$CO_2 + H_2O$ (1:2)	303	6-W/cm^2 Hg lamp	CH_4: 82/L	[84]

RT, room temperature.

TABLE 11.11 Studies on Different Metal Oxides

Number	Catalyst	Type of Reactor Used	Conditions		Light Source and Intensity	Product Formation (μmol/g·cat)	References
			Reactant	Temperature (K)			
1.	Cu/ZrO_2	Inner-irradiation quartz reactor	CO_2 in $NaHCO_3$ (pH 2.3)	308	400 W Hg	CO: 2.5	[85]
					UV Hg		
2.	TiO_2 (P25), ZrO_2	Circulating packed-bed quartz reactor	1.1 atm CO_2	316	UV/365 and 254 nm	TiO_2- CH_4: 0.21,	[86]
						CO: 0.28,	
						C_2H_6: 0.2,	
						ZrO_2, CO: 1.24	
						with $H_2 + H_2O$	
3.	Pd, Cu on mixed oxides of MgO, ZnO Al_2O_3, TiO_2, SiO_2	Batch reactor	CO_2 in 0.5M $KHCO_3$ (pH = 7.5)	308	250 mW Hg	C_1-C_3 product. Acidic supports: CH_4, HCHO, HCOOH. Basic supports: C_2H_6, C_2H_5OH, acetone	[87]
4.	MgO	Closed static system	CO_2 and H_2	308	500 W Hg	CH_4,CO	[88]
5.	Ga_2O_3	Closed circulating system–quartz	CO_2: H_2: (150:50 μmol)	308	200 W Hg Xe	CO: 2.1	[89]

Continued

TABLE 11.11 Studies on Different Metal Oxides—cont'd

			Conditions				
Number	Catalyst	Type of Reactor Used	Reactant	Temperature (K)	Light Source and Intensity	Product Formation (μmol/g·cat)	References
6.	ZrO_2	Quartz reactor	CO_2: H_2 or $CH_4 = 150:50$ (μmol)	308	500 W Hg	CO: 9.5	[90]
7.	Meso Ga_2O_3	Quartz window —closed chamber	CO_2 in H_2O	300	300 W Xe	CH_4: 2.9	[91]
8.	WO_3 nanosheets	Pyrex glass cell	$CO_2 + H_2O$ vapor	RT	300 W Xe arc, visible light λ 420 nm	CH_4: 16	[92]
9.	Rectangular sheet-like WO_3	Stainless-steel reaction chamber	$CO_2 + H_2O$ vapor	RT	UV Xe lamp	CH_4: 0.34	[93]
10.	$W_{18}O_{49}$ nanowires	–	$CO_2 + H_2O$ vapor	RT	Visible light λ 420 nm	CH_4: 4000 ppm/g	[94]
11.	Porous MgO/ TiO_2	Continuous-flow glass reactor	$CO_2 + H_2O$ vapor	423	400-W Xe lamp	CO: 30/h	[95]
12.	g-C_3N_4 nanosheet	Stainless-steel with quartz window	$CO_2 + H_2O$ vapor	RT	300-W Xe lamp	CH_4: 0.5, CH_3CHO: 0.05	[96]

13.	g-C_3N_4/ Bi_2WO_6	Glass reactor with quartz window	$CO_2 + H_2O$ vapor	RT	300-W Xe lamp	CO: 5.19/h	[97]
14.	CNTs–TiO_2	Quartz reactor	$CO_2 + H_2O$	RT	15-W energy-saving light bulb	CH_4: 1.2	[98]
15.	CuO nanowire/ ZnO	Cylindrical stainless-steel with quartz window	$(CO_2 + H_2O)$ vapor	RT	400-W Xe lamp	CO: 1980/h	[99]
16.	Fe_2O_3/Cu_2O	Cylindrical stainless-steel reactor	$(CO_2 + H_2O)$ vapor	RT	300-W Xe lamp	CO: 5	[100]
17.	Rh/Cr_2O_3/ GaN	Pyrex reaction cell	$(CO_2 + H_2O)$ vapor	RT	300-W Xe lamp	CH_4: 85, CO: 100, H_2: 3.8	[101]
18.	ZnS	Fused silica reactor	CO_2 in Na_2S	288	1-kW high-pressure Hg (Xe) arc lamp	$HCOO^-$: 440 μM	[102]
19.	g-CN	Schlenk flask	CO_2 in CO-$(bpy)_3C_{l2}$ and TEOA mixture	333	300-W Xe lamp	CO: 10.2, H_2: 2.8	[103]
20.	ZnO/ZnTe	Quartz reactor	$CO_2 + H_2O$	–	300-W Xe lamp	CH_4: 356.51	[104]

RT, room temperature.

TABLE 11.12 Studies on Mixed-Metal Oxide Catalysts

			Conditions				
Number	Catalyst	Type of Reactor Used	Reactant	Temperature (K)	Light Source and Intensity	Product Formation (μmol/g·cat)	References
1.	$SrTiO_3$ in aqueous suspension	UHV surface analysis chamber: batch	$CO_2 + H_2O$	Sunlight	Natural sunlight	HCOOH, HCHO, CH_3OH. QE = 0.011%	[105]
2.	$CaFe_2O_4$	Quartz beaker	CO_2/0.01 M NaOH (pH 5.8)	308	500-W Hg	CH_3OH, HCHO	[106]
3.	Pt/$K_2Ti_6O_{13}$ + Fe-Cu-K/day	Photo-reactor with quartz window	$CO_2 + H_2O$	308	300 W Xe and 150 W Hg	CH_4: 0.014	[107]
						HCOOH: 3.88	
						CH_3OH: 1.45	
						C_2H_5OH: 0.35	
4.	Pt/$K_2Ti_6O_{13}$ with Cu/ZnO	Photo-reactor with flat quartz window	$CO_2 + H_2O$	308	300 W Xe or 150 W Hg	CH_4: 0.86,	[108]
						HCOOH: 43.47,	
						CH_3OH: 18.93	
						HCHO: 10.97	
5.	$ATaO_3$ (A = Li, Na, K)	Closed circulating reactor	CO_2: H_2 = 150:50 (μmol)	303	200 W Hg–Xe	Order of activity: $LiTaO_3$> $NaTaO_3$> $KTaO_3$. CO: 0.42 on $LiTaO_3$	[109]

6.	Bi_2WO_6 square nanoplates	230-mL glass reactor	$CO_2 + H_2O$		300 W Xe	CH_4: 6	[110]
7.	Pt-$NaNbO_3$	Pyrex glass	$CO_2 + H_2O$	308	300 W Xe	CH_4: 653*	[111]
8.	Pt/$NaNbO_3$	Closed Pyrex cell	80 kPa CO_2 in H_2O	308	300 W Xe	CH_4: 0.486	[112]
9.	Pt/Meso Zn-Ge (ZGON)				($\lambda > 400$ nm)	CH_4: 26.8*	[113]
10.	Ag- $ALa_4Ti_4O_{15}$ (A = Ca, Sr, and Ba)	Inner-irradiation quartz cell	CO_2 flow system (15 mL/min)		400 W Hg	$BaLa_4Ti_4O_{15}$ > $SrLa_4Ti_4O_{15}$ > $CaLa_4Ti_4O_{15}$. CO, HCOOH	[114]
11.	Ni@NiO on $InTaO_4$	Pyrex glass (continuous flow)	CO_2(3 kg/ cm^2) in H_2O	298	100-mW Xe lamp	Methanol – $InTaO_4$ < $InTaO_4$-N<Ni@NiO/$InTaO_4$-N.	[115]
12.	NiO–$InNbO_4$, Co_3O_4–$InNbO_4$	Pyrex glass (continuous flow)	CO_2 in 0.2 M $KHCO_3$	308	500 W halogen H-500; 143 μm W/cm^2; λ500 –900 nm	CH_3OH	[116]
						NiO-$InNbO_4$ –1.577	
						CO_3O_4-$InNbO_4$ –1.503	
13.	HNb_3O_8 nanobelts	Quartz tubular reactor	$H_2O + CO_2$	318	350 W Xe	CH_4: 3.58	[117]
14.	Zn-Al lactate dehydrogenase (LDH), Zn-Cu-Ga	Quartz reactor	2.3 kPa CO_2 + 21.7 kPa H_2	305–313	500-W Xe arc lamp	CO: 0.62,	[118]
						CH_3OH: 0.17	

Continued

TABLE 11.12 Studies on Mixed-Metal Oxide Catalysts—cont'd

			Conditions				
Number	Catalyst	Type of Reactor Used	Reactant	Temperature (K)	Light Source and Intensity	Product Formation (μmol/g·cat)	References
15.	$FeTiO_3/TiO_2$	Pyrex glass	CO_2 in $NaHCO_3$	RT	500 W Xe (with cutoff)	CH_3OH: 0.46	[119]
16.	$CuGaO_2$, $CuGa_{1-x}Fe_xO_2$	SS cell	CO_2 in H_2O	RT	300 W Xe arc	CO: 9**	[120]
17.	SiO_2–pillared HNb_3O_8	Quartz tubular reactor	$H_2O/CO_2 = 0.14$	333	34.8 mW/cm^2 Xe arc	CH_4: 2.9	[121]
18.	$Pt/SrTiO_3–Rh/Pt/CuAlGaO_4WO_3$	H-twin reactor	2 mM $FeCl_2/FeCl_3$	RT	AM 1.5G	CH_3OH: 0.52 H_2: 0.12	[122]
19.	$Pt/SrTiO_3–Rh/Pt/CuAlGaO_4WO_3$	H-twin reactor	2 mM $FeCl_2/FeCl_3$	–	AM 1.5G	CH_3OH: 8 H_2: 1	[123]
20.	Pt/self-doped $SrTiO_{3-\delta}$	–	$CO_2 + H_2O$ water	–	300-W Xe lamp, visible light λ 400 nm	CH_4: 0.26/cm^2	[124]
21.	$Ag/SrTiO_3$	Batch slurry reactor	CO_2 in CH_3OH	298	Hg lamp	HCOOH	[125]
22.	Au/leaf-architectured $SrTiO_3$	Pyrex reaction cell	$CO_2 + H_2O$ vapor	RT	300-W Xe lamp	CO, CH_4	[126]

23.	$Pt/Fe_2V_4O_{13}$ nanoribbons		$CO_2 + H_2O$ vapor		Visible light λ 420-nm	CH_4	[127]
24.	Pt and RuO_2/ $Na_2\ V_6O_{16}$ nanoribbons	Circular glass reactor	$CO_2 + H_2O$ vapor		300-W Xe lamp Visible light λ 420 nm	CH_4: 2	[128]
25.	$Bi_6Mo_2O_{15}$ submicrowires	–	$CO_2 + H_2O$ vapor	RT	UV Xe lamp	CH_4 : 17***	[129]
						O_2: 250	
26.	Bi_2WO_6 hollow microspheres	–	(CO_2 + H_2O) vapor	RT	Visible light λ 420 nm	CH_3OH: 33	[130]
27.	Pt/mesoporous $In(OH)_3$	Gas closed circulation pump	(CO_2 + H_2O) vapor	RT	UV Xe lamp	CH_4: 25	[131]
28.	RuO_2/ mesoorous $ZnGa_2O_4$	Glass	(CO_2 + H_2O) vapor	RT	UV Xe lamp	CH_4: 145 ppm	[132]
29.	RuO_2/$ZnGa_2O4$ nanocubes	Glass reactor	(CO_2 + H_2O) vapor	–	UV Xe lamp	CH_4: 2, O_2: 21	[133]
30.	$Pt/ZnAl_2O_4$-modified mesoporous ZnGaNO	–	(CO_2 + H_2O) vapor	–	Visible light λ 420 nm	CH_4: 40	[134]
31.	Pt/ZIF-8/ Zn_2GeO_4 nanorods	–	CO_2 bubbled Na_2SO_3	RT	UV Xe lamp	CH_3OH: 3	[135]
32.	Pt and RuO_2 Sheaf-like, hyperbranched $Zn_{1.7}GeN_{1.8}O$	Pyrex glass	(CO_2 + H_2O) vapor	RT	Visible light λ 420 nm	CH_4: 55	[136]

Continued

TABLE 11.12 Studies on Mixed-Metal Oxide Catalysts—cont'd

			Conditions				
Number	Catalyst	Type of Reactor Used	Reactant	Temperature (K)	Light Source and Intensity	Product Formation (μmol/g·cat)	References
33.	4.5($ZnGa_2O_4$)(Zn_2GeO_4)	Glass reactor	(CO_2 + H_2O) vapor	RT	UV Xe lamp	CH_4: 2.75, O_2: 17	[137]
34.	Pt/$In_2Ge_2O_7$(En) nanowires	Pyrex glass cell	(CO_2 + H_2O) vapor	RT	300-W Xe lamp	CO: 4, O_2	[138]
35.	Pt/Zn_2SnO_4	Pyrex glass cell	(CO_2 + H_2O) vapor	RT	UV Xe lamp	CH_4: 350	[139]
36.	Layered Mg-In LDH	Closed circulation system	(CO_2 + H_2O) vapor	RT	UV Hg–Xe lamp	CO: 3, O_2	[140]
37.	Layered Zn–Cu–Ga LDH	Quartz reactor	(CO_2 + H_2) vapor	RT	500-W UV Xe lamp	CH_3OH, CO	[141]
38.	Ag/$KTaO_3$	Glass reactor	CO_2 + H_2O	RT	300-W Xe lamp	CO: 152.6, H_2: 1134, O_2: 499.9	[142]
39.	(MgAl)-LDO-grafted TiO_2	Quartz tube reactor	(CO_2 + H_2) vapor	423	100-W he/450-W Xe lamp	CO: 4/h	[143]

40.	ZnGaNO nanorod	Glass reactor	$CO_2 + H_2O$	RT	300-W Xe lamp	CH_4: 0.19	[144]
41.	$NiCo_2O_4$ with $[Ru(bpy)_3]Cl_2.6H_2O$	Schlenk reactor	TEOA/water/acetonitrile	303	300-W Xe lamp	CO: 35, H_2: 5	[145]
42.	$SnNb_2O_6$	Schlenk reactor	$CO_2 + H_2O$	RT	300-W Xe lamp	CH_4: 110 μL/h	[146]
43.	$MTaO_3$ (M = Li, Na, K)	Pyrex glass cell	$CO_2 + H_2O$	RT	200-W Hg–Xe arc lamp	CH_4: 1.9 nmol/h CO: 4	[147]
44.	$Ag/BaZrO_3$	Cylindrical quartz reactor	(CO_2 + H_2O) vapor	RT	300-W Xe lamp	CH_4: 4	[148]
45.	$MnCo_2O_4$	–	CO_2 in TEOA	303	300-W Xe lamp	CO: 27, H_2 -8	[149]
46.	Cu_xO_y-$ZrOCo^{II}$ SBA-15	Stainless infrared vacuum cell	CO_2 in diethylamine	RT	Laser	CO	[150]
47.	Cr-doped $SrTiO_3$	–	(CO_2 + H_2O) vapor	RT	300-W Xe lamp	CH_4: 8.8	[151]
48.	$Ca_xTi_yO_3$	Quartz reactor	$CO_2 + H_2O$	303	6-W/cm^2 UV lamp	CH_4: 17	[152]
49.	C, N/sodium titanate	Stainless steel	(CO_2 + H_2O) vapor	RT	100-W Xe solar simulator	CH_4: 9.75/h	[153]

RT, room temperature. *ppm/h/g; **ppm/g/h; ***ppm.

TABLE 11.13 Studies on Dye Sensitized Photo Catalysts

			Conditions				
Number	Catalyst	Type of Reactor Used	Reactant	Temperature (K)	Light Source/ Intensity	Product Formation (μmol/g cat)	References
1.	Metal porphyrin/ phthalocyanine on Nafion	Glass cell with an optical window	CO_2 in TEA	308	500 W tungsten-halogen	HCOOH	[154]
2.	Dye([Ru(B py)$_3$]$^{2+}$ perylene di-imide based) sensitized, Q- yield:Pt onTiO_2	Glass chamber connected to a vacuum line	0.09 MPa CO_2 inH_2O	298	75 W visible daylight lamp, 150 W UV	CH_4: 0.6	[155]
3.	Ruthenium dye (N_3)- sensitized Cu−Fe/TiO_2	Continuous circulating Pyrex glass	$CO_2 + H_2O$ (vapor)	348	Solar 20 mW/ cm^2, UV lamp 225 mW/cm^2	CH_4: 0.617	[156]
4.	CoPc/TiO_2	Pyrex glass cell (batch mode)	CO_2 in 0.1N NaOH	308	500W tungsten-halogen	HCOOH: 300, CH_3OH: 10, HCHO: 50	[157]
5.	0.7 wt.% in-situ CoPc/TiO_2	Pyrex glass cell	CO_2 in 0.1N NaOH	308	500W tungsten-halogen	HCOOH: 1487.6	[158]
						CH_3OH: 93	
						HCHO: 134.3	

6.	Ru/RuOx sensitized TiO_2	–	$CO_2 + H_2O$	–	Solar simulator	CH_4: 900 μL/h	[159]
7.	Dye sensitized (perylene diimide derivatives) Pt impregnated on TiO_2	–	$CO_2 + H_2O$	RT	75 W daylight lamp	CH_4: 0.74	[155]
8.	N719/TiO_2	H Type reactor	$CO_2 + H_2O$/ 2 M NaOH	RT	300 W Xe lamp	CH_3OH: 178.1/cm^2	[160]
						HCHO: 129.2/cm^2	
9.	IrOx/SiGe-jn/CC/ p-RuCP	Stainless steel Monolithic tablet shaped device	$CO_2 + H_2O$	RT	Simulated solar light	HCOOH: 50.2,	[161]
						H_2: 3.6, O_2 :26.4	
10.	Zinc porphyrin/ TiO_2	Pyrex glass reactor	$(CO_2 + H_2O)$ vapor	RT	300 W Xe-lamp	CO: 8.07/h	[162]
						CH_4: 1.01/h	

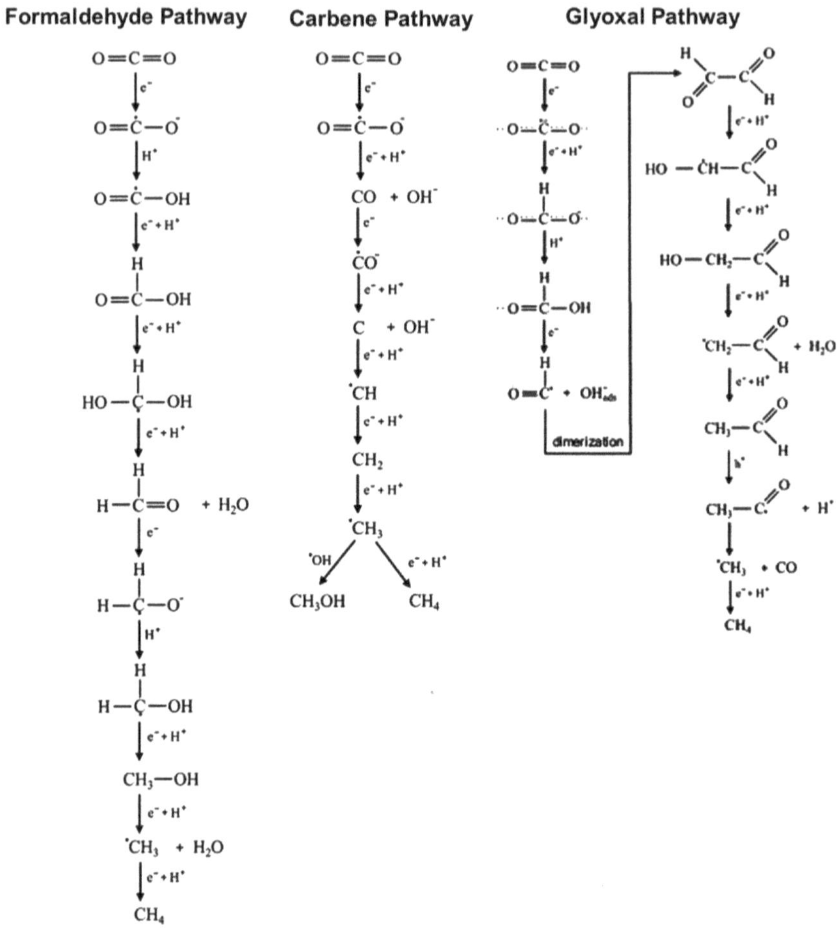

FIGURE 11.8 Proposed mechanisms for the reduction of CO_2 to methane.

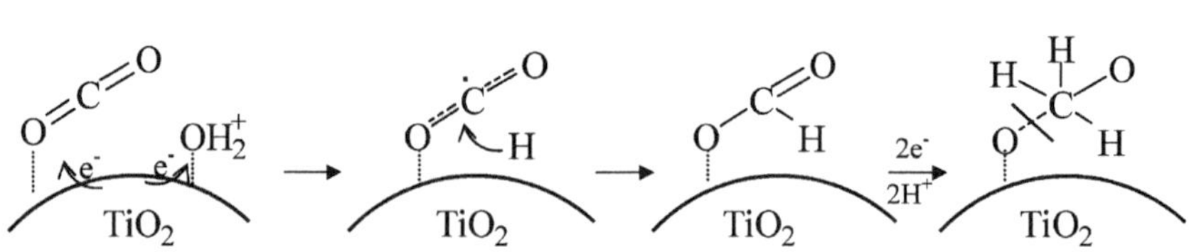

SCHEME 11.1 Proposed mechanism of photo-catalytic transformation of CO_2 to methoxy radical over TiO_2 in the presence of dissociated or bound water.

and traced the reaction pathway using isotopic labeling studies with $C^{18}O_2$, $^{13}CO_2$, and $^{13}CH_3OH$. HCOOH was identified as the primary product, which on pyrolysis gives CO. In the absence of methanol, water was confirmed as the electron donor.

A kinetic model for CO_2 photo-reduction with respect to the formaldehyde pathway was studied. The result showed that the formation of methane did not correspond to the consecutive reduction of formic acid and methanol. Hence the carbene pathway for CO_2 photo-reduction shown in Fig. 11.8(B) was proposed. In this pathway, anion radicals bind through carbon on the semiconductor surface; the formation of CO and O occurs in the first step. Then adsorbed CO is reduced with two electrons to form carbon residue on the semiconductor surface, which further combines with H to form carbene and methyl radicals and finally CH_4. The formation of different transient surface species and their transformations on TiO_2 during the photo-catalytic reduction of CO_2 by water was identified using electron spin resonance (ESR), extended X-ray absorption fine structure, and other spectroscopic techniques. Ti^{3+}, H, C, and CH_3 radicals were detected by ESR studies carried out at 77 K. The charge transfer excited state $(Ti^{3+}\text{-}O^-)^{3*}$ with Ti in tetrahedral coordination, formed by the absorption of light energy, was proposed as the active site on which CO_2 becomes reduced to CO and C, which upon addition of active surface hydrogen forms hydrocarbons. Further mechanistic pathways in this case are in Scheme 11.2.

The photo-catalytic reduction of CO_2 using Ti-SBA-15 deals with investigations into the effects of varying feed compositions on product patterns, with a view to understanding the reaction mechanism. In particular, addition of CO and H_2 in the place of H_2O, methanol, and formaldehyde was studied in detail. Whereas CH_4, C_2H_4, and C_2H_6 could be observed with CO + H_2O with reasonable rates, with the CO + H_2 mixture, the reaction rates were low. It was concluded that H_2O was a more efficient reductant with CO_2 or CO than was H_2. CH_3OH was not a significant product, and when added to the feed it did not increase the reaction rate. Similarly, formic acid also did not enhance product formation when added. These observations rule out methanol or formic acid as intermediates in the formation of hydrocarbons on Ti-SBA-15. However, added formaldehyde underwent facile conversion to hydrocarbon products. Based on

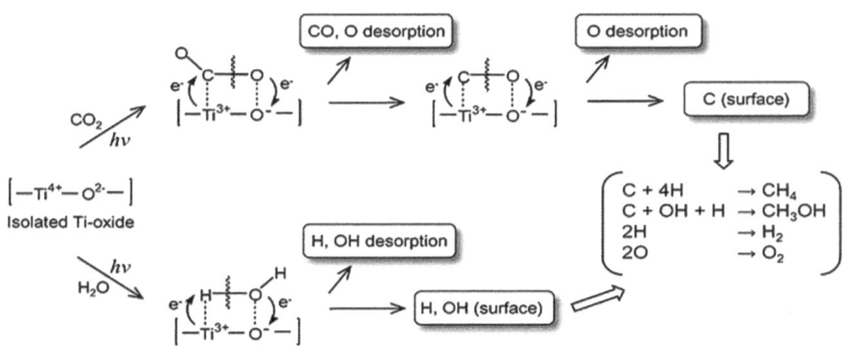

SCHEME 11.2 Schematic representation of the photo-catalytic reduction of CO_2 with H_2O on anchored titanium dioxide. *Reproduced from M. Anpo, H. Yamashita, Y. Ichihashi, S. Ehara, J. Electroanal. Chem. 396 (1995) 21.*

the detailed analysis of the products, it was proposed that CO_2 photo-reduction proceeds through the formation of CO in the initial stages followed by its conversion to formaldehyde, which in turn is converted to other hydrocarbon products. Ti-OH site, which upon photo-activation is converted to [Ti (III)-O]* (*denotes appropriate charge), is considered as the active site.

Kinetic models for the photo-catalytic reduction of CO_2 with water on titania have shown a single-site L-H−type kinetic model to simulate the process. CO is proposed as the primary intermediate. Hence the carbene reaction pathway is a more plausible reaction pathway for CO_2 reduction.

Glyoxal reaction pathway for CO_2 photo reduction by adopting electron paramagnetic resonance study, which includes both oxidation and reduction reactions. In the previous reaction pathway the authors were interested in reduction reactions (electrons), but highly oxidative holes are also present in the system. The formation of oxidized products from formaldehyde and methanol is more favorable compared with the reduced product form it. Hence the C2 complex reaction pathway is shown in Fig. 11.8(C), in which dimerization of the formyl radical gives glyoxal, further reduced to form acetaldehyde through the vinoxyl radical. CH_3CHO is easily oxidized to form an unstable acetyl radical that further undergoes decarbonylation to form a methyl radical, and then combines with H to form CH_4.

Although the formation of C_1 hydrocarbons could be explained to a reasonably acceptable level, understanding of the pathways leading to the formation of C_2 and $>C_2$ hydrocarbons is not clear. Dimerization of surface C_1 species was proposed as a possible route for the formation of C_2 hydrocarbon products, although direct experimental evidence to this effect is lacking.

The backward reaction, ie, oxidation of hydrocarbons back to CO_2 and water, proceeds to a significant extent. This implies that the oxygen produced by the splitting of water, which is responsible for the backward reaction, has to be removed effectively to increase hydrocarbon yields. The means of physically separating the oxidation and reduction sites would be crucial in this respect. It is clear that the reaction mechanism is complex and depends on the type of the catalysts and active sites.

DEACTIVATION OF PHOTO-CATALYSTS

In some cases the catalysts reported for the photo-reduction of CO_2 tend to become deactivated within a short period, whereas in a few cases activity beyond 20 hours has been reported. Transformation of photo-activated CO_2 to hydrocarbons proceeds through several hydrocarbon intermediates whose identity or nature has yet to be established. Accumulation of these intermediates on the catalyst surface and blocking of active sites are major reasons for catalyst deactivation. In the case of catalyst nanoparticles dispersed in liquid media, coagulation and reduction in light absorption power and changes in the oxidation state of metal ions may lead to catalyst deactivation.

Regeneration with air helps restore activity. However the phenomenon of catalyst deactivation in this process needs further study, especially when large-scale applications of this process are desired.

An in-depth understanding of the reaction mechanism and fast deactivation of the photo-catalyst would be helpful in the development of superior catalysts for this process, which has the potential to emerge as a sustainable route for the production of fuels and chemicals.

Photo-catalysis and photo-electrochemical routes can also be used to prepare and produce useful chemicals and fuel precursors. These aspects are beyond the scope of this compilation. Interested readers should refer to original literature in this respect.

REFERENCES

[1] O. Ishitani, C. Inoue, Y. Suzuki, T. Ibusuki, J. Photochem. Photobiol. A 72 (1993) 269.
[2] F. Saladin, L. Fross, I. Kamber, J. Chem. Soc. Chem. Commun. (1995) 533.
[3] F. Saladin, I. Alxneit, J. Chem. Soc. Faraday Trans. 93 (1997) 4159.
[4] S. Kaneco, Y. Shimizu, K. Ohta, T. Mizuno, J. Photochem. Photobiol. A Chem. 115 (1998) 223.
[5] K. Ikeue, H. Yamashita, M. Anpo, T. Takewaki, J. Phys. Chem. B 105 (2001) 8350.
[6] I.H. Tseng, W.C. Chang, J.C.S. Wu, Appl. Catal. B Environ. 37 (2002) 37.
[7] Y. Ku, W.-H. Lee, W.-Y. Wang, J. Mol. Catal. A Chem. 212 (2004) 191.
[8] P. Pathak, M.J. Meziani, L. Castillo, Y.P. Sun, Green Chem. 7 (2005) 667.
[9] J. Hwang, J. Chang, S. Park, K. Ikeuen, M. Anpo, Top. Catal. 35 (2005) 311.
[10] S.S. Tan, L. Zou, E. Hu, Sci. Technol. Adv. Mater. 8 (2007) 89.
[11] X.-H. Xia, Z.-J. Jia, Y. Yu, Y. Liang, Z. Wang, L.-L. Ma, Carbon 45 (2007) 717.
[12] G. Li, S. Ciston, Z.V. Saponjic, L. Chen, N.M. Dimitrijevic, T. Rajh, K.A. Gray, J. Catal. 253 (2008) 105.
[13] K. Kocí, M. eli, O. Kozák, Z. Lacny, D. Plachá, P. Praus, L. Obalov, Catal. Today (2010). DOI:10.1016/j.cattod.2010.12.054.
[14] J.C.S. Wu, T.H. Wu, T. Chu, H. Huang, D. Tsai, Topic Catal. 47 (2008) 131.
[15] J.C.S. Wu, H.M. Lin, C.-H. Lai, Appl. Catal. A Gen. 296 (2005) 194.
[16] J.C.S. Wu, H.M. Lin, Int. J. Photoenergy 7 (2005) 115.
[17] J.C.S. Wu, Catal. Surv. Asia 13 (2009) 30.
[18] T. Mizuno, K. Adachi, K. Ohta, A. Saji, J. Photochem. Photobiol. A Chem. 98 (1996) 87.
[19] S. Kaneco, H. Kurimoto, K. Ohta, T. Mizuno, Akira, J. Photochem. Photobiol. A Chem. 109 (1997) 59.
[20] S. Kaneco, H. Kurimoto, Y. Shimizu, K. Ohta, T. Mizuno, Energy 24 (1999) 21.
[21] G.R. Dey, A.D. Belapurkar, K. Kishore, J. Photochem. Photobiol. A Chem. 163 (2004) 503.
[22] S.S. Tan, L. Zou, E. Hu, Catal. Today 115 (2006) 269.
[23] K. Koč, L. Obalova, L. Matějová, D. Plachá, Z. Lacný, J. Jirkovský, O. Sǒlcová, Appl. Catal. B Environ. 89 (2009) 494.
[24] K. Rajalakshmi, V. Jeyalakshmi, R. Mahalakshmy, K.R. Krishnamurthy, B. Viswanathan, Indian J. Chem. Sect. A Inorg. Bio-Inorg. Phys. Theor. Anal. Chem. 51A (2012) 411.
[25] J. Pan, X. Wu, L. Wang, G. Liu, G.Q.(M.) Lu, H.-M. Cheng, Chem. Comm. 47 (2011) 8361.

[26] W.-N. Wang, W.-J. An, B. Ramalingam, S. Mukherjee, D.M. Niedzwiedzki, S. Gangopadhyay, P. Biswas, J. Am. Chem. Soc. 134 (2012) 11276.
[27] H. Xu, S. Ouyang, P. Li, T. Kako, J. Ye, ACS Appl. Mater. Interfaces 5 (2013) 1348.
[28] T. Wang, X. Meng, P. Li, S. Ouyang, K. Chang, G. Liu, Z. Mei, J. Ye, Nano Energy 9 (2014) 50.
[29] M. Halmann, E. Katzir Borgarello, J. Kiwi, Sol. Energy Mater. 10 (1984) 85.
[30] Z. Goren, I. Willner, A.J. Nelson, J. Phys. Chem. 94 (1990) 3784.
[31] K. Adachi, K. Ohta, T. Mizuno, Solar Energy 53 (1994) 187.
[32] H. Yamashita, H. . Nishigushi, N. Kamada, M. Anpo, Res. Chem. Interm 20 (1994) 815–823.
[33] I.-H. Tseng, J.C.S. Wu, H.-Y. Chou, J. Catal. 221 (2004) 432.
[34] H.W. Slamet Nasution, E. Purnama, S. Kosela, J. Gunlazuardi, Catal. Commun. 6 (2005) 313.
[35] T.-V. Nguyen, J.C.S. Wu, Appl. Catal. A Gen. 335 (2008) 112.
[36] O.K. Varghese, M. Paulose, T.J. Latempa, C.A. Grimes, Nano Lett. 9 (2009) 731.
[37] Q.-H. Zhang, W.-D. Han, Y.-J. Hong, J.-G. Yu, Catal. Today 148 (2009) 335.
[38] K. Koci, K. Mateju, L. Obalova, S. Krejcikova, Z. Lacny, D. Placha, L. Capek, A. Hospodkova, O. Solcova, Appl. Catal. B Environ. 96 (2010) 239.
[39] J.C.S. Wu, C.W. Huang, Front. Chem. Eng. China 4 (2010) 120.
[40] Y. Liu, B. Huang, Y. Dai, X. Zhang, X. Qin, M. Jiang, M. Whangbo, Catal. Commun. 11 (2009) 210.
[41] J. Fan, E.Z. Liu, L. Tian, X.Y. Hu, Q. He, T. Sun, Am. Soc. Civil Eng. (2010). http://dx.doi.org/10.1061/(ASCE)EE.1943-7870.0000311.
[42] S.T. Hussain, K. Khan, R. Hussain, J. Nat. Gas Chem. 18 (2009) 383.
[43] Z., Zhao, J. Fan, J. Wang, R. Li, Catal. Commun. 21 (2012) 32.
[44] X. Li, Z. Zhuang, W. Li, H. Pan, Appl. Catal. A Gen. 429–430 (2012) 31.
[45] Q. Zhang, Y. Li, E.A. Ackerman, M. Gajdardziska-Josifovska, H. Li, Appl. Catal. A Gen. 400 (2011) 195–202.
[46] B. Michalkiewicz, J. Majewska, G. Ka̧dziołka, K. Bubacz, S. Mozia, A.W. Morawski, J. CO_2 Util. 5 (2014) 47.
[47] S. Zhou, Y. Liu, J. Li, Y. Wang, G. Jiang, Z. Zhao, D. Wang, A. Duan, J. Liu, Y. Wei, Appl. Catal. B Environ. 158–159 (2014) 20.
[48] T. Phongamwong, M. Chareonpanich, J. Limtrakul, Appl. Catal. B Environ. 168–169 (2015) 114.
[49] P. Pathak, M.J. Meziani, Y. Li, L.T. Cureton, Y.-P. Sun, Chem. Commun. (2004) 1234.
[50] C. Wang, R.L. Thompson, J. Baltrus, C. Matranga, J. Phys. Chem. Lett. 1 (2010) 48.
[51] Y.T. Liang, B.K. Vijayan, K.A. Gray, M.C. Hersam, Nano Lett. 11 (2011) 2865.
[52] X. Li, H. Liu, D. Luo, J. Li, Y. Huang, Y. Fang, Y. Xu, L. Zhu, Chem. Engg. J. 180 (2012) 151–158.
[53] W. Kim, T. Seok, W. Choi, Energy Environ. Sci. 5 (2012) 6066.
[54] C.L. Zhang, Q.Y. Zhang, S.F. Kang, B. Li, X. Li, Y.G. Wang, ECS Solid State Lett. 2 (2013) M49.
[55] W.J. Ong, M.M. Gui, S.P. Chai, A.R. Mohamed, RSC Adv. 3 (2013) 4505.
[56] L.J. Liu, H.L. Zhao, J.M. Andino, Y. Li, ACS Catal. 2 (2012) 1817.
[57] W.G. Tu, Y. Zhou, Q. Liu, S.C. Yan, S.S. Bao, X.Y. Wang, M. Xiao, Z.G. Zou, Adv. Funct. Mater. 23 (2013) 1743.
[58] B.J. Liu, T. Torimoto, H. Matsumoto, H. Yoneyama, J. Photochem, Photobiol. A Chem. 108 (1997) 187.

[59] M. Anpo, H. Yamashita, Y. Ichihashi, Y. Fujii, M. Honda, J. Phys. Chem. B 101 (1997) 2632.

[60] H. Yamashita, Y. Fujii, Y. Ichihashi, S.G. Zhang, K. Ikeue, D.R. Park, K. Koyano, T. Tatsumi, M. Anpo, Catal. Today 45 (1998) 221.

[61] N. Ulagappan, H. Frei, J. Phys. Chem. A 104 (2000) 7834.

[62] K. Ikeue, S. Nozaki, M. Ogawa, M. Anpo, Catal. Lett. (2002) 111.

[63] H. Ulagappan, J. Frei, Phys. Chem. A 104 (2000) 7834. W. Lin, H. Han, H. Frei, J. Phys. Chem. B 108 (2004) 18269.

[64] N. Sasirekha, S.J.S. Basha, K. Shanthi, Appl. Catal. B 62 (2006) 169.

[65] T.-V. Nguyen, J.C.S. Wu, Solar Energy Mat. Solar Cells 92 (2008) 864.

[66] Y. Li, W. Wang, Z. Zhan, M. Woo, C. Wu, P. Biswas, Appl. Catal. B Environ. 100 (2010) 386.

[67] H. Li, Y. Lie, Y. Huang, Y. Fang, Y. Xu, L. Zhu, Z. Li, J. Nat. Gas Chem. 20 (2011) 145.

[68] C.-C. Yang, J. Vernimmen, V. Meynen, P. Cool, G. Mul, J. Catal. 284 (2011) 1.

[69] W. Wang, J. Park, P. Biswas, Catal. Sci. Technol. 1 (2011) 593.

[70] H.C. Yang, H.Y. Lin, Y.S. Chien, J.C.S. Wu, H.H. Wu, Catal. Lett. 131 (2009) 381.

[71] K. Kŏci, V. Matějka, P. Kovar, Z. Lacny, L. Obalova, Catal. Today 161 (2011) 105.

[72] C. Zhao, L. Liu, Q. Zhang, J. Wang, Y. Li, Catal. Sci. Technol. 2 (2012) 2558.

[73] P. Akhter, M. Hussain, G. Saracco, N. Russo, Nanoscale Res. Lett. 9 (2014) 1.

[74] M. Hussain, P. Akhter, G. Saracco, N. Russo, Appl. Catal. B Environ. 170–171 (2015) 53.

[75] J. Mao, T. Peng, X. Zhang, K. Li, L. Ye, L. Zan, Catal. Sci. Technol. 3 (2013) 1253.

[76] Y. Liu, G. Ji, M.A. Dastageer, L. Zhu, J. Wang, B. Zhang, X. Chang, M.A. Gondal, RSC Adv. 4 (2014) 56961.

[77] O. Ola, M.M. Maroto-Valer, Catal. Sci. Technol. 4 (2014) 1631.

[78] L. Yuan, C. Han, M. Pagliaro, Y-J. Xu, J. Phys. Chem. C, doi:10.1021/acs.jpcc.5b08893.

[79] I. Shown, H.-C. Hsu, Y.-C. Chang, C.-H. Lin, P. Kumar Roy, A. Ganguly, C.-H. Wang, J.-K. Chang, C.-I. Wu, L.-C. Chen, K.-H. Chen, Nano Lett. 14 (2014) 6097.

[80] M.M. Gui, W.M. Ping Wong, S.-P. Chai, A.R. Mohamed, Chem. Eng. J. 278 (2015) 272.

[81] B. Tahir, M. Tahir, N. Saidina Amin, Appl. Surf. Sci. 338 (2015) 1.

[82] L.L. Tan, W.-J. Ong, S.-P. Chai, A.R. Mohamed, Appl. Catal. B Environ. 166–167 (2015) 251.

[83] J.-J. Wang, Y.-H. Jing, T. Ouyang, Q. Zhang, C.-T. Chang, Catal. Commun. 59 (2015) 69.

[84] S. Won Jo, B.S. Kwak, K.M. Kim, J.Y. Do, N.-K. Park, S.O. Ryu, H.-J. Ryu, J.-I. Baek, M. Kang, Appl. Surf. Sci. 355 (2015) 891.

[85] K. Sayama, H. Arakawa, J. Phys. Chem. 97 (1993) 531.

[86] C.C. Lo, C.H. Hung, C.S. Yuan, J.F. Wu, Sol. Energy Mater. Sol. Cells 91 (2007) 1765.

[87] M. Subrahmanyam, S. Kaneco, N. Alonso-Vante, Appl. Catal. B Environ. 23 (1999) 169.

[88] K. Teramura, T. Tanaka, H. Ishikawa, Y. Kohno, T. Funabiki, J. Phys. Chem. B 108 (2004) 346.

[89] K. Teramura, H. Tsuneoka, T. Shishido, T. Tanaka, Chem. Phys. Lett. 467 (2008) 191.

[90] S. Yoshida, Y. Kohno, Catal. Surv. Jpn. 4 (2000) 107.

[91] H. Park, J.H. Choi, K.M. Choi, D.K. Lee, J.K. Kang, J. Mater. Chem. 22 (2012) 5304.

[92] X.Y. Chen, Y. Zhou, Q. Liu, Z.D. Li, J.G. Liu, Z.G. Zou, ACS Appl. Mater. Interfaces 4 (2012) 3372.

[93] Y.P. Xie, G. Liu, L.C. Yin, H.M. Cheng, J. Mater. Chem 22 (2012) 6746.

[94] G.C. Xi, S.X. Ouyang, P. Li, J.H. Ye, Q. Ma, N. Su, H. Bai, C. Wang, Angew. Chem. Int. Ed 51 (2012) 2395.

[95] L. Liu, C. Zhao, D. Pitts, H. Zhao, Y. Li, Catal. Sci. Technol. 4 (2014) 1539.

[96] P. Niu, Y. Yang, J.C. Yu, G. Liu, H.-M. Cheng, Chem. Commun. 50 (2014) 10837.
[97] M. Li, L. Zhang, X. Fan, Y. Zhou, M. Wu, J. Shi, J. Mater. Chem. A 3 (2015) 5189.
[98] M.M. Gui, S.-P. Chai, B.-Q. Xu, A.R. Mohamed, RSC Adv. 4 (2014) 24007.
[99] W.-N. Wang, F. Wu, Y. Myung, D.M. Niedzwiedzki, H.S. Im, J. Park, P. Banerjee, P. Biswas, ACS Appl. Mater. Interfaces 7 (2015) 5685.
[100] J.-C. Wang, L. Zhang, W.-X. Fang, J. Ren, Y.-Y. Li, H.-C. Yao, J.-S. Wang, Z.-J. Li, ACS Appl. Mater. Interfaces 7 (2015) 8631.
[101] B. AlOtaibi, S. Fan, D. Wang, J. Ye, Z. Mi, ACS Catal. 5 (2015) 5342.
[102] R. Zhou, M.I. Guzman, J. Phys. Chem. C 118 (2014) 11649.
[103] J. Lin, Z. Pan, X. Wang, ACS Sustainable Chem. Eng. 2 (2014) 353.
[104] F. Ehsan, T. He, Appl. Catal. B Environ. 166–167 (2015) 345.
[105] M. Halmann, M. Ulman, B.A. Blajeni, Sol. Energy 31 (1983) 429.
[106] Y. Matsumoto, M. Obata, J. Hombo, J. Phy. Chem. 98 (1994) 2950.
[107] G. Guan, T. Kida, A. Yoshida, App. Catal. B Environ. 41 (2003) 387.
[108] G. Guan, T. Kida, T. Harada, M. Isayama, A. Yoshida, App. Catal. A Gen. 249 (2003) 11.
[109] K. Teramura, H.S.-I. Okuoka, H. Tsuneoka, T. Shishido, T. Tanaka, Appl. Catal. B Environ. 96 (2010) 565.
[110] Y. Zhou, Z. Tian, Z. Zhao, Q. Liu, J. Kou, X. Chen, J. Gao, S. Yan, Z. Zou, ACS Appl. Mater. Interfaces 3 (2011) 3594.
[111] H. Shi, T. Wang, J. Chen, C. Zhu, J. Ye, Catal. Lett. 141 (2011) 525.
[112] P. Li, S. Ouyang, G. Xi, T. Kako, J. Ye, J. Phys. Chem. C 116 (2012) 7621.
[113] N. Zhang, S. Ouyang, T. Kako, J. Ye, Chem. Commun. 48 (2012) 1269–1271.
[114] K. Iizuka, T. Wato, Y. Miseki, K. Saito, A. Kudo, J. Am. Chem. Soc. 133 (2011) 20863.
[115] C.W. Tsai, H.M. Chen, R.S. Liu, K. Asakura, T.S. Chan, J. Phys. Chem. C 115 (2011) 10180.
[116] D.S. Lee, H.J. Chen, Y.W. Chen, J. Phys. Chem. Solids 73 (2012) 661.
[117] X. Li, H. Pan, W. Li, Z. Zhuang, Appl. Catal. A Gen. 413–414 (2012) 103.
[118] A. Ahmed, Y. Shibata, T. Taniguchi, T., Izumi, J. Catal. 279 (2011) 123.
[119] Q.D. Truong, J.Y. Liu, C.C. Chung, Y.C. Ling, Catal. Commun. 19 (2012) 85.
[120] J.W. Lekse, M.K. Underwood, J.P. Lewis, C. Matranga, J. Phys. Chem. C 116 (2012) 1865.
[121] X. Li, W. Li, Z. Zhuang, Y. Zhong, Q. Li, L. Wang, J. Phys. Chem. C (2012). http://dx.doi.org/10.1021/jp303365z.
[122] W.-H. Lee, C.-H. Liao, M.-F. Tsai, C.-W. Huang, J.C.S. Wu, Appl. Catal. B Environ. 132 (2013) 445.
[123] Y.-H. Cheng, H.-Y.C. V-H Nguyen, J.C.S. Wu, W.-H. Wang, Appl. Energy 147 (2015) 318.
[124] K. Xie, N. Umezawa, N. Zhang, P. Reunchan, Y. Zhang, J. Ye, Energy Environ. Sci. 4 (2011) 4211.
[125] D.D. Sui, X.H. Yin, H.Z. Dong, S.Y. Qin, J.S. Chen, W.L. Jiang, Catal. Lett. 142 (2012) 1202.
[126] H. Zhou, J.J. Guo, P. Li, T.X. Fan, D. Zhang, J.H. Ye, Sci. Rep. 3 (2013) 1667.
[127] P. Li, Y. Zhou, W.G. Tu, Q. Liu, S.C. Yan, Z.G. Zou, Chem. Plus Chem. 78 (2013) 274.
[128] S. Feng, X. Chen, Y. Zhou, W. Tu, P. Li, H. Li, Z. Zou, Nanoscale 6 (2014) 1896.
[129] P. Li, Y. Zhou, W.G. Tu, R. Wang, C.F. Zhang, Q. Liu, H.J. Li, Z.D. Li, H. Dai, J.J. Wang, S.C. Yan, Z.G. Zou, Cryst. Eng. Comm. 15 (2013) 9855.
[130] H.F. Cheng, B.B. Huang, Y.Y. Liu, Z.Y. Wang, X.Y. Qin, X.Y. Zhang, Y. Dai, Chem. Commun. 48 (2012) 9729.
[131] J.J. Guo, S.X. Ouyang, T. Kako, J.H. Ye, Appl. Surf. Sci. 280 (2013) 418.

[132] S.C. Yan, S.X. Ouyang, J. Gao, M. Yang, J.Y. Feng, X.X. Fan, L.J. Wan, Z.S. Li, J.H. Ye, Y. Zhou, Z.G. Zou, Angew. Chem. Int. Ed 49 (2010) 6400.

[133] S.C. Yan, J.J. Wang, H.L. Gao, N.Y. Wang, H. Yu, Z.S. Li, Y. Zhou, Z.G. Zou, Adv. Funct. Mater. 23 (2013) 758.

[134] S. Yan, H. Yu, N. Wang, Z. Li, Z. Zou, Chem. Commun. 48 (2012) 1048.

[135] Q. Liu, Z.X. Low, L.X. Li, A. Razmjou, K. Wang, J.F. Yao, H.T. Wang, J. Mater. Chem. A 1 (2013) 11563.

[136] Q. Liu, Y. Zhou, Z.P. Tian, X.Y. Chen, J. Gao, Z.G. Zou, J. Mater.Chem 22 (2012) 2033.

[137] S.C. Yan, J.J. Wang, H.L. Gao, N.Y. Wang, H. Yu, Z.S. Li, Y. Zhou, Z.G. Zou, Adv. Funct. Mater. 23 (2013) 1839.

[138] Q. Liu, Y. Zhou, Y. Ma, Z.G. Zou, RSC Adv. 2 (2012) 3247.

[139] Z.D. Li, Y. Zhou, J.Y. Zhang, W.G. Tu, Q. Liu, T. Yu, Z.G. Zou, Cryst. Growth Des. 12 (2012) 1476.

[140] K. Teramura, S. Iguchi, Y. Mizuno, T. Shishido, T. Tanaka, Angew. Chem. Int. Ed. 51 (2012) 8008.

[141] N. Ahmed, M. Morikawa, Y. Izumi, Catal. Today 185 (2012) 263.

[142] K. Li, A.D. Handoko, M. Khraisheh, J. Tang, Nanoscale 6 (2014) 9767.

[143] C. Zhao, L. Liu, G. Rao, H. Zhao, L. Wang, J. Xua, Y. Li, Catal. Sci. Technol. 5 (2015) 3288.

[144] P. Zhou, S.C. Yan, Z.G. Zou, CrystEngComm 17 (2015) 992.

[145] Z. Wang, M. Jiang, J. Qin, H. Zhou, Z. Ding, Phys. Chem. Chem. Phys. 17 (2015) 16040.

[146] S. Zhu, S. Liang, J. Bi, M. Liu, L. Zhou, L. Wu, X. Wang, Green Chem. (2015). http://dx.doi.org/10.1039/C5GC02308D.

[147] H. Zhou, P. Li, J. Guo, R. Yan, T. Fan, D. Zhang, J. Ye, Nanoscale 7 (2015) 113.

[148] X. Chen, J. Wang, C. Huang, S. Zhang, H. Zhang, Z. Li, Z. Zou, Catal. Sci. Technol. 5 (2015) 1758.

[149] S. Wang, Y. Hou, X. . Wang, ACS Appl. Mater. Interfaces 7 (2015) 4327.

[150] W. Kim, H. Frei, ACS Catal. 5 (2015) 5627.

[151] Y. Bi, M. Fahad Ehsan, Y. Huang, J. Jin, T. He, J. CO_2 Util. 12 (2015) 43.

[152] B.S. Kwak, M. Kang, Appl. Surf. Sci. 337 (2015) 138.

[153] S.K. Parayil, A. Razzaq, S.-M. Park, H.R. Kim, C.A. Grimes, Appl. Catal. A Gen. 498 (2015) 205.

[154] J. Premkumar, R. Ramaraj, J. Photochem. Photobiol. A Chem. 110 (1997) 53.

[155] O. Ozcan, F. Yukruk, E.U. Akkaya, D. Uner, Appl. Catal. B Environ. 71 (2007) 291.

[156] T.-V. Nguyen, J.C.S. Wu, C.-H. Chiou, Catal. Commun. 9 (2008) 2073.

[157] Z. Zhao, J. Fan, S. Liu, Z. Wang, Chem. Eng. J. 151 (2009) 134.

[158] Z. Zhao, J. Fan, M. Xie, Z. Wang, J. Cleaner Prodn. 17 (2009) 1025.

[159] K.R. Thampi, J. Kiwi, M. Graetzel, Nature 327 (6122) (1987) 506.

[160] G. Qin, Y. Zhang, X. Ke, X. Tong, Z. Sun, M. Liang, S. Xue, Appl. Catal. B Environ. 129 (2013) 599.

[161] T. Arai, S. Sato, T. Morikawa, Energy Environ. Sci. 8 (2015) 1998.

[162] K. Li, L. Lin, T. Peng, Y. Guo, R. Li, J. Zhang, Chem. Commun. 51 (2015) 12443.

Chapter 12

Batteries

Chapter 12.1: Primary Batteries

INTRODUCTION

"The term battery comes from the old French word meaning action of beating relating to a group of cannons in battle. Energy storage devices, scientists in the 1900s adopted the term battery to represent multiple electrochemical cells connected together." As a device, a battery consists of two electrodes that are soaked in an electrolyte and are separated by a membrane or divider. New active materials are introduced now and then and improvements have been advocated regarding the specific energy levels of batteries much beyond achievable limits.

The main limitation of a battery is that it delivers for a certain period of time and then requires recharging, and the time for recharging can be the same level as the delivery time in some batteries. Let us first compare the battery with other power sources. Compared with fossil fuels, the battery has a low storage capability. The energy on a mass basis is 12,000 Wh/kg whereas a typical battery such as lithium ion carries about 200 Wh/kg, but the energy delivery is more effective in batteries than in a thermal engine. The batteries have short response times compared with other power sources and the bandwidth in terms of load is wide. Batteries run clean and stay favorably cool. Batteries are highly efficient (nearly >90%) compared with a thermal internal combustion engine, which is 25–30%.

Terminology in Batteries

Batteries are specified by three main characteristics: chemistry taking place inside the battery, voltage, and its specific energy, usually identified as capacity. The most common batteries are lead, nickel, and lithium; each of these has its own chemistry, which has to be taken into account while shipping

Energy Sources. http://dx.doi.org/10.1016/B978-0-444-56353-8.00012-5

and disposing because regulatory requirements differ for each battery. The open circuit voltage on a fully charged battery is 5–7% higher than the nominal voltage marked on each battery. The closed circuit voltage is the operating voltage. Capacity is the specific energy in ampere-hours (Ah). Ah is the discharge current a battery can deliver over a specified time. The specific power or gravimetric power density denotes the loading capability. Watts are the power that is metered. Volt-amps (VA) are the apparent power that is affected by the reactive load.

A brief history of the development of the battery is appropriate. Table 12.1.1 summarizes battery advancements and lists inventors when available (this table was adopted from BU-102 Battery Developments: batteryuniversity.com/learn/article/sharing_battery_knowledge).

TABLE 12.1.1 History of Modern Battery Development

Year	Inventor	Activity
1600	William Gilbert (United Kingdom)	Start of electrochemistry study
1745	Ewald Georg von Kleist (The Netherlands)	Invention of Leyden jar (stores static electricity)
1791	Luigi Galvani (Italy)	Discovery of animal electricity
1800	Alessandro Volta (Italy)	Invention of voltaic cell (zinc copper disks)
1802	William Cruickshank (United Kingdom)	First electric battery capable of mass production
1820	Andre-Marie Ampere (France)	Electricity through magnetism
1833	Michael Faraday (United Kingdom)	Discovery of Faraday's law
1836	John F. Daniell (United Kingdom)	Daniel cell invented
1839	William Robert Grove (United Kingdom)	Concept of fuel cell (H_2/O_2)
1859	Gaston Plante (France)	Introduction of lead acid battery
1868	Georges Leclanché (France)	Invention of Leclanche cell (carbon-zinc)
1899	Waldemar Jungner (Sweden)	Invention of nickel cadmium battery
1901	Thomas A. Edison (United States)	Invention of nickel-iron battery
1932	Schlecht & Ackermann (Germany)	Invention of sintered pole plate

TABLE 12.1.1 History of Modern Battery Development—cont'd

Year	Inventor	Activity
1947	Georg Neumann (Germany)	Sealing of nickel cadmium battery
1949	Lewis Urry (Eveready)	Invention of alkaline manganese battery
1970s	Group effort	Invention of valve-regulated lead acid battery
1990	Group effort	Invention of nickel-metal hydride battery
1991	Sony (Japan)	Invention of lithium-ion battery
1994	Bellcore (United States)	Invention of lithium-ion polymer battery
1996	Molienergy (Canada)	Invention of Li-ion with manganese cathode
1996	University of Texas (United States)	Identification of Li-phosphate ($LiFePO_4$)
2002	University of Montreal Quebec Hydro, MIT, others	Improvements of Li-phosphate, nanotechnology, commercialization

Primary Batteries: Brief Comparison

Primary batteries are commonly referred to as dry cells. They are also known as nonchargeable batteries. The dry cell is a modified version of the Leclanche cell and was successfully introduced in 1888 by Gassner. It is **zinc-carbon**, also known as carbon-zinc or the Leclanché battery, is one of the earliest and least expensive primary batteries. It delivers 1.5 V and often comes with consumer devices. The first zinc-carbon invented by Georges Leclanché in 1859 was wet. The term "dry cell" is a misnomer. A totally dry battery will not function under ordinary temperatures. The term only conveys that the contents of the battery are unspillable whatever the position of the battery, even in motion. The electrolyte is kept in the form of a gel or in some absorbent material to achieve unspillability. Primary dry cells must be distinguished from sealed rechargeable batteries, which are also unspillable. Some reserve batteries remain dry until they are activated for use. Solid electrolyte batteries are also unspillable but they function only at high temperatures.

Alkaline-manganese, also known as alkaline, is an improved version of the zinc-carbon battery and also delivers 1.5 V. Lewis Urry invented alkaline cells in 1949 while working with the Eveready Battery Company laboratory in, Ohio, United States. Alkaline has low self-discharge and does not leak electrolyte when depleted as did the old zinc-carbon, but it is not totally leak-proof.

Lithium iron disulfide (Li-FeS_2) is a newcomer to the primary battery family and offers improved performance compared with an alkaline battery. It has higher capacity and lower internal resistance (IR) than do alkaline batteries. They have advantage of improved low-temperature performance, superior leak resistance, and low self-discharge.

Lithium-thionyl chloride ($LiSOCl_2$ or LTC) is one of the most rugged lithium-metal batteries. The ability to withstand high heat and strong vibration enables horizontal drilling, also known as fracking. Some LTCs are said to operate from 0 to 200°C (32–392°F). With a specific energy of over 500 Wh/kg, LTC offers twice the capacity of the best Li-ion. The nominal voltage is 3.60 V/cell; the end-of-discharge cutoff voltage is 3.00 V.

Like alkaline, LTC has a relatively high resistance and can be used only for moderate discharge loads. If stored for a time, a passivation layer forms between the lithium anode and the carbon-based cathode that dissipates when applying a load. This layer protects the battery by granting low self-discharge, and a long shelf-life LTC is one of the most powerful and potent battery chemistries and should be used only by trained workers.

Lithium manganese dioxide (MnO_2 or Li-M) is similar to LTC but has a lower specific capacity and is safe for public use. The voltage is 3.0–3.30 V and the specific energy is about 280 Wh/kg. Li-M is economically priced and has a long life. The operational temperature ranges from −30 to 60°C (−22 to 140°F). Typical uses are meter sensing, medical devices, road toll sensors, and cameras.

Lithium sulfur dioxide ($LiSO_2$) is a primary battery with a voltage of 2.8 V and an energy density of up to 330 Wh/kg. It offers a wide temperature range of −54 to 71°C (−65 to 160°F) with a projected shelf life of nearly 10 years at room temperature. $LiSO_2$ is inexpensive to make and is commonly used by the military.

Tables 12.1.2 and 12.1.3 list data on potentially available anode and cathode materials.

Among these materials successful anode cathode combinations are given in Table 12.1.4.

The Leclanche Dry Cell

The battery can be represented as:

$$Zn/ZnCl_2,\ NH_4Cl/MnO_2,\ C$$

The zinc container also serves as the anode. The electrolyte is in the form of jelly. A central carbon rod serves as the electrical conductor surrounded by the depolarizer MnO_2. The container is made of 99.9% pure zinc with 0.03% cadmium (for mechanical strength) and 0.3% lead for ductility. The presence of other metals (especially Cu) is harmful and can lead to galvanic corrosion of the zinc. The cylinder is occupied by the bobbin, which is a mixture of MnO_2 (60%) carbon black (20%), NH_4Cl (10%), and water (10%). The positive electrode is a sintered carbon rod centrally placed with a brass cap for external

TABLE 12.1.2 Potential Anode Materials for Primary Batteries

Anode	Open Circuit (OC) Potential Developed (V)		Weight Capacity (mAh/g)	Volumetric Capacity (Ah/mL)
	Acid	Alkali		
Mg	−2.35	−2.69	2204	3.84
Al	−1.87	−2.35	2982	8.05
Zn	−0.76	−1.25	820	5.68
Fe	−0.44	−0.88	960	7.55
Cd	−0.40	−0.81	477	4.13
Sn	−0.14	−0.91	452	3.30
Pb	−0.13	−0.54	259	2.90

TABLE 12.1.3 Potential Cathode Materials for Batteries

Cathode Material	Open Circuit Potential Developed (volts)		Weight Capacity (mAh/g)	Volume Capacity (Ah/mL)
	Acid	Alkali		
MnO_2	+0.80	+0.29	307	1.54
PbO_2	+1.46	+0.25	248	2.26
HgO	+0.85	+0.10	248	2.76
CuO	+0.34	−0.36	670	4.32
Ag_2O	+0.80	+0.35	232	1.67
Ag_2O_3	+1.90	+0.57	432	1.92
AgCl	+0.22	+0.22	187	1.64

electrical contact. The bobbin is surrounded by a muslin bag and the space between the bag and zinc is filled with a paste of electrolyte of NH_4Cl (26%) and $ZnCl_2$ (9%) gelled with wheat flour or wheat starch. Corrosion inhibitors such as chromates and $HgCl_2$ are added to the electrolyte (Fig. 12.1.1).

Excess zinc (nearly four times the theoretical requirement) must be provided for mechanical strength because zinc is consumed anodically.

TABLE 12.1.4 Successful Anode and Cathode Combinations

Name	System	Open Circuit Resistance (Ω)	Power Density (W/lb)	Maximum Current Drawn (A)
Dry cell (Leclanche)	$Zn/ZnCl_2NH_4Cl/MnO_2$, C	1.55	26	0.3–0.2
Alkaine-MnO_2	$Zn/KOH/MnO_2$, C	1.5	30	1.3
Ruben-Mallory	Zn/KOH/HgO, Hg	1.35	35	1.0
Silver-Zinc	$Zn/KOH/Ag_2O$	1.60	75	0.8
Zinc-Air	$Zn/NaOH/O_2$(air), C	1.40	20	0.9–2.0

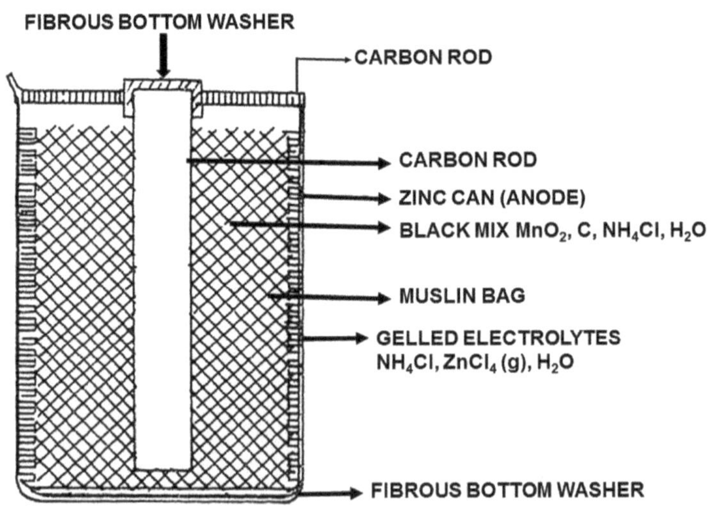

FIGURE 12.1.1 Schematic diagram of a dry Leclanche primary cell.

Several varieties of MnO_2 are possible, but γ-MnO_2 is the best suitable and ensures a long enough shelf life. To increase the conductivity of MnO_2, carbon or acetylene black is added to lower the resistance. A high degree of porosity is desirable because the depolarizing power depends on the surface area. The central carbon rod must be porous to vent hydrogen and ammonia produced by the cell reaction.

The electrolyte is gelled ammonium chloride with sufficient water. $ZnCl_2$ is added to NH_4Cl so that (1) drying of the electrolyte is slowed, (2) the electrolyte remains acidic and the open circuit voltage (OCV) is maintained; (3) the starch flour is well-gelatinized; (4) the freezing point is low; and (5) the ammonia evolved under a high current drain is absorbed by $ZnCl_2$.

Cell reactions may be represented as:

$$Zn \rightarrow Zn^{2+} + 2e$$

$$MnO_2 + H^+ + e^- \rightarrow MnOOH$$

$$\text{or } MnO_2 + H_2O + e^- \rightarrow MnOOH + OH^-$$

The pH of the solution near the cathode increases with the time and intensity of discharge. The cathode potential and rate of discharge change with time. The overall electromotive force (emf) decreases with time owing to the accumulation of zinc ions in the analyte and OH^- ions in the catholyte. The accumulation of zinc ions can lead to precipitation and thus increase IR.

Other possible secondary reactions include:

$$Zn^{2+} + 2OH^- \rightarrow Zn(OH)_2$$

$$3Zn(OH)_2 + ZnCl_2 \rightarrow ZnCl_2 3Zn(OH)_2$$

$$Zn(OH)_2 + 2MnOOH \rightarrow ZnO \cdot Mn_2O_3 + 2H_2O$$

$$NH_4^+ + OH^- \rightarrow NH_3 + H_2O$$

$$ZnCl_2 + 2NH_3 \rightarrow ZnCl_2 \cdot 2NH_3 \text{ (unstable complex)}$$

Cell Performance

The emf falls from 1.5 to 1.45 V for moderate loads of small discharge currents and remains constant for most of the useful life of the battery. If the discharge current is very low, natural deterioration will set in before the full capacity is used. On the other hand, concentration polarization sets in early with high discharge currents. Low temperature will increase shelf life because the corrosion reactions become slow and the hydrogen overpotential is increased. However performance efficiency also decreased because the cell reaction rates decrease and diffusion rates become small, leading to concentration polarization effects and increases in IR. The IR of a typical cell is 0.25 Ω at 298 K, 0.4 Ω at 273 K, and 1 Ω at 252 K (the freezing point of the electrolyte). The electrolyte can be maintained in a liquid state at low temperatures by adding antifreeze agents such as alcohol or ethylene glycol. Adding LiCl not only decreases the freezing point but provides high conductivity. The service life is not a fixed quantity but depends on the discharge current, operating schedule, cutoff voltage, temperature, and storage period.

Alkaline MnO_2 Batteries

These cells were patented as Crown cells in 1882 but were manufactured only in 1949. Development of these cells was delayed owing to the failure of porous pots under alkaline conditions. MnO_2 shows extremely high depolarizing efficiency under these conditions. The battery is represented as Zn/KOH/MnO_2, C.

The cell is constructed using a central zinc cylinder anode or a gelled powder anode connected to an outer steel container. The cathode is high-density MnO_2 mixed with conducting carbon (graphite or acetylene black) contained in a steel case. Binder materials such as Portland cement and fibers are added to increase strength to the cathode material. The cathode is lined with a sandwich separator of two materials, one with good absorption for KOH and the other to prevent penetration by zinc. A liquid alkali reservoir supplies electrolyte to the gelled powder anode during discharge and storage. Assembly of a D-size battery is made as follows: First the cathode mix is pressed into the steel can and molded. The separator sandwich is assembled and inserted. The anode cylinder is then placed into the separator basket, followed by the current collector assembly. After sealing the jacket and cell cover, plates are fitted around the cell to conform to the polarity of a Leclanche flashlight battery with the cap being the positive terminal. Gassing in this battery is low so that the cell can be hermetically sealed, which increases the shelf life. A release vent valve is provided to withstand pressures developed by misuse of the battery.

Performance

The OCV developed is 1.5 V. High efficiencies are reported for conditions of continuous as well as high current draw (300 mA). The voltage drops steadily from 1.2 to 0.6 V over a period of 12 h for 1 A current discharge. Table 12.1.5 compares alkaline and Leclanche cells.

One should be able to get 70 Wh/kg from an alkaline battery. The efficiency of an alkaline cell is 47% at 0.65 V and 0.5 Ω compared with 12% of a Leclanche cell and is 85% with 10 Ω compared with 70% of a Leclanche cell. The theoretical capacity is 10.9 Ah compared with 6.0 Ah of a Leclanche cell. Main advantages of alkaline batteries are:

1. higher efficiency than for acid batteries
2. low resistance and low IR drop

TABLE 12.1.5 Comparison of Alkaline Batteries With Leclanche Cells

	Capacity (Ah)		Service in Use for Flashlight (min)	
Temperature (°C)	Alkaline	Leclanche	Alkaline	Leclanche
18	6.93	0.89	1200	150
0	4.45	0.58	835	100
−18	1.39	0.26	275	50

3. higher short-circuit currents
4. higher energy per unit weight than for that of a Leclanche cell
5. shelf life on the order of 2 years
6. good performance in the temperature range 0–60°C
7. recharging a few times (two to five cycles) is more efficient than that for a Leclanche cell; however, alkaline cells are more costly than Leclanche cells.

Magnesium Dry Cell

The magnesium dry cell can be represented as:

$$Mg|MgBr_2gel|MnO_2, C$$

A cell reaction involves the dissolution of Mg as Mg^{2+} and the formation of MnOOH at the cathode. Advantages of Mg over Zn are that:

1. The anode is more active. E for Mg is $= -2.38$ V whereas for Zn it is $= -0.76$ V. The OCV of a magnesium battery is 1.9 V.
2. Mg has a lower equivalent weight (12; whereas for Zn it is 32.69).
3. Magnesium batteries have higher energy density.
4. The shelf life of magnesium batteries is on the order of 2 years.
5. Magnesium batteries can withstand higher humidity and temperature conditions than can zinc batteries.

The specific gravity of Mg is 1.7 whereas that for Zn is 7.1. The natural corrosion of Mg is higher than that of Zn. The cost is also higher. Hence magnesium batteries are employed in military applications and emergency equipment.

The construction of a magnesium dry cell is similar to that of a Leclanche cell. A magnesium alloy AZ 31 containing 3% Al, 1% Zn, 2% Mn and 15% Ca is extruded as the outer container can. The thickness of Mg must be greater than that of Zn because the anode efficiency of Mg is only 60%, since 40% is consumed by the corrosion reaction.

The cathode is a combination of MnO_2 and acetylene black with 3%$BaCrO_4$. The electrolyte is 19.25% gelled $MgBr_2$ containing 0.02% lithium chromate. Lithium chromate acts as a corrosion inhibitor whereas $MgBr_2$ provides the necessary conductivity. A paper separator between the cathode and electrolyte is employed. Limited entry of air is permitted through a vent.

The battery has an OCV of 1.9 V and is stabilized at 1.3 V during discharge. The ampere-hour capacity of magnesium battery is 2.5 times that of a zinc battery of the same size. A magnesium battery can undergo corrosion and the shelf life can be increased by alloying, but the anode efficiency and potential developed will also vary. Characteristics of some materials used as anode materials are given in Table 12.1.6.

TABLE 12.1.6 Comparison of Anode Materials for Magnesium Batteries

Alloy	Percentage of Al	Mn	Ca	Zn	Efficiency	Anode Potential (V)
A210A	0.8–1.2	0.35	0.15	0.1–0.25	63.3	1.327
A231	2.5–3.5	0.7–1.3	0.15	0.1–0.25	67.9	1.287
A280	7.9–9.2	0.20–0.8		0.15	67.9	1.232
ZK60		4.8–6.5		0.45	58.8	1.233
M1A		1.2	0.8		52.6	1.328

As the content of aluminum or calcium increases, the anode efficiency decreases. However the decrease does not change appreciably over a wide range of discharge current densities. Heat can be liberated and it can be used if the battery is properly designed.

Hydrogen evolution requires gas escape valves to be incorporated into the battery. There are two other forms of magnesium battery: $Mg|Mg(ClO_3)_2|MnO_2$, C and $Mg|Mg(ClO_3)_2|$dinitrobenzene, M. Cell reactions of a magnesium battery are:

$$Mg + H_2O + 2MnO_2 \rightarrow Mn_2O_3 + Mg(OH)_2$$

$$\text{And, } 6\,Mg + 8H_2O + C_6H_4(NO_2)_2 \rightarrow 6Mg(OH)_2 + C_6H_4(NH_2)_2$$

Ten percent acetylene black, 3% barium chromate, and 1% magnesium hydroxide are incorporated with MnO_2 into the cathode mix. The electrolyte is a mixture of 3M $Mg(ClO_3)_2$ with 2% lithium chromate, which acts as a corrosion inhibitor. A kraft paper separator is generally used for absorption of the electrolyte from the bobbin, and also to permit the migration of water.

Mercury Oxide-Zinc Battery (Ruben–Mallory Battery)

The use of HgO as a cathode material was introduced in 1884 but the commercial battery was made in 1942, when Ruben succeeded in reducing the evolution of gas during battery discharge. The cell is represented as:

$$Zn(s)(amalgam)|KOH(aq)34\text{–}40\%;\ 5\%ZnO|HgO(s),\ C(s),\ steel$$

The battery is made in the conventional cylindrical form or in the form of a button cell (Fig. 12.1.2).

The anode is a pellet of amalgamated zinc powder whereas the cathode is a mixture of HgO and finely divided graphite powder pressed into pellets. HgO is taken in excess of the theoretical requirement, to avoid profuse gassing toward the end of the battery life. The cathode is consolidated at the bottom of a nickel-plated steel can. The compartments are separated by alkali-resistant

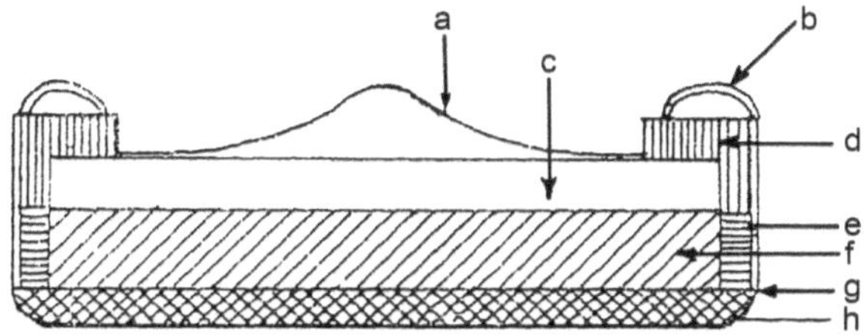

FIGURE 12.1.2 Design of Ruben–Mallory dry cell: (a) Steel closure as anode connector; (b) outer steel container (cathode connector); (c) compressed powdered zinc (anode); (d) insulating seal insulator; (f) cellulose material impregnated with alkaline electrolyte; (g) alkali-resistant parchment paper disc; (h) mercuric oxide depolarizer mixed with graphite (cathode).

parchment paper. Carboxymethyl cellulose is used with a powdered anode and α-cellulose with foil anodes. The electrolyte is contained in absorbing cotton. The cell is sealed by compiling the outer can over plastic insulation. A steel cap can be provided as the anode connector and separated from the outer can by a gasket. The cell reactions are:

$$HgO + H_2O + 2e \rightarrow Hg + 2OH^-$$

$$Zn + 2OH^- \rightarrow ZnO + H_2O + 2e$$

This battery has an OCV of 1.34 V drops to 1.145 V and even to 0.9 V when the discharge current is on the order of 250 mA. Nearly 90% of the energy stored can be recovered during discharge. Main advantages of this battery are:

1. longer shelf life up to 24 months at 298 K
2. more charge yielded than a Leclanche battery
3. flat discharge curve
4. excellent mechanical strength
5. nearly 90% use is possible
6. good performance up to 65°C
7. can withstand high momentary discharge rates although it has a high cost and poor performance at low temperatures

Zinc-CuO Leclanche Battery

The zinc-CuO Leclanche battery is represented as:

$$Zn|KOH \text{ or } NaOH|CuO, \text{ metal grid}$$

The cathode is made of CuO compacted with KOH in the form of flat plates and then baked and encased on a metal grid. The anode is made of amalgamated zinc plates. The electrolyte is a 20% solution of NaOH. The choice is based on maximum conductivity and a minimum freezing temperature. ZnO is added to the electrolyte to maintain a minimum zincate ion concentration.

The sequence of cathode reactions is:

$$CuO + H_2O + e \rightarrow CuO..H + OH^-$$

$$2CuO..H \rightarrow Cu_2O + H_2O$$

$$Cu_2O + 2H_2O + 2e \rightarrow Cu_2O..H_2(ad) + 2OH^-$$

$$Cu_2O..H_2(ad) \rightarrow 2Cu + 2H_2O$$

$$\text{Anode } Zn + 4OH^- \rightarrow ZNO_2^- + 2H_2O + 2e$$

Fifty percent more than theoretical requirement of zinc is used to allow for local corrosion and prevention of mechanical failure. An excess of NaOH is used to prevent the crystallization of $Zn(OH)_2$. The theoretical OCV is 1.0 V but it usually delivers 0.7 at 298 K and remains constant over the period of cell discharge.

Silver Oxide–Zinc Battery

A silver oxide battery was introduced in 1902 by Junger, who used a cadmium anode. Andre perfected this battery in 1943 using zinc. The cell representation is:

Zn|KOH|silver oxide, C, and the cell reaction is $Ag_2O + 2Zn \rightarrow 2ZnO + 2Ag$

The cell is made of gray spongy zinc plates and silver oxide as a cathode material. Cellophane or wetted polypropylene is employed as a separator material. High concentrations of KOH are used as an electrolyte for heavy duty and NaOH is used for long reliability. Additions of Ag_2O and MnO_2 to the cathode mix result in flat discharge conditions. The battery delivers 1.3–1.5 V during discharge. The battery is costly and is used in ballistic missiles and in the button form in digital watches.

Metal–Air Batteries

In these batteries metal is the anode and oxygen in air is the cathodic depolarizer. High-capacity batteries can be made with a low overall weight. For use in electric vehicles, there is a required energy density of 50–150 Wh/lb and it needs to be able to deliver power at the rate of 35–90 W/lb. Many metal–air systems seem to be capable of achieving this requirement. Table 12.1.7 lists data on known metal air systems and their OCV and energy density.

Data on silver-Zn and lead acid systems are included for comparison. Only systems that develop more than 1 V are included in Table 12.1.7. The metal anode must be stable in contact with the electrolyte. The cost of the metals is in the following order: Fe $<$ Zn $<$ Al $<$ Ni $<$ Cu. The relative cost of power that can be recovered ensuring 100% conversion efficiency of the metal–air cell is in the order: Al $<$ Mg $>$ Zn $>$ Cd $>$ Fe.

The high OC potential (0.35 V in alkaline solutions), energy density, and relatively low cost make Al an attractive candidate. The formation of thin

TABLE 12.1.7 Known Metal–Air Systems

Metal–O_2	Open Circuit Voltage	Energy Density (Wh/lb)
Li	3.44	6020
Al	2.70	3645
Mg	3.09	3090
Zn	1.65	614
Fe	1.28	555
Cd	1.59	186
Ag-Zn battery	1.86	254
Pb acid battery	2.14	112

aluminum oxide under battery conditions makes this battery unsuitable for the value of OCV. Corrosion inhibitors have been tried but the current situation produces a satisfactory aluminum–air battery. Cadmium can be stabilized without using inhibitors whereas Zn and iron require corrosion inhibitors. Zinc–air batteries are in use for military radio receivers and in railways. The zinc–air cell can be represented as Zn|NaOH 20%|air, C. The outer container is made of glass or ebonite casing. Rectangular flat plates of Zinc are placed on either side of the cathode. Nongraphitic porous carbon plates activated for better reduction of O_2 are treated with water repellents and used as cathodes. The density of carbon is between 0.6 and 0.8 and the porosity is 30–65%. Twenty percent NaOH is used as an electrolyte and has a prominent part in the cell reactions. The zinc ions formed as a result of the anodic reaction are converted to zincate ions. The electrolyte becomes weak in caustic soda during the discharge and the direct reaction of zinc with NaOH. This can be reduced by amalgamation. The concentration of NaOH has to be maintained to avoid IR. Addition agents such as $Cu(OH)_2$ are used to keep zinc ions in solution. When the solution is saturated with zincate ions then an oxide layer on the electrode surface, the cell ceases to function. The porous anode has both advantages and reduces life and conductivity.

The cell's cathode reaction is: $O_2 + 2H_2O + 4e \rightarrow 4OH^-$. The reduction mechanism suggested is: $2C + 2H_2O + 2e \rightarrow 2CH + 2OH^-$:

$$2CH + O \rightarrow H_2O + 2C$$

$$H_2O + O + 2e \rightarrow 2OH^-$$

The solution the cathode will become increasingly alkaline during discharge.

Performance

1. The OCV is 1.45 and the voltage developed is 1.1 V for a 10-mA current discharge. The voltage remains constant within 0.1 V throughout the active life of the cell.
2. Hydrogen evolution occurs whenever O_2 supply is reduced.
3. When high currents are drawn, the carbon cathode becomes clogged with nitrogen from air.
4. If the carbon used is porous, the electrolyte is better absorbed. However with continuous discharge, the cathode becomes increasingly viscous. CO_2 absorption also increases and the precipitation of carbonates can clog the pores and increase IR.
5. The shelf life is limited with activated cells.

Cadmium–Air Cells

The US Army laboratories have developed a cell that can deliver 40–50 Wh/lb at the 5-h rate of discharge. A long life cycle of the cadmium electrode is retained in the cadmium–air configuration in the absence of CO_2, and the high capacity of the cadmium electrode is used to take advantage of the infinite capacity of the air electrode. The air electrode is similar to that in the zinc–air system. Compact cadmium oxide is reduced to sponge cadmium and used.

Lithium-Based Cells

The chemistry of the lithium battery is composed of a number of cell designs that use lithium as the anode. Lithium is gaining a lot of popularity as an anode for a number of reasons. Note that lithium, the lightest of metals, also has the highest standard potential of all metals, at over 3 V. Some lithium cell designs have a voltage of nearly 4 V. This means that lithium has the highest energy density. Many different lithium cells exist because of its stability and low reactivity with a number of cathodes and nonaqueous electrolytes. The most common electrolytes are organic liquids, with the notable exceptions of $SOCl_2$ (thionyl chloride) and SO_2Cl_2 (sulfuryl chloride). Solutes are added to the electrolytes to increase conductivity. Lithium cells have only recently become commercially viable because lithium reacts violently with water as well as nitrogen in air. This requires sealed cells. High-rate lithium cells can build up pressure if they short-circuit and cause temperature and pressure to rise. Thus, the cell design needs to include weak points, or safety vents, that rupture at a certain pressure to prevent explosion.

Lithium cells can be grouped into three general categories: liquid cathode, solid cathode, and solid electrolyte. Let us look at some specific lithium cell designs within the context of these three categories.

These cells tend to offer higher discharge rates because the reactions occur at the cathode surface. In a solid cathode, the reactions take longer because the lithium ions must enter the cathode for discharge to occur. Direct contact between the liquid cathode and the lithium forms a film over the lithium, called the solid electrolyte interface (SEI). This prevents a further chemical reaction when not in use, thus preserving the cell's shelf life. One drawback, though, is that if the film is too thick it causes an initial voltage delay. Usually, water contamination is the reason for the thicker film, so quality control is important.

Lithium–Iron Disulfide Cell

Lithium iron disulfide is a newcomer to the primary battery family. Normally Li batteries deliver 3 V or higher but $LiFeS_2$ has 1.5 V. It has a higher capacity and lower IR. These cells have improved performance at low temperatures and exhibit low self-discharge. The disadvantage of these cells is high cost and transportation issues as a result of the lithium metal content in the anode.

Lithium–Thionyl Chloride

LTC is one of the rugged cells in this family. This cell can withstand heat and strong vibration. Some LTC cells can operate from 273 to 473 K. These cells can be used in medical and sensor applications. With a specific energy of over 500 Wh/kg, LTC offers twice the capacity of the best Li-ion cells. If stored for a time, a passivation layer forms between the lithium anode and carbon-based cathode that dissipates when a load is applied. This passivation layer helps to reduce the self-discharge rate and thus imparts a longer life. LTC is a powerful and potent battery but requires an experienced user.

Lithium–Sulfur Dioxide

Lithium-sulfur dioxide ($LiSO_2$) is a primary battery with a voltage of 2.8 V and an energy density of up to 330 Wh/kg. The temperature range of operation is 219–344 K. The shelf life is 5–10 years. This is costly and is commonly used in the military.

Solid Cathode Lithium Cells

These cells cannot be used in high-drain applications and do not perform as well as liquid cathode cells at low temperatures. However they do not have the same voltage delay and the cells do not require pressurization. They are used generally for memory backup, watches, and portable electronic devices.

$LiMnO_2$

$LiMnO_2$ batteries account for about 80% of all primary lithium cells; one reason is their low cost. The cathode used is a heat-treated MnO_2 and the electrolyte is a mixture of propylene carbonate and 1,2-dimethoyethane. The half reactions are:

$$Li \rightarrow Li^+ + e$$

$$MnO_2 + Li^+ + e \rightarrow Mn^{III}\,O_2\,(Li^+)$$

$$\text{Overall reaction: } Li + Mn^{IV}\,O_2 \rightarrow Mn^{III}\,O_2\,(Li^+)$$

At lower temperatures and at a high rate of discharge, the $LiSO_2$ cell performs better than the $LiMnO_2$ cell. At a low rate of discharge and higher temperatures, the two cells perform equally well, but the $LiMnO_2$ cell has an advantage because it does not require pressurization.

Li(CF)*n* Lithium Polycarbon Monofluoride

The cathode in this cell is carbon monofluoride, a compound formed through high-temperature intercalation. This is the process in which foreign atoms (in this case fluorine gas) are incorporated into some crystal lattice (graphite powder), with the atoms of the crystal lattice retaining their positions relative to one another. A typical electrolyte is lithium tetrafluorobate ($LiBF_4$) salt in a solution of propylene carbonate (PC) and dimethoxyethane. These cells also have a high voltage (about 3.0 V open voltage) and a high energy density (around 250 Wh/kg). All this and a 7-year shelf life make them suitable for low to moderate drain use, eg, watches, calculators, and memory applications.

Solid Electrolyte Lithium Cells

All commercially manufactured cells that use a solid electrolyte have a lithium anode. They perform best in low-current applications and have a very long service life. For this reason, they are used in pacemakers.

LiI_2—Lithium iodine cells use solid LiI as their electrolyte and also produce LiI as the cell discharges. The cathode is poly-2-vinylpyridine (P2VP) with the following reactions:

$$2Li \rightarrow 2Li^+ + 2e$$

$$2Li^+ + 2e + P2VP\cdot nI_2 \rightarrow P2VP\cdot(n-1)I_2 + 2LiI$$

$$2Li + P2VP\cdot nI_2 \rightarrow P2VP\cdot(n-1)I_2 + 2LiI$$

LiI is formed in situ by the direct reaction of the electrodes.

Lithium—Iron Cells

Lithium—iron chemistry deserves a separate section because it is one of a handful of lithium metal systems that have a 1.5-V output (others are lithium—lead bismuthate, lithium—bismuth trioxide, lithium—copper oxide, and lithium—copper sulfide). Consumer cells that use Li-Fe have reached the market, including the Energizer. These have the advantage of having the same voltage as alkaline batteries with more energy storage capacity, so they are called "voltage compatible" lithium cells. They are not rechargeable. They have about 2.5 times the capacity of an alkaline battery of the same size, but only under high current discharge conditions (digital cameras, flashlights, motor driven toys, etc.). For small currents they have no advantage. Another advantage is the low self-discharge rate: 10 years' storage is quoted by the manufacturer. Discharge reactions are:

$$2FeS_2 + 4Li \rightarrow Fe + 2Li_2S \quad 1.6\ V$$

$$FeS + 2Li \rightarrow Fe + Li_2S \quad 1.5\ V$$

Both iron sulfide and iron disulfide are used; FeS_2 is used in the Energizer. Electrolytes are organic materials such as propylene carbonate, dioxolane, and dimethoxyethane.

Selection Criteria for Battery Systems

A set of criteria that illustrate the characteristics of the materials and reactions for commercial battery systems are listed (these criteria are applicable to any battery system although they are given under primary batteries):

1. Mechanical and chemical stability: The materials must maintain their mechanical properties and their chemical structure and composition over the course of time and temperature as much as possible and during operation. Mechanical and chemical stability limitations arise from reactions with the electrolyte, irreversible phase changes and corrosion, isolation of active materials, and poor conductivity of materials in the discharged state.
2. Energy storage capability: The reactants must have sufficient energy content to provide a useful voltage and current level, measured in Wh/L or Wh/kg. In addition, the reactants must be capable of delivering useful rates of electricity, measured in terms of W/L or W/kg. This implies that the kinetics of the cell reactions is fast and without significant kinetics hindrances. The carbon–zinc and Ni-Cd systems set the lower limit of storage and release capability for primary and rechargeable batteries, respectively.
3. Temperature range of operation: For military applications, the operational temperature range is from −50 to 85°C. Essentially the same temperature range applies to automotive applications. For a general purpose consumer battery, the operating temperature range is 0–40°C and the storage temperatures range from −20 to 85°C. These temperatures are encountered when using automobiles and handheld devices in the winter in northern areas and in the hot summer season in southern areas.
4. Self-discharge: Self-discharge is the loss of performance when a battery is not in use. An acceptable rate of loss of energy in a battery depends somewhat on the application and the chemistry of the system. People expect a battery to perform its intended task on demand. Li-MnO_2 primary

cells will deliver 90% of their energy even after 8 years on the shelf; that is, their self-discharge is low. Some military batteries have a 20-year storage life and still deliver their rated capacity.

5. Cost: The cost of the battery is determined by the materials used in its fabrication and the manufacturing process. The selling price must be in keeping with its perceived value.
6. Safety: All consumer and commercial batteries must be safe in the normal operating environment and not present any hazard under mild abuse conditions. The cell or battery should not leak, vent hazardous materials, or explode.

Chapter 12.2: Secondary Batteries

INTRODUCTION

The storage battery, secondary battery, or charge accumulator is a cell or combination of cells in which the cell reactions are reversible. This means that the original chemical conditions within the cell can be restored by passing current to flow into it: that is, by charging from an external source. These batteries find application in transport, telephone exchange, and emergency lighting sources. The first secondary battery, today's lead–acid battery, was designed in 1859 by Plante. Other secondary batteries include the Edison nickel–cadmium battery (1919), and the Andre silver–zinc system introduced in 1960.

Ideal battery storage has low resistance, simplicity, and ease and strength of construction, durability and low cost. The lead–acid battery is cheap, reliable, and durable. In addition electrochemical reactions in this battery are reversible and the deteriorative side reactions have little effect. However it does not have high capacity per unit weight; the mechanical strength of the lead plates are also low. The nickel–cadmium battery has a high capacity per unit weight and is durable and strong but expensive. In addition not all of the reactions are always reversible. Despite their high cost, silver–zinc batteries provide high energy density.

PERFORMANCE

The performance of battery is usually expressed in terms of ampere-hour efficiency or voltage or watt-hour efficiency, defined as:

Ah efficiency = economic Ah output during discharge/all input to fully charged state

V efficiency = average V during discharge/average V during charge

WH efficiency = Ah efficiency × V efficiency

Ampere-hour efficiency indicates how much stored electricity is used during discharge and how much is lost in gas evolution, heating, and other electrode processes. Voltage efficiency is a measure of polarization in either direction; when there is overvoltage (η_a or $\eta_c = 0$) and if IR is minimum, this efficiency will be one for a reversible process.

Normal charge–discharge curves are shown in Fig. 12.2.1. Average discharge voltage will be smaller with increasing discharge rate and so will be the quantity of total charge extracted. The arrows indicate the limit of optimum discharge. The longest flat region is observed with silver–zinc cells. All

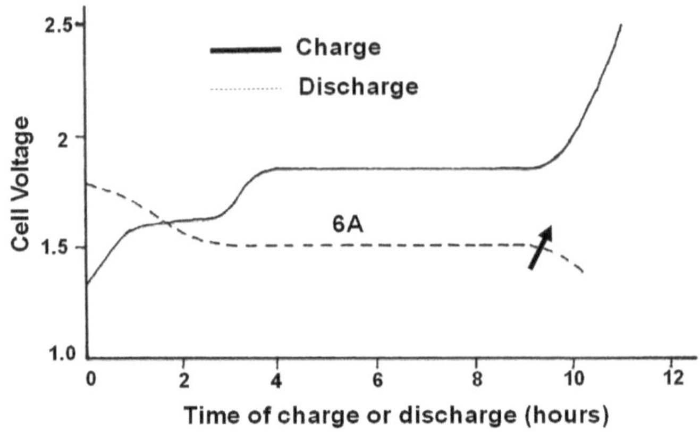

FIGURE 12.2.1 Charge–discharge curves for 60-Ah silver–zinc accumulator.

secondary cells eventually show a voltage drop to zero but the discharge must not be stopped when (1) only a small portion of the charge remains, (2) the equipment may not operate at these low voltages, and (3) if permanent damage to electrode takes place at low voltages (an example is that sulfating of the lead electrodes can occur below 1 V in a lead acid battery).

Similarly, charging should be stopped when the emf starts climbing rapidly and gassing sets in. Thermodynamically, electrolysis of water takes place at the working potentials of these batteries. However the high overvoltages prevent them from taking place. It is necessary to add water to maintain the concentration of the electrolyte.

TYPES AND CLASSIFICATION

Classification of secondary batteries can be based on whether they are construction sealed or vented, electrolyte acid or alkaline, stationary (specific gravity of 1.3) or portable (specific gravity of lead acid, 1.2), and end use. Ampere-hour capacities can vary from less than 1 to several thousand.

LEAD–ACID BATTERY

A lead–acid battery is composed of: anode: sponge metallic lead; cathode: lead dioxide (PbO_2); electrolyte: dilute mixture of aqueous sulfuric acid.

Applications are motive power in cars, trucks, forklifts, construction equipment, recreational water craft, and standby/backup systems. They are used mainly for engine batteries. These cells represent over half of all battery sales. Some advantages are their low cost, long lifecycle, and ability to withstand mistreatment. They also perform well in high and low temperatures and in high-drain applications. The battery is represented as $^{-}Pb/H_2SO_4/PbO_2$,

Pb^{+} or $^{-}Pb/PbSO_4$, $H_2SO_4/PbSO_4/PbO_2/Pb^{+}$. The chemistry of lead acid battery in terms of half-cell reactions is:

$$\text{Anode: } Pb \rightarrow Pb^{2+} + 2e$$
$$Pb^{2+} + SO_4{}^{2-} \rightarrow PbSO_4$$

$$\text{Cathode: } PbO_2 + 4H^{+} + 2e^{-} \rightarrow Pb^{2+} + 2H_2O$$
$$Pb^{2+} + SO_4{}^{2-} \rightarrow PbSO_4$$

The overall reaction is: $Pb + PbO_2 + 2H_2SO_4 \rightarrow$ (for 2F) $2PbSO_4 + 2H_2O$

The stepwise mechanisms are not yet completely understood but the overall reaction is confirmed. Because $PbSO_4$ is formed in both of the half-cell reactions it is called the double sulfate theory. The OCV is given by $E_{rev} = E_0 - (RT/nF)\ \ln\ (a_{H_2O})^2/(a_{H_2SO_4})^2$, where E_0 is the algebraic sum of the two standard potentials of -0355 and $+1.68$ V is equal to 2.042 V at 298 K. A maximum of 2.2 V is attained upon proper charging and the battery is to be discharged only up to 1.8 V.

DESIGN

There are a few problems with the design. If the cell voltages exceed 2.39 V, the water breaks down into hydrogen and oxygen (this so-called gassing voltage is temperature dependent). This requires replacing the cell's water. Also, because the hydrogen and oxygen vent from the cell, too high a concentration of this mixture will cause an explosion. Another problem arising from this system is that fumes from the acid or hydroxide solution may have a corrosive effect on the area surrounding the battery. These problems are mostly solved by sealed cells, made commercially available in the 1970s. In the case of lead acid cells, the term "valve-regulated cells" is more accurate, because they cannot be sealed completely. If they were, the hydrogen gas would cause the pressure to build up beyond safe limits. Catalytic gas recombination does a great deal to alleviate this problem. It converts hydrogen and oxygen back into water, achieving about 85% efficiency at best. Although this does not entirely eliminate the hydrogen and oxygen gas, the water lost becomes so insignificant that no refill is needed for the life of the battery. For this reason, these cells are often referred to as maintenance-free batteries. Also, this cell design prevents corrosive fumes from escaping. These cells have a low cycle life, a quick self-discharge, and low energy densities (normally between 30 and 40 Wh/kg). However, with a nominal voltage of 2 V and power densities of up to 600 W/kg, the lead–acid cell is an adequate, if not perfect, design for car batteries.

A rectangular design is adopted for convenience. The anodes are called negative plates, made of rectangular plates of spongy lead, whereas the cathodes (positive electrode) are made of lead antimony alloy grids coated with PbO_2. The electrodes are dipped in 27–39% by weight sulfuric acid. A porous insulator separates the positive electrode from the negative plates.

Acid corrosion-resistant materials are employed for the containers, cover, and vent. The interior working volume is divided into three parts called the bottom, element, and gas spaces. The element space is occupied by the bulk volumes and holds electrodes and separators and supporting plates. The positive and negative electrodes are interleaved with the separators. The electrolyte fills the space between the electrodes and the bottom space.

ELECTRODE PLATES

Lead alloyed with antimony or calcium is cast as grids and coated with litharge or red lead and dilute sulfuric acid. This is dried to form a basic sulfate and further electrolyzed in dilute sulfuric acid. The grids, which are cathodes in this electrolysis, are gradually covered with spongy lead owing to the reduction of the oxide paste. These form the anodes of the battery. The anodes in the electrolysis are covered by PbO_2 as a result of oxidation of the paste. Two percent $PbSO_4$ may also be incorporated in these grids. These form the cathodes of the battery. In the process of electroforming, the porosity of the plates increases considerably. The porosity helps the acid to penetrate better and helps retain $PbSO_4$ formed during discharge of the battery. $BaSO_4$, lampblack, or graphite is also incorporated as an expander for heavy current discharge batteries.

The positive place, lead dioxide, is supported on narrow vertical ebonite tubes with portholes for the entry of electrolyte. The life of the plates is determined by (1) shedding of PbO_2 and (2) parasitic corrosion of the grid material. The shedding is minimized by providing robust plates. The corrosion is caused by local galvanic action with the grid material; lead becomes anodic and hydrogen evolution occurs at the cathodic sites where antimony is present. In the presence of silver, copper can also promote corrosion. Incorporation of calcium (0.08%) as an alloying material instead of antimony is beneficial.

Alternatively the plates may be formed by a cycle of the charging and discharging of pure lead sheets in dilute sulfuric acid. The introduction of 8–12% antimony improves the castability of lead by reducing the freezing point, increasing hardness, and lowering the creep point. Antimony increases the self-discharge of the plates. If there is less than 5% antimony, casting defects cause brittleness and cracking increases. Heat treatment can improve the strength of the grid. Beneficial grain refinement can be achieved by incorporating silver, silicon, tellurium, selenium, and tin (0.075%). Precipitation-hardened alloys of lead and calcium have a low corrosion rate and longer life.

SEPARATORS

It is necessary to keep the positive and negative plates as close as possible without contact so that the IR is reduced. Very close spacing is not possible because (1) lead tends to grow in dendritic form on the substrate during charge and is likely to cause a short-circuit by coming into contact with the adjacent plate; (2) the plates tend to buckle under pressure developed by the $PbSO_4$ formed by buckling when the plates are overcharged or undercharged; and (3) the plates are displaced by mechanical and thermal shocks to the supports, so some form of porous separator that fills the intervening space completely is required. Several materials have been tried for this.

ELECTROLYTE

The emf developed in the battery depends on the activity of sulfuric acid. The open circuit potential at 298 K varies from 1.88 V with 5% acid to 2.3 V with 40% acid. The acid is diluted during discharge. The specific gravity in a charged battery is that of a dilute acid. The dilution becomes faster as $i_{discharge}$ is increased. Hence, there should be excess sulfuric acid to start with to maintain low resistance. In Table 12.2.1 the change of specific gravity with concentration is given.

Variations of resistance and freezing point with specific gravity are shown in Fig. 12.2.2. The conductivity is maximum when the specific gravity is 1.2.

One has to choose an optimum concentration for a specific application. These data are provided in Table 12.2.2.

TABLE 12.2.1 Acid Concentration and Specific Gravity

Acid concentration (%)	0	7.5	14.5	21	27.5	33.5	39
Specific gravity	1	1.05	1.1	1.15	1.2	1.25	1.3

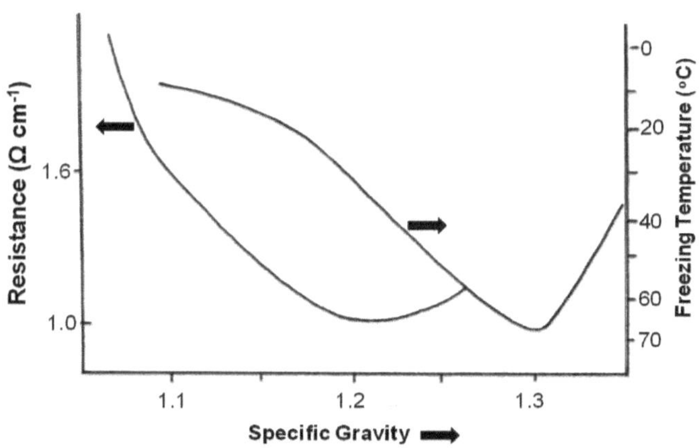

FIGURE 12.2.2 Variations in resistance and freezing point of sulfuric acid with concentration.

TABLE 12.2.2 Recommended Specific Gravity for Lead–Acid Batteries for Different Applications

Type	Acid Concentration (%)	Specific Gravity	Container
Stand-by batteries	27–30	1.2–1.225	Steel + lead-based wood
Automobiles	29–31	1.21–1.230	Ebonite
Submarine			
Portable	30	1.23	Glass, polystyrene
Aircraft	36–37.5	1.275–1.285	Unspillable polystyrene
Very heavy duty	36–39		

CAPACITY OF LEAD–ACID BATTERIES

The total capacity of the lead–acid batteries depends on the amount of active material loaded in the system. All of the lead used is converted into lead sulfate, a nonconducting material. Simultaneously 3.65 g of sulfuric acid is converted into water per ampere-hour. Thus the IR will increase. Lead sulfate is the product of the cathodic reduction of PbO_2. Hence IR increases during the discharge of the battery. In practice, the amount of active material used is three to four times the theoretical requirement.

Experience has given us the following information:

1. Lead sulfate clogs pores in the plates and obstructs the penetration of acid, preventing it from interacting with the active material.
2. The accumulation of lead sulfate results in stress on the positive plate (cathode), causing it to bulge and buckle.
3. Plates with a large area retain more charge but have lower mechanical strength.
4. High currents cannot be sustained with thick plates by diluting the acid.

TEMPERATURE EFFECT

Lowering the temperature results in reduced capacity. This mainly happens because of the lowering of diffusion rates. Expanders are essential for battery plates for low-temperature operations. A reduction in temperature causes an increase in viscosity and a decrease in conductivity. It also leads to a reduction in the rates of the electrochemical reactions. It is conventional to quote the capacity with a temperature label. For example, 60 Ah with an 8-h rating at 15–25°C means that the accumulator will deliver 60 Ah of charge upon being discharged continuously for 8 hours until the emf falls to 1.75 V. The ampere-hour capacity quoted is applicable for a freshly charged battery. The actual capacity will depend on the time of standing. The trickle charging for standby systems has to be accounted for.

EFFICIENCY

The efficiency of a lead–acid battery in the absolute sense is low because only one-third of active material is used up during the discharge. However the active material is wholly reformed during charging. The ampere-hour efficiency can be as high as 90–99% and is normally 80%. The lower value has to be associated with the generation of heat, gassing during charge, and discharge at a rate higher than that quoted or while operating at a lower temperature.

Deterioration of the lead–acid may occur because of:

- physical deterioration of the plates, such as buckling and warping
- excess gassing if overcharged
- overoxidation, causing the lead grid to weaken
- filling of the bottom space with lead sulfate and shed products, causing a reduction in active volume
- Sulfating from overdischarge (the formation of hard crystalline lead sulfate, which is nonconducting and clogs pores)
- self-discharge (estimated to be around 1% per day)
- inverse or reverse charging (accidental), forming lead and lead dioxide on the wrong plates

Dry Charged Batteries

Dry charged battery systems are similar to wet lead–acid batteries except for the additional processing steps of washing with water and drying in an inert atmosphere. The battery consists of formed positive plates of lead oxide and formed negative plates of spongy lead along with dry separators. These components are activated by adding battery-grade sulfuric acid of appropriate strength. The major advantage of using dry charged batteries is during storage, when wet batteries deteriorate. Deterioration is reduced by frequent charging. Nonspillable lead–acid batteries are also manufactured.

ALKALINE STORAGE BATTERIES

Alkaline batteries were introduced in 1919. There are two types of Edison cells using nickel oxide combined with iron or with cadmium. The cathodes are a higher grade of nickel with an admixture of nickel supported on steel grids.

In the nickel–iron battery the anodes are iron powder pasted onto a steel support. The electrolyte is a 20–25% KOH solution contained in an inert steel container. The cell is conventionally represented by:

$$\mathrm{Fe|FeO|KOH|NiO,\ Ni_2O_3|Ni}$$

The cell reactions are complicated. Some representative reactions at the left-hand side electrode of the cell can be:

$$\mathrm{Fe \rightarrow Fe^{2+} + 2e}$$

$$\mathrm{Fe^{2+} + 2OH^- \rightarrow FeO(S) + H_2O}$$

$$\text{or}\quad \mathrm{Fe + 2OH^- \rightarrow FeO + H_2O + 2e}$$

Reactions at the right-hand side electrode can be represented as:

$$\mathrm{Ni_2O_3 + 3H_2O = 2Ni^{3+} = 6OH^-}$$

$$\mathrm{2Ni^{3+} + 2e \rightarrow 2Ni^{2+}}$$

$$\mathrm{2Ni^{2+} + 4OH^- \rightarrow 2NiO + 2H_2O}$$

$$\mathrm{Ni_2O_3 + H_2O + 2e \rightarrow 2NiO + 2OH^-}$$

The overall reaction can be written as:

$$\mathrm{Fe + Ni_2O_3 \rightarrow FeO + 2NiO}$$

$$E_{\mathrm{LHS}} = E^0{}_{\mathrm{Fe,FeO}} + (\mathrm{R}T/2\mathrm{F})\ln(\mathrm{OH^-})^2 = -0.88\ \mathrm{V}$$

$$E_{\mathrm{RHS}} = E^0{}_{\mathrm{Ni_2O_3,NiO}} + (\mathrm{R}T/2\mathrm{F})\ln(\mathrm{OH^-})^2 = 0.55\ \mathrm{V}$$

The emf is thus independent of the concentration of OH ions and is observed in the range of 1.33–1.35 V.

NICKEL–CADMIUM CELLS

The anode is composed of: cadmium cathode: nickel oxyhydroxide $Ni(OH)_2$; electrolyte: aqueous potassium hydroxide (KOH).

Applications are as calculators, digital cameras, pagers, laptop computers, tape recorders, flashlights, medical devices (eg, defibrillators), and electric vehicles, and also space applications.

The cathode is nickel-plated woven mesh, and the anode is a cadmium-plated net. Because the cadmium is just a coating, this cell's negative environmental impact is often exaggerated. (Incidentally, cadmium is also used in television tubes and some semiconductors, and as an orange-yellow dye for plastics.) The electrolyte, KOH, acts only as an ion conductor and does not contribute significantly to the cell's reaction. This is why not much electrolyte is needed, and so this keeps the weight down. (NaOH is sometimes used as an electrolyte, which does not conduct as well, but also does not tend to leak out of the seal as much.)

The cell reactions can be written as:

$$Cd + 2OH^- \rightarrow Cd(OH)_2 + 2e$$

$$NiO_2 + 2H_2O + 2e^- \rightarrow Ni(OH)_2 + 2OH^-$$

$$\text{Overall reaction: } Cd + NiO_2 + 2H_2O \rightarrow Cd(OH)_2 + Ni(OH)_2$$

Advantages of these cells include good performance in high-discharge and low-temperature applications. They also have long shelf and use life.

Disadvantages are that they cost more than the lead–acid battery and have lower power densities. Possibly the most well-known limitation is a memory effect, in which the cell retains the characteristics of the previous cycle. This term refers to a temporary loss of cell capacity, which occurs when a cell is recharged without being fully discharged. This can cause cadmium hydroxide to passivate the electrode, or the battery to wear out. In the former case, after a few cycles of discharging and charging, the cell will help correct the problem, but it may shorten the lifetime of the battery. The true memory effect comes from experience with a certain style of Ni-Cd in space use, which was cycled within a few percent of discharge each time. An important thing to know about "conditioning" an Ni-Cd battery is the deep discharge.

Construction of the Electrodes

$Ni(OH)_2$ is precipitated by reacting $NiSO_4$ with NaOH, and dried. This is then compressed with nickel plaques. About 300 layers of alternating Ni and $Ni(OH)_2$ are formed and incorporated into perforated steel cylinders. Active cathodes are formed during the first charging when $Ni(OH)_2$ changes into Ni_2O_3 or $NiO_2 + Ni(OH)_3$, NiO_2 decomposes to produce oxygen.

Anodes (the negative electrode): Ferric oxide is reduced by H_2 to a powdered mixture of Fe and Fe_2O_3. The powder is packed into perforated flat steel grids.

Twenty-two percent KOH with a specific gravity of 1.21 is recommended for a nickel–iron cell and 20% KOH for a nickel–cadmium cell. The capacity of the cell is independent of the volume of KOH taken. The volume of electrolyte should be sufficient to submerge the active area of the plate. Because $Fe(OH)_2$ and $Cd(OH)_2$ are the products of oxidation and are nonconductors, the iron powder must be taken in excess to ensure high conductivity during discharge. The nickel oxide and hydroxide are also poor conductors. Hence incorporation of metallic nickel into the cathode is essential.

No clogging is observed in the acid cell as the volute decreases upon discharge. The variation of emf with capacity for a 100-Ah cell is shown in Fig. 12.2.3.

Factors that affect the capacity of these cells are that:

1. The emf falls with the rate of discharge (I_{amp}) ampere-hour capacity independently of i_{dis}. This is a result of the superior access of KOH to the active surface and to the interior of the plates.

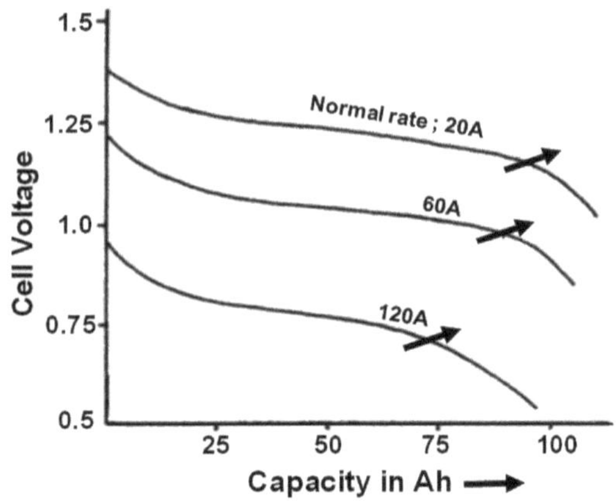

FIGURE 12.2.3 Discharge characteristics of 100-Ah Edison alkaline accumulator.

2. The capacitance is reduced as the temperature of operation is lowered because the iron anodes tend to become passive. This is not so in cadmium anodes.
3. The cell has a low rate of self-discharge (20% in a year) and hence does not require trickle charging.

MERITS OF THE SYSTEM

- Even after a long idling time, although charge-cycling is not efficient for the first five cycles, the damage is never permanent.
- Overcharging does not affect performance.
- Low maintenance cost offsets the initial cost of the product.
- It can be charged with alternating current.

EFFICIENCY

Efficiency is 25% in terms of the metal put in, but the ampere-hour efficiency is normally 80%. The smaller value is due to gassing during the charge step. The Watt-hour efficiency is 55–65%. The voltage efficiency is 70% because the charging and discharging curves are different.

SILVER–ZINC BATTERY

The silver–zinc battery is depicted as:

$$Zn|KOH|Ag_2O_2, Ag$$

The overall reaction is:

$$Ag_2O_2 + 2Zn \rightarrow 2Ag + 2ZnO$$

$$\text{or } Ag_2O_2 + 4KOH + 2Zn \rightarrow 2Ag + 2K_2ZnO_3 + 2H_2O$$

The anode reaction is:

$$Zn + 2OH \rightarrow Zn(OH)_2 + 2e$$

The cathode reactions are:

$$4Ag + 4H_2O + 4e \rightarrow 4AgH + 4OH^-$$

$$2AgH + Ag_2O_2 \rightarrow Ag_2O + 2Ag + H_2O$$

$$2AgH + Ag_2O \rightarrow 4Ag + H_2O$$

The overall reaction is $Ag_2O_2 + 2H_2O + 4e \rightarrow 2Ag + 4OH^-$

The OH produced in the catholyte is consumed in the anode. However H_2O is consumed in the cathode reaction. Hence the electrolyte can become saturated with KOH. The Ag_2O is more stable and has higher reactivity. Ag_2O can also be formed by the reaction $2Ag + Ag_2O_2 \rightarrow 2Ag_2O$.

Performance

The discharge curve shows a single stage whereas charging curves show two steps. The two stages correspond to the two different oxides of Ag. The emf stays constant during discharge. IR decreases when the oxides are converted to the highly conducting metal.

Advantages

- During discharge lowering in capacity toward the end is small (smaller than for the lead–acid cell).
- IR is very small; it can be discharged rapidly to 1 V and yet remain unchanged.
- There is little self-discharge.
- High temperature does not increase the corrosion rate but life is reduced above 323 K. Capacity is reduced below 273 K without damage. It works well up to 253 K.
- A low-temperature operation is better than in the nickel iron cell even with high current discharge.
- The battery is compact and delivers up to 3 Ah/cm^2.
- The theoretical capacity is 33.6 Ah/lb and has nearly achieved been in practice.
- The charge retention is 50% after 2 months of storage for small cells and 75% for large cells.

Silver plates are made by pasting sintering or electrolyte deposition, depending on the end use and type of battery. Silver powder or Ag_2O powder is sintered; the former can be sintered more easily even at low temperatures and then converted to Ag_2O_2. Zinc plates are made by the deposition of porous dendritic zinc and are further compressed mechanically.

SEALED STORAGE BATTERIES

Sealing or non-spillability can be achieved in principle by avoiding or reducing the formation of gases produced during discharge or by converting them electrochemically into other products. The electrolyte may be maintained as a gel absorbed into a conducting medium. Such batteries have the following advantages:

- They can be used in any position.
- There is no leakage and hence operating instruments are not affected by the battery chemicals.
- There is no escape of vapors or sprays.
- No topping is generally required.

It is possible to make an unspillable lead–acid battery. This is achieved by:

- using a separator, which will absorb the electrolyte;
- providing vent traps to prevent leakage in any position. This does not prevent acid spray;
- using spongy lead to absorb oxygen during charging; and
- tripping the charger to avoid gassing at high voltages.

The sealed nickel–cadmium penlight cell was introduced in 1951. Sealing requires:

- removal of gases on one of the cell electrodes or on a third electrode material,
- a decrease in the rate of evolution of gas during charging and idling, and
- control of the final cell voltage during charging.

These are achieved by:

1. using special valve closures;
2. making evolved oxygen diffuse through the cell and combining it with the active materials of the negative electrode. An excess of cadmium oxide is present. Oxygen from the positive electrode diffuses and reacts with spongy cadmium. Cadmium oxide does not react with KOH to evolve H_2;
3. using small amounts of electrolytes to decrease the rate of gas evolution; and
4. employing a controlled charging voltage unit.

Sealed cells are of the following types: flat button cells (50–400 mAh), cylindrical cells (AA) (450 mAh to 8 or even 50 Ah), and rectangular cells (5–25 Ah).

Sealed silver–zinc cells were introduced in 1960. Distinctive features regarding their operation are that they are provided with controlled charging voltage to prevent the voltage from exceeding 1.97 V during charging, heavily amalgamated zinc is used to prevent hydrogen evolution, and a third electrode is provided in some cells that can oxidize H_2 to H^+ or reduce O_2 to OH^- ions.

NICKEL–METAL HYDRIDE CELLS

In nickel–metal hydride (NiMH) cells, the anode is a rare-earth or nickel alloy with many metals. The cathode is nickel oxyhydroxide. The electrolyte is potassium hydroxide.

Applications are cellular phones, camcorders, emergency backup lighting, power tools, laptops, portable, and electric vehicles.

The battery may be conventionally represented as:

$$^{-}MH|KOH\ (30\%)|Ni(OH)_2,\ NiOOH^{+}$$

where MH is a metal hydride and is a hydrogen storage material. Half-cell reactions are:

$$C + M + H_2O \leftrightarrow MH + OH^-$$

$$\text{And, } Ni(OH)_2 + OH^- \leftrightarrow NiOOH + H_2O + e$$

$$\text{The overall reaction is: } NiOOH + MH \rightarrow Ni(OH)_2 + M;\ E = 1.35\ V$$

There is no net change in the concentration of the electrolyte over the charge discharge cycle. The nominal cell voltage is 1.2 V. A specific advantage of this battery over an Ni-H_2 battery is that hydrogen does not need to be stored at high pressures.

This sealed cell is a hybrid of the Ni-Cd and Ni-H_2 cells. Previously this battery was not available for commercial use because although hydrogen has wonderful anodic qualities, it requires cell pressurization. Fortunately, in the late 1960s scientists discovered that some metal alloys (hydrides such as $LiNi_5$ or $ZrNi_2$) could store hydrogen atoms, which then could participate in reversible chemical reactions. In modern NiMH batteries, the anode consists of many metals alloys, including V, Ti, Zr, Ni, Cr, Co, and Fe.

Except for the anode, the NiMH cell closely resembles the Ni-Cd cell in construction. Even the voltage is virtually identical at 1.2 V, making the cells interchangeable in many applications.

The anodes used in these cells are complex alloys containing many metals, such as an alloy of V, Ti, Zr, Ni, Cr, Co, and Fe. The underlying chemistry of these alloys and reasons for superior performance are not clearly understood, and the compositions are determined by empirical testing methods. An interesting fact about these alloys is that some metals absorb heat when absorbing hydrogen, and some give off heat when absorbing hydrogen. Both of these are bad for a battery, because one would like the hydrogen to move easily in and out with no energy transfer. Successful alloys are all mixtures of exothermic and endothermic metals to achieve this. The electrolyte of commercial NiMH batteries is typically 6 M KOH. The NiMH cell costs more and has half the service life of the Ni-Cd cell, but it also has 30% more capacity, increased power density (theoretically 50% more, practically 25% more). The memory effect, which at one time was thought to be absent from NiMH cells, is present if the cells are treated correctly. To avoid the memory effect, fully discharge it once every 30 or so cycles. There is no clear winner between the two. The better battery depends on what characteristics are crucial for a specific application.

LITHIUM ION CELLS

The anode for lithium ion cells is a carbon compound, with a graphite cathode of lithium oxide.

Applications are laptops, cellular phones, and electric vehicles.

Lithium batteries that use lithium metal have safety disadvantages when used as secondary (rechargeable) energy sources. For this reason a series of cell chemistries have been developed using lithium compounds instead of lithium metal.

Cathodes consist of a layered crystal (graphite) into which the lithium is intercalated. Experimental cells have also used lithiated metal oxide such as $LiCoO_2$, $NiNi_{0.3}Co_{0.7}O_2$, $LiNiO_2$, LiV_2O_5, LiV_6O_{13}, $LiMn_4O_9$, $LiMn_2O_4$, and $LiNiO_{0.2}CoO_2$. Electrolytes are usually $LiPF_6$, although this has a problem with aluminum corrosion, and so alternatives are being sought. One such possibility is $LiBF_4$. The electrolyte in current production batteries is liquid and uses an organic solvent. Membranes are necessary to separate the electrons from the ions. Currently batteries in wide use have microporous polyethylene membranes. Intercalation is a long-studied process that has finally found a practical use. It has long been known that small ions (such as lithium, sodium, and the other alkali metals) can fit in the interstitial spaces in a graphite crystal. Not only that, these metallic atoms can go further and force the graphitic planes apart to fit two, three, or more layers of metallic atoms between the carbon sheets. One can imagine what a great way this is to store lithium in a battery—the graphite is conductive, dilutes the lithium for safety, is reasonably cheap, and does not allow dendrites or other unwanted crystal structures to form.

HYDROGEN–NICKEL BATTERY

$$Pt(s)H_2(g)|KOH(aq)\ 30\%|NiO(OH),\ (s);\ \text{OCV } 1.5\text{–}1.6\ \text{V}$$

$$1/2H_2 + NiOOH + H_2O \leftrightarrow Ni(OH)_2 \cdot H_2O(s)$$

Hydrogen pressure rises from 0.5 MPa in the fully discharged state to 3–10 MPa when charged. The pressure is the monitor of the state of charge. Self-discharge is reported to be in the range of 6–12% per day. SS or Inconel pressure vessels are to be used. Batteries (10–60 Ah) with 1000-cycle capabilities have been constructed. Practical energy density has been 40–50 Wh/g. The practical application has been to replace Ni-Cd cells in some space applications because of the available high energy densities and longer life. Safety during operation and the high cost are the limiting factors.

METAL HALOGEN CELLS

Zinc–chlorine and zinc–bromine cells are two systems that have reached an advanced stage of development and production. Chlorine is abundantly available and inexpensive and will have a high positive equilibrium reduction potential. A high energy density is theoretically possible. The gas is highly soluble in aqueous solutions (~0.1 mol/dm at ambient temperature). Hence three-phase electrode technology of the type experienced with hydrogen fuel electrodes is not required. Furthermore, few dendritic deposits of zinc will be observed in acidic solutions for normal discharge. The storage of chlorine with high corrosivity is the major difficulty. Chlorine can be stored as a solid hydrate (Cl_26H_2O) that is stable below 9.6°C. The cell may be represented as:

$$^{-}Zn|ZnCl_2(aq)|Cl_2(aq)\ M(s)^{+}$$

The cell reaction is represented as $Zn(s) + Ck_2(g) \leftrightarrow ZnCl_2(aq)$ and the emf is 2.12 V at 298 K. The negative electrode is a solid graphite substrate upon which zinc is deposited. The other electrode is porous graphite or porous platinized or ruthenized titanium. No separator is necessary. A 4-mol/dm^3 solution of $ZnCl_2$ is used as electrolyte. When a charging current is passed, zinc is deposited on the graphite electrode and chlorine is evolved at the positive electrode. The chlorine is removed and stored as the hydrate. For discharge the chlorine is released from the hydrate by heating and the chlorine-saturated electrolyte is pumped through the pores of the positive electrode.

Zinc corrosion with accompanying hydrogen evolution must be minimized. Porous graphite with necessary surface treatments is preferred to ruthenium-treated titanium substrates. The graphite electrodes undergo slow oxidative degradation not at a significant rate. Energy densities of 80–100 Wh/kg have been reported. A 1000-cycle deep discharge has been demonstrated.

Zinc–bromine can be represented as $Zn(s)|ZnBr_2(aq)|Br_2(g)$, M(s) and the overall cell reaction is $Zn(s) + Br_2(g) \leftrightarrow ZnBr_2(aq)$, but the energy density is less than that for a zinc–chlorine battery. The emf is 1.83 V at 298 K and the theoretical energy density is 430 Wh/kg. KCl is added to maintain conductivity because the concentration of $ZnBr_2$ decreases during discharge. Titanium and ruthenium dioxide perform well as a bromine electrode. With graphite, intercalation of bromine occurs and hence it cannot be used. The rate of self-discharge is high in this system.

MANGANESE–TITANIUM (LITHIUM) CELLS

The anode of manganese–titanium (lithium) cells is lithium–titanium oxide. The cathode is lithium intercalated manganese dioxide.

Applications are watches and other ultra-low discharge uses.

This technology might be called manganese–titanium but it is just another lithium coin cell. It has "compatible" voltage, 1.5–1.2 V, like the lithium–iron cell, which makes it convenient for applications that formerly used primary coin cells. It is unusual for a lithium-based cell because it can withstand a continuous overcharge at 1.6–2.6 V without damage. Although rated for 500 full discharge cycles, it only has a 10% per year self-discharge rate, and so it is used in solar-charged watches with an expected life of more than 15 years with shallow discharging. The amp-hour capacity and available current output of these cells are extremely meager. The range of capacities from Panasonic is 0.9–14 mAh. The maximum continuous drain current is 0.1–0.5 mA.

RECHARGEABLE ALKALINE MANGANESE CELLS

The anode for rechargeable alkaline manganese (RAM) cells is zinc. The cathode is manganese dioxide. The electrolyte is a potassium hydroxide solution.

Applications are in consumer devices.

This is the familiar alkaline battery, specially designed to be rechargeable. In the charging process, direct-current electrical power is used to reform the active chemicals of the battery system to their high-energy charge state. In the case of the RAM battery, this involves oxidation of manganese oxyhydroxide (MnOOH) in the discharged positive electrode to manganese dioxide (MnO_2), and of zinc oxide (ZnO) in the negative electrode to metallic zinc. Care must be taken not to overcharge, to prevent electrolysis of the KOH solution electrolyte, or to charge at voltages higher than 1.65 V (depending on temperature), to avoid the formation of higher oxides of manganese.

REDOX (LIQUID ELECTRODE) CELLS

Redox (liquid electrode) cells consist of a semipermeable membrane with different liquids on either side. The membrane permits ion flow but prevents mixing of the liquids. Electrical contact is made through inert conductors in the liquids. As the ions flow across the membrane an electric current is induced in the conductors. These cells and batteries have two ways of recharging. The first is the traditional way of running current backward. The other is replacing the liquids, which can be recharged in another cell. A small cell can also be used to charge a great quantity of liquid, which is stored outside the cells. This is an interesting way to store energy for alternative energy sources that are unreliable, such as solar, wind, and tides. These batteries have low volumetric efficiency but are reliable and very long-lived. Electrochemical systems that can be used are $FeCl_3$ (cathode) and $TiCl_3$ or $CrCl_2$ (anode).

VANADIUM REDOX CELLS

Vanadium redox cells are a particularly interesting cell that uses vanadium oxides of different oxidation states as the anode and cathode. These solutions will not be spoiled if the membrane leaks, because the mixture can be charged as either reducing or oxidizing components.

BATTERY ECONOMICS

Secondary batteries are expensive sources of electricity. Other factors must considered:

- The cost of ac-dc converters for charging the battery should be included in the overall cost.
- The cost of dc-ac converters has to be included when the operating instruments are designed to work on ac.
- Dissipation of energy cannot be avoided but can be minimized.
- When using a multicell assembly, a large quantity of material is required.
- Pollution hazards caused by the spent battery materials have to be taken into account.

Chapter 12.3: Other Batteries

Reserve batteries are known as deferred action, one-shot, or dry-charged batteries. Reserve batteries are inactive when not in use and may be activated in several different ways before application. The principle is to assemble the battery in all respects except for one vital component. When this component is added the battery is activated. Main advantages of reserve batteries are: (1) unlimited shelf life; (2) high reliability of performance, (3) freedom of the choice of system, permitting the use of highly reactive chemistry leading to high rates of discharge and efficient performance at low temperatures; and (4) a variety of design options. The reserve battery is designed for reliable performance but not efficiency of performance. These batteries are classified as liquid-, gas-, and thermally activated systems, which may include fused salt batteries. Activation can be manual or automatic.

LIQUID-ACTIVATED BATTERIES

Water-activated batteries: There are essentially two systems activated by the addition of water. These batteries find use in meteorology and signaling in midsea. The two systems are:

(1) Mg|seawater|AgCl, Ag; and (2) Mg|seawater|Cu_2Cl_2, Cu

The cathode reactions are: $AgCl + e \rightarrow Ag + Cl^-$ and

$$Cu_2Cl_2 + 2e \rightarrow Cu + 2Cl^-$$

The OCV developed for the first cell is 1.6 V. The IR of the battery decreases during discharge. The performance is good even at 223 K. The second system develops 1 V under open circuit conditions and is less stable. The active life of these batteries is only a few hours.

ACID-ACTIVATED BATTERIES

These batteries use Zn, Cd, or Pb as an anode and PbO_2 as a cathode with an acid electrolyte. They can be activated with the addition of acid. They function at high discharge rates and high temperatures. OCV developed in a lead battery using sulfuric acid or perchloric acid ranges from 1.3 to 1.9 V.

Alkali-Activated Batteries

The system that is used is Ag_2O|KOH|Zn. The upper limit of requirement can be in the range of 10 kW/min in a missile. A multiple battery system with a voltage of 28–100 V is used. The merits of the alkali-activated batteries are: (1) short activation time; (2) operation over a wide range of temperature, automatic or chemical heating; (3) electrical stability and reliability. The OCV is 1.9 V but the operating value can be 1.3 V. Zinc anodes are preferred over cadmium anodes for more rapid discharge.

GAS-ACTIVATED BATTERIES

The system can be of a hydrogen oxygen fuel cell type. These batteries are activated by fuel gas or a type in which the electrolyte is produced by a chemical reaction initiated with a gas. The anode employed is Mg, Zn, or lead; the cathode can be PbO_2 Ag_2SO_4, or any other heavy metal sulfate. The electrolyte is an ammoniacal salt produced by the action of ammonia vapor on the thiocyanates of Li^+, K^+, or NH^+. The OCV developed varies from 1.5 to 2.5 V depending on the electrode and the electrolyte.

HEAT-ACTIVATED BATTERIES

These systems are unique in the sense that all chemicals and metals are assembled and activated by heat. The metal used is zirconium and barium chromate, which is hermetically sealed and ignited by percussion or flaring. The anodes are made of calcium or magnesium sheets. The electrolyte is an LiCl–KCl eutectic and the cathode is a mixture of chromate and heavy metal sulfate or oxide of iron, tungsten, or vanadium. Thermally stable material (300–600°C) can be used as cathode. The OCV developed varies 1 to 3 V. Characteristics of these batteries are: (1) a long shelf life, (2) a wide range of temperature of operation (273–873 K), (3) rugged construction and quick activation, and (4) wide-range electrical capacity but a short active life and low energy output.

SOLID-STATE AND MOLTEN SOLVENT BATTERIES

Although liquid-state batteries provide good contact with electrodes and high conductivity, they face two major limitations: corrosion of the electrodes and materials and solvent consumption. A fully solid state configuration is appealing in advanced technology.

A force is something that pushes against something else, such as gravity. If it succeeds, work gets done. If a 1-lb weight is lifted 1 ft, 1 ft-lb of work has been done on the weight itself. Both force and distance are needed before work gets done. Energy is just the capacity to do work or the ability to employ a force that moves something through a distance or performs some exact electrical, thermal, chemical, or whatever equivalent to mechanical work. Power is the time rate of doing work.

Thus, energy is "how much" and power is "how fast." An energy source is a substance or a system that may be capable of delivering net kilowatt hours of energy. An energy carrier is some means of moving energy from one location to another. Batteries, flywheels, utility pumped storage, and terrestrial hydrogen are examples. They are carriers or "energy transfer systems" because one first has to "fill" them with energy before one can "empty" them. Without fail, all energy carriers consume significantly more existing old energy than they can return as new.

All batteries operate on the principles of electrochemistry. An electrochemical reaction is one in which electrons are transferred from one chemical species to another as the chemical reaction is taking place. In a battery these reactions take place at the electrodes of the battery. At the battery electrode known as the anode, a reaction takes place known as oxidation. During oxidation a chemical species loses electrons. The other electrode in a battery is known as the cathode. Reduction reaction takes place at the cathode where electrons are combined with ions to form stable, electrically neutral chemical species. Batteries take advantage of these reactions by making the electrons formed by oxidation on the anode flow through a wire in the external circuit to the cathode, where they are used in the reduction reaction. A load can be attached along this circuit to take advantage of the current of electrons, to power a device. Electrons move through the wire from the anode to cathode because the conductive nature of the wire connecting the two makes that path the easiest way for the electrons to get there.

The rechargeable, or secondary, batteries can be distinguished according to the following parameters: voltage, current (maximum, steady-state, and peak), energy density (Watt hours per kilogram and per liter), power density (Watts per kilogram and per liter), and service life (cycles to failure) and cost (per kilowatt hour). The energy density per unit volume (Wh/L) and per unit weight (Wh/kg) of various rechargeable batteries is shown in Fig. 12.3.1.

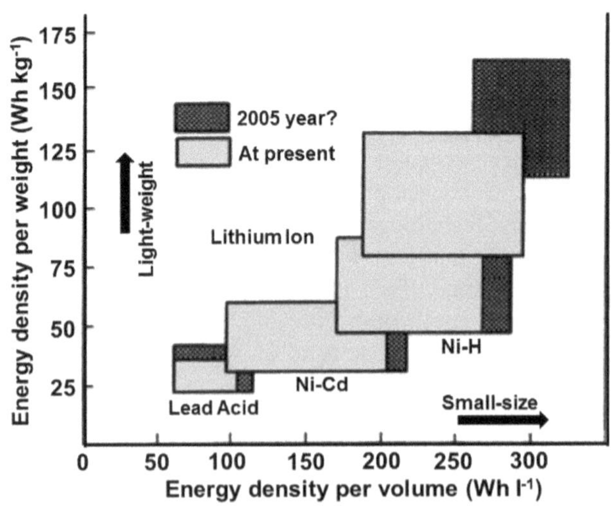

FIGURE 12.3.1 Energy density in selected secondary batteries.

In the case of conventional batteries, for instance, these systems contain a liquid electrolyte, generally a concentrated aqueous solution of potassium hydroxide or sulfuric acid. The use of aqueous battery electrolytes theoretically limits the choice of electrode reactants to those with decomposition voltages less than that of water, that is, 1.23 V at 25°C, although because of the high overpotential normally associated with the decomposition of water, the practical limit is 2.0 V. The liquid state offers good contacts with the electrodes and high ionic conductivities, but anion and cation mobilities are of the same order of magnitude and their simultaneous flow results in two major problems: (1) corrosion of the electrodes; and (2) consumption of the solvent (water) by electrolysis during recharging and by corrosion during storage, which makes periodic refilling necessary. In addition, these two processes give off gases, thereby prohibiting the design of totally sealed systems. The resulting problems include leakage of the corrosive electrolyte and air entries which, even when kept to a minimum, deteriorate the electrolyte and the electrodes. A further drawback is the risk of electrode passivation; the formation of insulating layers of $PbSO_4$ and $Zn(OH)_2$.

SOLID-STATE ELECTROLYTES

The demand for batteries with high energy densities has inevitably led to research and the development of systems using electrolytes that are thermodynamically more stable to aqueous environments. Essential requirements of an electrolyte are that: (1) it is ionically but not electronically conducting; (2) it is neither a solvent for the reactants nor, preferably, for the reaction product; and (3) it has a decomposition potential greater than that of the chosen reaction product. It is advantageous for the electrolyte to be inexpensive and nontoxic and to have a low vapor pressure. In general these requirements can be met in three classes of compounds: (1) molten salts, (2) ionically conducting solids, and (3) organic liquids and low-melting solids.

The concept of an all solid-state battery is appealing because such a system would possess a number of desirable characteristics, eg, the absence of possible liquid leakage or gassing, the likelihood of an extremely long shelf life, and the possibility of operation over a wide temperature range. Solid-state batteries could be constructed with excellent packaging efficiency for active components without separators and using simple lightweight containers. The opportunities for extreme miniaturization and simple fabrication techniques are of obvious importance in applications, and reliability is a key factors, as, for example, in implantable electronic instrumentation such as cardiac pacemakers and physiologic monitoring/telemetry packages. A solid electrolyte is a phase that has an electric conductance wholly due to ionic motion within the solid. Furthermore, the only mobile charge carrier is the cation associated with an anion immobilized in a crystal lattice. Such phases have been known for over a century, but until recently all known materials of this type had high

TABLE 12.3.1 Solid-State Batteries as of 1960

System	Cell Potential	Organization Involved in Development
$Ag/AgI/V_2O_5$	0.46	National Carbon
$Ag/AgBr/CuBr_2$	0.74	General Electric
$Ag/AgBr\text{-}Te/CuBr_2$	0.80	Patterson-Moos Research
$Ag/AgCl/KICl_4$	1.04	Sprague Electric
$Ni\text{-}Cr/SnSO_4/PbO_2$	1.2–1.5	P. R. Mallory & Rayovac

resistivities at ambient temperatures. This high IR of the cells is a direct result of the lack of any ambient temperature solid with fast ion conduction. The most ionic conducting material at that time was AgI with a conductivity value of about 10^{-6} S/cm at 25°C. Table 12.3.1 shows the five solid electrolyte batteries that were under development; the very high IR ranged from 50 to 40 MΩ. This restricted the development of solid electrolyte devices in a number of laboratory cells used for thermodynamic studies, and they were of little interest in power sources. At room temperature solid electrolytes did not conduct current well. A value of 10^{-6} S/cm was a high value of conductivity for a solid electrolyte. A striking development occurred toward the end of 1960 with the discovery of a series of solids of general formula MAg_4I_5 (M = Rb, K), which had exceptionally high ionic conductivity (>10 S/m at room temperature).

A number of structural features have been found to characterize solids with high ionic conductivity and to distinguish them from the more usual ionic crystals. Ionic conductivities of some solid-state electrolytes are shown in Fig. 12.3.2.

The electrolyte is a solid fast ion conductor. The blocking of the anions prevents passivation, corrosion, and solvent electrolysis reactions. Consequently there is no gas formation. It is therefore possible to design totally sealed batteries, which eliminates the deterioration of the electrolyte and the electrodes by the outside environment. Under these conditions, the electrolyte can coexist with couples that are highly reducing at the negative electrodes and highly oxidizing at the positive electrode. In such systems higher energy densities can be achieved.

IONIC CONDUCTIVITY IN SOLID ELECTROLYTES (FAST ION CONDUCTORS)

Point defects are responsible for possible movements of atoms or ions through the lattice. If a crystal structure is perfect it would be difficult to envisage how

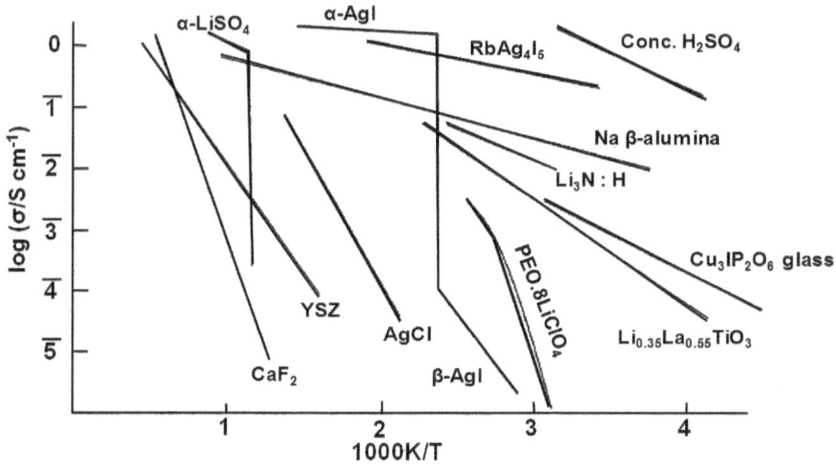

FIGURE 12.3.2 Ionic conductivity of some solid electrolytes.

the atoms move by either diffusion though the lattice or ionic conductivity (ion transport under the influence of an external electric field). There are two possible mechanisms for the movement of ions through a lattice: vacancy mechanism (it can be described as the movement of a vacancy rather than the movement of the ion) or an interstitial mechanism in which an interstitial ion jumps or hops to an adjacent equivalent site. This simple picture of movement in an ionic lattice is known as the hopping model. In Fig. 12.3.3 the movement of vacancy or interstitial is shown.

Ionic conductivity σ is defined in the same way as electronic conductivity:

$$\sigma = n\, q\, \mu$$

where n is the number of charge carriers per unit volume, q is their charge, and μ is their mobility, which is a measure of the drift velocity in a constant electric field. This equation is a general equation defining conductivity in all conducting materials. To understand why some ionic solids conduct better than

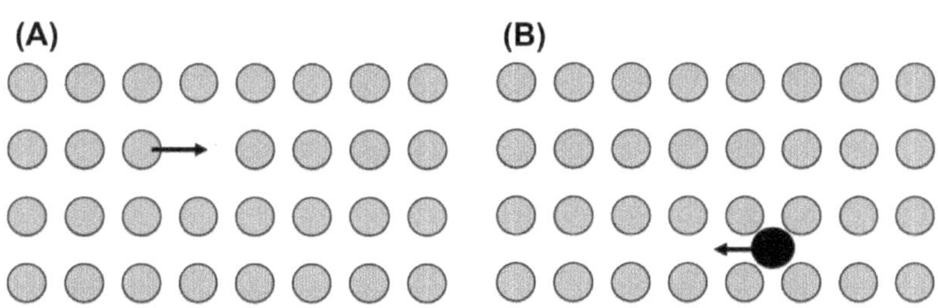

FIGURE 12.3.3 Ion motion via point defects: (A) mobile vacancy; (B) mobile interstitial.

others, it is useful to look at the definition more closely in terms of the hopping model. In the case of crystals in which the ionic conductivity is carried by vacancy or an interstitial mechanism, the concentration of charge carrier n will be closely related to the concentration of defects in the crystal, and μ will thus refer to the mobility of these defects in such cases. Fast ion transport in crystalline solids appears to be limited to compounds in which either group IA or IB cations or group VI-A or VII-A anions are mobile, with cation conductors being far more numerous. Typical examples of compounds in each of these categories include α-AgI, Naβ-Al_2O_3, cubic stabilized ZrO_2, and β-PbF_2 respectively.

SOLID-STATE BATTERIES

A solid-state battery is an energy converter transforming chemical energy into electrical energy by means of internal electron exchange. The electron transfer is mediated by mobile ions released from an ion source, the anode, and neutralized in the electron exchanger, the cathode. The positive ion is transmitted through a dielectric, which is a good electronic insulator, the separator. The ideal solid-state battery should be based on one unique material in which three regions, corresponding to the ion source, the separator, and the electron exchanger, are separated only by internal homo-junctions. The conventional structure of the battery currently available is shown in Fig. 12.3.4.

Materials constituting the electrochemical cell are the ion source (anode), the separator, and the electron exchanger (cathode). The anode emits positive ions into the separator and supplies the external circuit with electrons obtained from the oxidation process. The ion-conducting separator is permeable only to the positive ions. The electron exchanger allowing the reduction process accepts electrons from the external circuit and positive ions through intercalation. The discharge of the battery occurs when the battery is connected to an external load with the metal ion source as negative and the intercalation compound as positive. An electrochemical cell is then formed and the spontaneous oxidation–reduction reaction is a source of electrical energy.

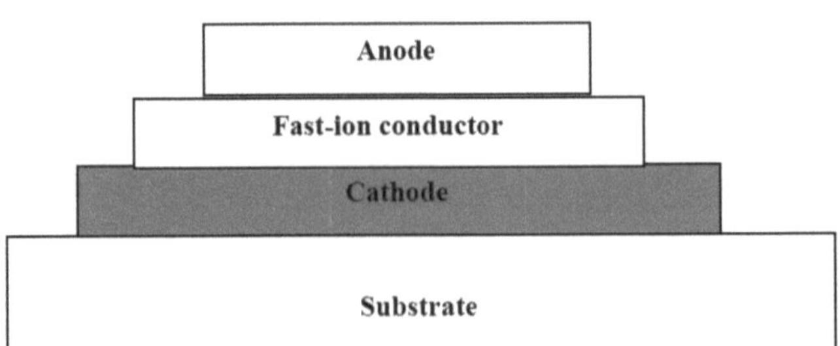

FIGURE 12.3.4 Schematic representation of construction of solid-state micro battery.

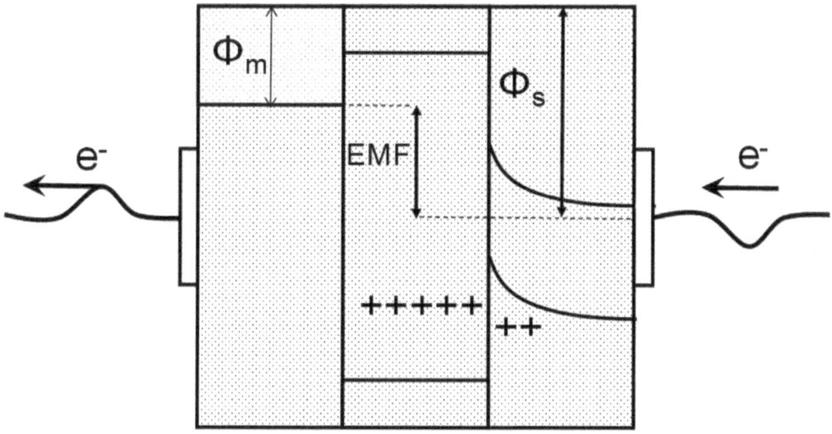

FIGURE 12.3.5 Schematic representation of the energy band diagram of a solid-state battery.

When the cell is connected to an external load, electrons are extracted from the metal and flow into the external circuit. Positive ions are injected into the separator and diffuse toward the insertion material cathode. Once transferred into the cathode, the positive ions are distributed near the surface to form a space charge layer. The quasi-Fermi level now depends on the distribution of charges in each material. A very thin layer of negative charge is formed at the metal-insulator surface to compensate for the positive charges distributed throughout the insulator. A space charge layer is formed in the semiconductor interface to account for the ion injection into the intercalation compound. An energy band diagram for a solid-state battery is represented in Fig. 12.3.5. It is convenient to classify solid-state batteries into four classes: high temperature, polymeric, lithium, and silver. A summary of the chronology of solid-state electrolytes and ambient temperature solid-state batteries investigated in 1950–1990 is given in Table 12.3.2.

HIGH-TEMPERATURE CELLS

The alkali metals lithium and sodium are attractive as battery anodes because of their high electrode potentials and low atomic masses, which together result in excellent values for battery-specific energy. Batteries that consist of solids (fast ion conductors) or fused salts as electrolytes and which operate at temperatures of 200–500°C are considered.

SILVER ION BATTERIES

AgI exhibits unusually high ionic conductivity at elevated temperatures that decrease ~20% upon melting. Silver iodide is known to go through a phase transition at 146°C to the high conducting phase, accompanied by an increase

TABLE 12.3.2 Chronology of Solid Electrolyte Batteries (1950–1990)

Date	Electrolyte	Log (S/cm)	Typical Cell System
1950–60	AgI	–5	Ag/V_2O_5
1960–65	Ag_2Si	–2	Ag/I_2
1965–70	$RbAg_4I_5$	–0.5	Li/Me_4Ni_5
1970–75	LiI	7	Li/I_2(P2VP)
1970–75	LiI (Al_2O_3)	5	Li/PBI_2
1970–75	β-Alumina	1.5	$Na\text{-}Hg/I_2PC$
1980–85	$LiI_aO_bS_cP_d$	–3	$Li/TiS_{2=}$
1978–85	LiX–polyethylene oxide	–7	Li/V_2O_5
1983–87	Poly{bis-[(methoxyethoxy) ethoxy] phosphazene}	–4	Li/TiS_2
1985–90	90 Plasticized SPE	–3	Li/V_6O_3

in conductivity of three orders of magnitude. Attempts to stabilize the high temperature α-AgI phase at room temperature by substituting foreign ions or complexes for either silver or iodine have been successful. These modified AgI conductors are classified into the following categories: (1) anion substituted, eg, S^{2-}, $PO_4{}^{3-}$, $P_2O_7{}^{4-}$, $SO_4{}^{2-}$; (b) cation substituted, eg, K^+, Rb^+, or $NH_4{}^+$ ions to produce the MAg_4I_5 class of compounds; and (c) mixed ion substituted, eg, the ternary system $AgI\text{-}HgI_2\text{-}Ag_2S$. Other Ag-conducting compounds based on the silver chalcogenides (Ag_2X, X = S, Se, and Te) were developed in a similar manner. The first commercial solid-state battery was manufactured at the end of the 1960s in the United States by Gould Ionics. This was a silver–iodine battery using $RbAg_4I_5$. Silver halides and rubidium silver iodide provide high Ag^+ ion conductivity. $RbAg_4I_5$ exhibits a conductivity of 27 Ω^{-1} cm^{-1} at 25°C, which is the highest value for all solid electrolytes at room temperature. A schematic diagram of the cell providing power to an external circuit is shown in Fig. 12.3.6.

SOLID-STATE PRIMARY LITHIUM BATTERIES

A major shortcoming of silver-based solid electrolytes, which limits galvanic cell voltages, is their low decomposition potentials. An electrolyte with room-temperature conductivity approaching that of silver compounds and possessing a high decomposition potential would open up a wide range of applications.

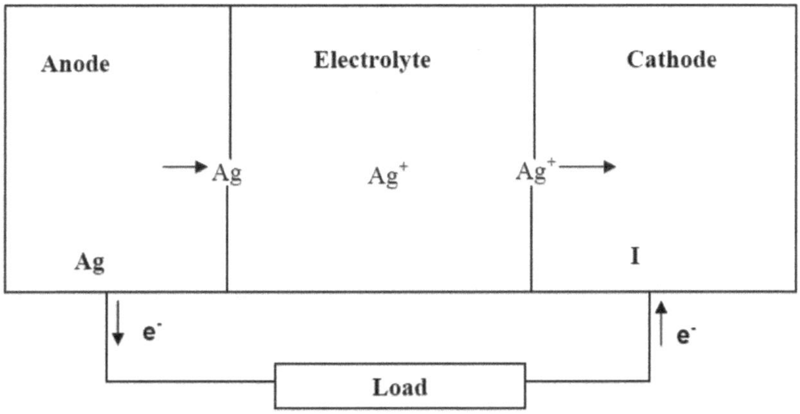

FIGURE 12.3.6 Schematic diagram of a silver ion, solid-state battery.

Many compounds have been studied with that goal in mind. One such material is lithium iodide. With its low density (0.53 g/cm^3), low electronegativity, and high electron–atom mass ratio, lithium has become the preferred choice for the active element of the anode, which upon discharge functions as an electron donor according to anode: $x\text{Li} \rightarrow x\text{Li}^+ + xe^{-1}$ discharge, in which Li enters the electrolyte and the electron exits the anode to the external circuit to power the load. The elemental lithium is typically present in a host insertion material, most commonly a lithiated carbon such as Li_xC_6. Fig. 12.3.7 shows a schematic representation of a lithium battery in discharge mode. The lithium–iodine battery has been used to power millions of cardiac pacemakers since its introduction in 1972. Lithium–iodine has established a record of reliability and performance unsurpassed by any other electrochemical power source. This battery has a solid anode of lithium and a polyphase cathode of poly-2-vinylpyridine,

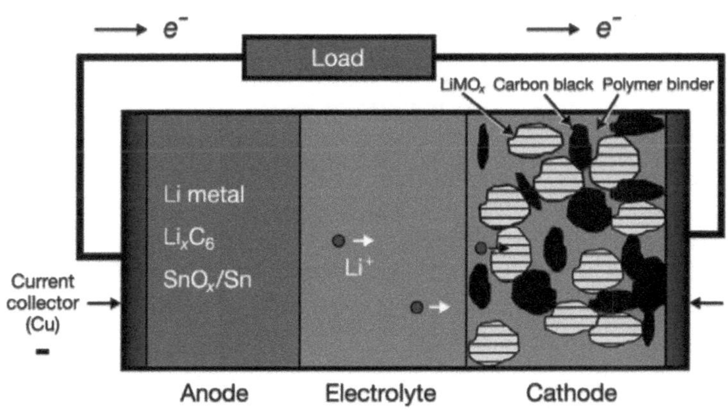

FIGURE 12.3.7 Schematic representation of a rechargeable lithium battery in discharge mode.

which is largely iodine (at 90% by weight). The solid electrolyte is a thin LiI film. The cell has an open circuit voltage of 2.8 V and the energy density is 100–200 Wh/kg. These batteries have extended lifetime of 10 years for 150- to 250-mAh capacities.

The main problem areas in primary solid-state batteries have been identified as: (1) volume changes, (2) electrolyte impedance, (3) discharge product impedance, (4) materials compatibility, and (5) manufacturability. Solid-state primary batteries can provide generally very long life at low currents. Another example of such batteries is the lithium–glass batteries, whose envisaged applications are mainly as power sources in electronic computers for complementary metal oxide semiconductor (CMOS) memory backup.

SODIUM BATTERIES

Sodium is most attractive as a negative electrode reactant because of its high electrochemical reduction potential of 2.271 V. Coupled with an appropriate electropositive material, it is capable of giving a cell of voltage greater than 2 V. Moreover, sodium is abundant in nature, cheap, and nontoxic. It is also of low atomic mass (23.0) and the combination of high voltage and low mass leads to the possibility of a battery of high specific energy. The realization of a practical battery based on sodium depended on identifying a suitable nonaqueous electrolyte. The sodium sulfur battery is the best developed solid electrolyte battery. It is composed of a molten sodium negative electrode and a molten sulfur positive electrode separated by a sodium ion conducting solid. Sodium β- and β''-alumina are nonstoichiometric aluminates that typically are synthesized from NaO and alumina. Sodium β-alumina is highly conductive toward Na^+ ions at 300°C although it is a good electronic insulator. This results in the possibility of a solid ceramic electrolyte. The cell discharges in two steps as Na^+ ions pass through the β-alumina to the sulfur electrode:

$$\text{Step 1: } 2Na + 5S \rightarrow Na_2S_5; \quad E^0 = 2.076 \text{ V}$$

$$\text{Step 2: } 2xNa + (5-x)Na_2S_5 \rightarrow 5Na_2S_{5-x} \ (0 < x < 2);$$
$$E^0 = 2.076 \rightarrow 1.76 \text{ V}$$

In the first step, sodium polysulfide (Na_2S_5) is formed as a liquid immiscible with liquid sulfur. The open circuit voltage is therefore invariant at 2.076 V. Once all of the sulfur has been converted to Na_2S_5, an ionic conducting liquid, further discharge to Na_2S_3 takes place in a single phase and therefore the voltage declines linearly to 1.78 V at the composition Na_2S_3. These reactions are fully reversible upon recharge. The sodium aluminum oxide phase diagram is complex and a great deal of work has gone into optimizing sodium β-alumina ceramic for this application.

The sodium sulfur cell, shown schematically in Fig. 12.3.8, can store several times more energy per weight and volume than a conventional lead acid battery can. Typical characteristics at 300°C are an open circuit voltage of

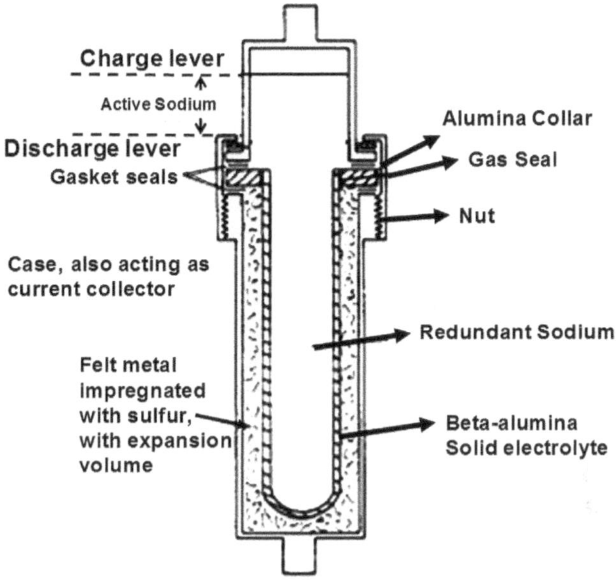

FIGURE 12.3.8 Schematic of sodium–sulfur cell.

2.18 V, an energy density of 200 Wh/kg, and a capacity of 15 Ah. When cells are assembled into a battery, this figure will be degraded and a value of 150 Wh/kg for a completed battery should be attainable. Despite their intensive development, sodium sulfur batteries are essentially dead because of the inability to maintain a consistent quality. It is possible to replace β-alumina by an Na-conducting borate glass, and other materials such as Nasicon of composition Na_{1-x}, $Si_xZr_2P_{3-x}O_{12}$ (with $1.8 < x < 2.4$) have been crystal engineered to maximize conductivity and ease of fabrication.

SOLID-STATE SECONDARY LITHIUM BATTERIES

Lithium–Iron Sulfide Batteries

One of the few examples of a commercial solid-state battery is the lithium heart pacemaker power source, but many systems of potential applicability have been proposed over the past 20 years. Solid-state rechargeable batteries have low capacity, generally on the order of milliampere hours, and yet research and development efforts are engaged; the target is the CMOS memory backup. Various solid electrolytes and intercalation cathode materials are used. The general structure of the cell is a solid lithium anode, fast-ion conducting glass electrolyte, and layered intercalation compound cathode. An example is the cell $Li/LiBF_4$–PC/CFx with a cell voltage of 2.8 V and energy density of 320 Wh/kg.

The lithium–iron sulfide battery operates at 400–500°C using a fused halide eutectic electrolyte immobilized in the pores of a suitable separator.

The most commonly used electrolytes are the LiCI-KC1 binary eutectic and the LiF-LiCI-LiI ternary lithium halides. With Li-Al alloy anodes, two distinct voltage plateaus are observed at 1.6 and 1.35 V. Most development work has concentrated on the LiA1−FeS couple, and the Varta Battery Company has produced a series of 140-Ah cells with a specific energy of 100 Wh/kg at low discharge rates of 80 mA/cm^2, falling to 50 Wh/kg at high rates of 250 mA/cm^2. There are still a number of unresolved scientific questions about the chemistry of LiAl−FeS cells and the mechanisms of degradation and failure. In this system the separator is clearly a crucial component that must not only keep the electrode materials apart but must also allow good permeation of the electrolyte, and the most suitable materials are boron nitride and zirconia in the form of woven cloths. However, there are obviously expensive options.

Polymeric Batteries

Polymer electrolyte batteries have been under development for cells operable at elevated temperatures (60−140°C). An experimental battery uses a thin electrolyte film made from a polymer such as polyethylene oxide to separate the lithium electrode from an ion-insertion−type electrode such as V_6O_{13}, TiS_2,or V_3O_8.

In many developmental cells, the positive electrode (cathode) is a composite and consists of small particles of insertion compound bound together with polymer electrolyte and carbon, which improves its electronic conductivity (Fig. 12.3.9). The composite electrode (50−75 μm thick) is deposited on a thin copper or nickel current collector less than 25 μm and a film (25−50 μm) of $[(C_2H_4O)_9\ LiF_3CSO_3]_n$. Polymer electrolyte completes

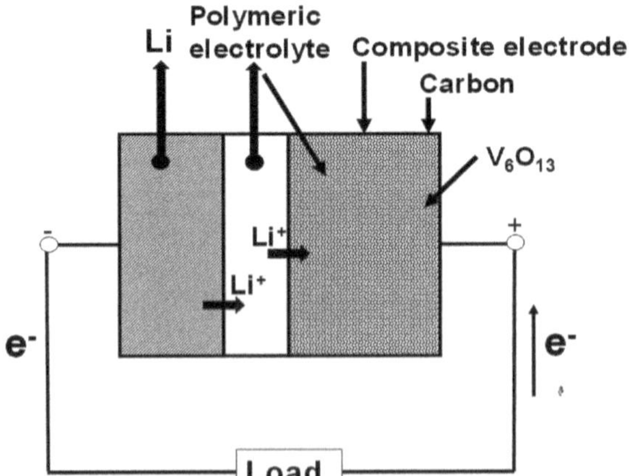

FIGURE 12.3.9 Schematic representation of a polymeric lithium cell using a composite electrode.

the lithium cell. Another possibility is to roll the cell assembly. A cell of this type using V_6O_{13} as the composite positive electrode would have a potential of about 2.8 V. During discharge, at currents of around 1 mA/cm^2, the voltage stabilizes around 2 V. The practical energy density is on the order of 200 Wh/kg, with the power density reaching 0.1 Wg^{-1}. The main application envisaged is storage batteries for all electric vehicles. Immediate advantages expected from polymer technology in electrochemistry are good mechanical properties, easy processing, and lightweight materials. The polymer salt phases transform to resistive phases at lower temperatures, and consequently little development has been reported at normal ambient temperatures. Cells have been reported, operated at 26°C with MoO_2 cathodes from which it is projected that energy densities in the range 0.1–0.2 Wh/cm^3 may be achieved.

Lithium Halogen Batteries

The theoretical specific energy for the $Li/LiI/I_2$ (P2VP) cell is 1.9 Wh/cm^3. The commercially available lithium–iodide batteries have a solid anode of lithium and a polyphase cathode (poly-2-vinyl pyridine) that is largely iodide (at 90% by weight). The discharge reaction is

$$2Li + P2VP\ nI_2 = P2VP(n-1)\ I_2 + 2LiI$$

This cell has an open circuit voltage of 2.8 V. The electrolyte ionic conductivity is 6.5×10^{-7} S/cm at 25°C, and the energy density is 100–200 Wh/kg.

Batteries of medium capacities (up to around 1 Ah) can be used for random access memory power supplies in electronics. Similar batteries using Li-Br have also been built. The greater electronegative bromine results in voltages on the order of 3.5 V and energy densities are as high as 1.25 Wh/cm^3. However, their practical application is held back by the limited conductivity of the Li-Br films formed.

Thin Film Batteries Using Copper Ion Conductors

Although copper ion–conducting materials are well known, applications in thin-film micro batteries are rare. One example is a battery with bimetal electrodes sandwiching a copper conductor. The latter was prepared by an electrolytic deposition method giving a CuI–CuCl mixture on a copper anode surface. Many other metals could be used on the other side. In all cases a low voltage was obtained, between 0.26 V (chromium) and 1.3 V (magnesium).

LEAD-CUPRIC FLUORIDE THIN-LAYER BATTERIES

This is a system that is interesting mainly because of its concentration and its method of fabrication. The cell is based on the fluoride ion solid conductor, β-PbF_2. The ionic conductance of this material is low at room temperature

($\sim 5 \times 1{-}5$ S/m). In this cell lead metal used is the negative electrode, PbF_2 is the electrolyte, and CuF_2 is the positive electrode. Because of the high resistivity of CuF_2, it was co-deposited with PbF_2 to form a more conductive mixed layer. The reactions lead to a voltage of 0.7 V. A current density of up to 10 mA/cm^2 could be obtained. This cell was not rechargeable. Other works are reported on micro batteries using PbFe (or derived structures) or other F conductors as the electrolyte:

$$\text{Anode: } Pb + 2F^- \rightarrow PbF_2 + 2e^-$$

$$\text{Cathode: } CuF_2 + 2e^- \rightarrow Cu + 2F^-$$

Lead was always the negative electrode and BiF_3 was often the positive electrode. In these cases the system was rechargeable. Current densities of 40 μA/cm^2 at 0.2 V were obtained. More recently the use of an evaporated mixture of PbF_2, and LiF as the electrolyte was mentioned.

So far, the diverse research activities that led to the development of cathode and anode active materials, separator, electrolyte, current collector (metal foil; cathode is aluminum foil and anode Cu foil), material optimization, and possible materials for use in a lithium ion battery have been described.

MANUFACTURING PROCESS

Manufacturing processes consist of (1) mixing the cathode or anode materials with binder and conductive additive, (2) painting on the current collector (metal foil), (3) drying, and (4) pressing. The next step involves (1) assembly

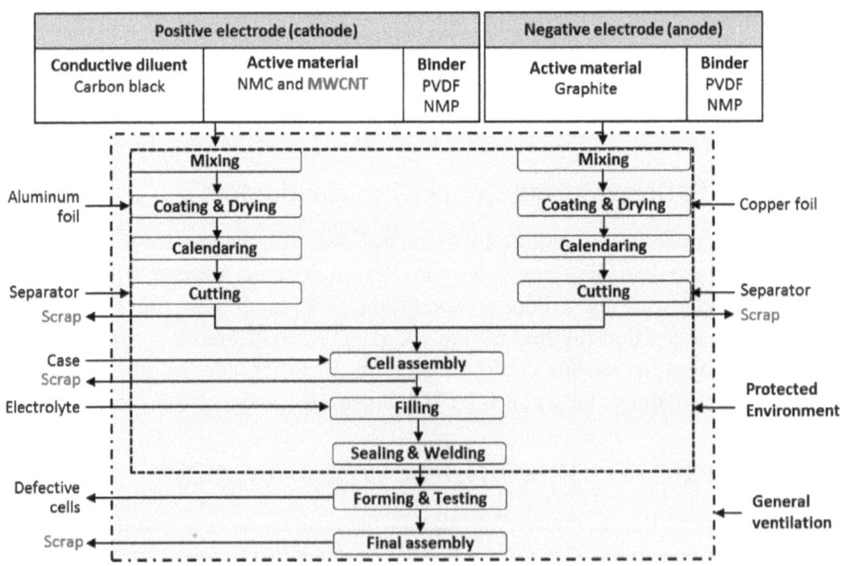

FIGURE 12.3.10 Schematic manufacture processing of lithium polymer battery.

of the cathode and anode and simultaneous rolling-up with the separator, (2) electrode insertion, (3) electrolyte injection into the battery case, and (4) sealing.

In the electrode manufacturing process, a mixture of active materials with conductive additive such as acetylene black or Ketjenblack, and a binder such as Teflon or polyvinylidene fluoride dissolved in *n*-methyl-2-pyrrolidinone are made in the form of a paste. The paste is painted over both sides of metal foil, dried, and roll-pressed. Then it is cut to the desired width. In the case of polymer lithium batteries, after carrying out direct polymerization of the gel electrolyte sheet, roll-press is carried out over the cathode and anode sheets, which are then cut a suitable size, wrapped in aluminum lamination film, and heat welded at the edges (Fig. 12.3.10).

Finally, the battery is checked for short-circuit over 2–4 weeks. A protective thin film (SEI) will form at the anode–carbon interface during this period. Shipment inspection has to be conducted and the products are shipped. Many battery manufacturing companies worldwide have announced their involvement in the large-scale production of lithium secondary batteries. However, the level of information released is limited, and thus it is impossible to evaluate the status of the different batteries at this stage.

Chapter 13

Supercapacitors

INTRODUCTION

In many modern application areas some of the most effective and practical appliances for electrochemical energy conversion and storage are batteries, fuel cells, and electrochemical supercapacitors (ES). Supercapacitors have attracted attention mainly because of their high power density and long lifecycle, and the possibility of bridging the gap power–energy gap. Supercapacitors possess higher energy density compared with common capacitors. Supercapacitors are electrical storage devices that can deliver a high amount of energy in a short time. Hybrid electric and fuel cell–powered vehicles need such a surge of energy to start, more than can be provided by regular batteries. Supercapacitors are also needed in a wide range of electronic and engineering applications, whenever a large, rapid pulse of energy is required. Capacitors that store energy within the electrochemical double layer at the electrode–electrolyte interface are known under various trademarks or established colloquial names such as "double-layer capacitors," "supercapacitors," "ultracapacitors," "power capacitors," "gold capacitors," and "power cache." "Electrochemical double-layer capacitor" describes the fundamental charge storage principle of such capacitors. However, because there are generally additional contributions to the capacitance other than double-layer effects, these capacitors are termed electrochemical capacitors (EC).

EC have been known since the first patent filed in 1957, when a capacitor based on high–surface area carbon was described by Becker. Later, in 1969, the first attempts to market such devices were undertaken by SOHIO. In the 1990s electrochemical capacitors became famous within the context of hybrid electric vehicles. A Department of Energy ultracapacitor development program was initiated in 1989, and short-term as well as long-term goals were defined for 1998–2003 and after 2003, respectively. The EC was supposed to boost the battery or the fuel cell in the hybrid electric vehicle to provide the necessary power for acceleration and to allow for the recuperation of brake energy. Today several companies such as Maxwell Technologies, Siemens Matsushita (now EPCOS), NEC, Panasonic, ELNA, and TOKIN invest in electrochemical capacitor development. The applications envisaged principally boost components supporting batteries or replacing batteries primarily in

Energy Sources. http://dx.doi.org/10.1016/B978-0-444-56353-8.00013-7

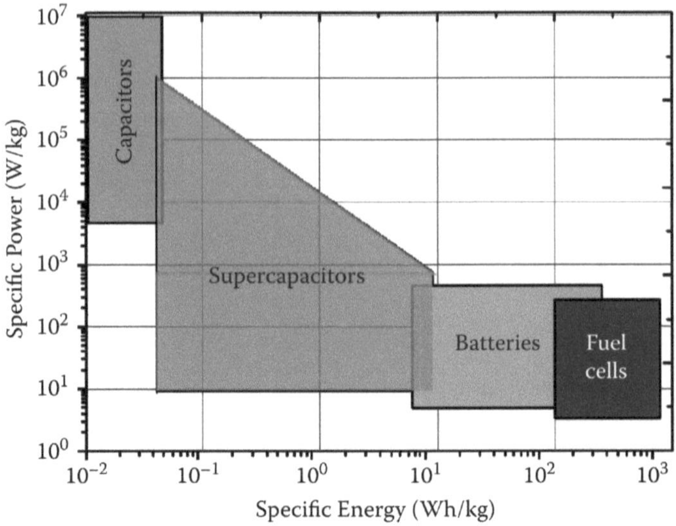

FIGURE 13.1 Sketch of Ragone plot for various energy conversion and storage devices. The indicated areas are only general guidelines.

electric vehicles. In addition, alternative applications of EC not competing with batteries but with conventional capacitors are coming up and show considerable market potential. The reason why electrochemical capacitors were able to raise considerable attention is visualized in Fig. 13.1. In this figure typical energy storage and conversion devices are presented in the so-called "Ragone plot" in terms of their specific energy and power. Electrochemical capacitors fill in the gap between batteries and conventional capacitors, such as electrolytic capacitors or metallized film capacitors. In terms of specific energy as well as specific power, this gap covers several orders of magnitude.

An ES is a charge storage device similar to batteries in its design and manufacture. An ES consists of two electrodes and a separator that electrically isolates the two electrodes. It can be seen from Fig. 13.2 that charges can be stored and separated at the interface between the conductive solid particles and the electrolyte. The interface can be treated as a capacitor with electrical double-layer capacitance, which is given by:

$$C = [A\varepsilon/4\pi d]$$

where A is the area of the electrode surface, ε is the dielectric constant of the medium (electrolyte), and d is the thickness of the electrical double layer.

Two types of ES exist. One is electrostatic supercapacitors (EDLS), in which the electrode material is not electrochemically active. There is no electrochemical reaction on the electrode material. During charging and discharge and pure physical charge, accumulation occurs at the electrode–electrolyte

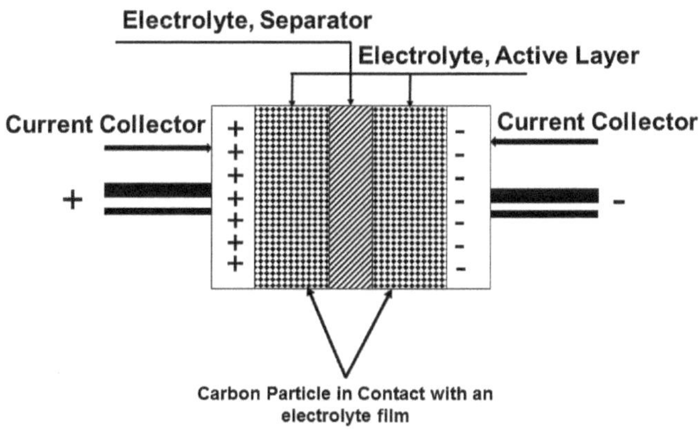

FIGURE 13.2 Principles of a single cell double-layer capacitor.

interface. The other type is the faradaic supercapacitors (FS) in which the electrode material is electrochemically active, which can directly store charge during the charging and discharging processes.

The charging and discharging cycles in the case of EDLS can be represented as follows: E_{s1} and E_{s2} are the designations of the two electrodes and the electrode–electrolyte interface is designated as: //. On one electrode (say, a positive one):

$$E_{s1} + A^- \rightarrow \text{charging } E_{s1}{}^{+}//A^- + e^-$$

$$E_{s1}{}^{+}//A^- + e^- \rightarrow (\text{discharging}) \; E_{s1} + A^-$$

On the other electrode (say, a negative one):

$$E_{s2} + C^+ + e^- \rightarrow (\text{charging}) \; E_{s2}{}^{-}//C^+$$

$$E_{s2}{}^{-}//C^+ \rightarrow (\text{discharging}) \; E_{s2} + C^+ + e^-$$

The overall charging and discharging process in EDLS can be expressed as:

$$E_{s1} + E_{s2} + A^- + C^+ \rightarrow E_{s1}{}^{+}//A^- + E_{s2}{}^{-}//C^+$$

$$E_{s1}{}^{+}//A^- + E_{s2}{}^{-}//C^+ \rightarrow E_{s1} + E_{s2} + A^- + C^+$$

FARADAIC SUPERCAPACITORS

FS or pseudocapacitors are different from electrostatic or EDLS capacitors. When a potential is applied to an FS, fast and reversible Faradaic reactions (redox reactions) take place on the electrode materials and involve the passage of charge across the double layer, similar to the charging and discharging processes that occur in batteries, resulting in Faradaic current passing through the supercapacitor cell. Materials undergoing such redox reactions include conducting polymers (CP) and several metal oxides, including RuO_2, MnO_2, and Co_3O_4. Three types of faradaic processes occur at FS electrodes: reversible adsorption (for example, adsorption of hydrogen on the surface of platinum or gold), redox reactions of transition metal oxides (eg, RuO_2), and reversible electrochemical doping−dedoping in conductive polymer based electrodes. In these, the electrochemical processes occur both on the surface and in the bulk near the surface of the solid electrode. An FS exhibits larger capacitance values and energy density than an EDLS. FS usually suffers from relatively lower power density than does an EDLS because faradaic processes are normally slower than nonfaradaic processes. Moreover, because redox reactions occur at the electrode, an FS often lacks stability during cycling, similar to batteries. In addition, hybrid ES with an asymmetrical electrode configuration (eg, one electrode consists of electrostatic carbon material and the other consists of faradaic capacitance material) have been extensively studied to capitalize on both electrode materials' advantages in improving overall cell voltage, energy, and power densities.

Compared with conventional capacitors in which capacitance is on the order of picofarads and microfarads, the capacitance and energy density stored in the supercapacitor by the electrochemical double layer is higher. To achieve a higher capacitance, the surface area of the electrode is additionally enlarged by using porous electrodes.

There are several techniques to determine the specific capacitance, such as a unit cell test (two-electrode system), a half-cell test (three-electrode system), and an impedance test. The unit cell and half-cell tests are mainly used to determine the specific capacitance of the supercapacitor. The specific capacitances reported in the literature are not consistent, mainly owing to the experimental methods used to determine them.

The important component in a supercapacitor, in addition to the two electrodes, is the electrolyte insider the separator as well as inside the active material layers. These electrolytes should provide a wide voltage window, electrochemical stability, high ionic concentration, and low solvated ionic radius, low resistivity, low viscosity low volatility, low toxicity, low cost, and high purity. The electrolyte used in supercapacitors can be classified into three types: aqueous electrolyte, organic electrolyte, and ionic liquids.

Aqueous electrolytes such as H_2SO_4, KOH, Na_2SO_4, and NH_4Cl can provide higher ionic concentration and lower resistance. These electrolytes are limited in terms of improving both energy and power densities because of their narrow voltage window.

Compared with aqueous electrolytes, organic electrolytes have a wider voltage window, up to 3.5 V. Commonly used solvents are acetonitrile and propylene carbonate. One limitation of these electrolytes water content is that they must be kept below 3−5 parts per million.

Ionic liquid electrolytes have low vapor pressure, high thermal and chemical stability, low flammability, and a wide electrochemical stability window (around 4.5 V) and conductivity (10 mS/cm).

DIFFERENCES BETWEEN A SUPERCAPACITOR AND A BATTERY

The most important difference between a supercapacitor and a battery is the principle of electrochemical energy storage. Electrochemical energy can be stored in two fundamentally different ways. In a battery, the potentially available chemical energy storage requires faradaic oxidation and the reduction of electrochemically active reagents to release charges that can perform electric work when they flow between two electrodes with different potentials; that is, the charge storage is achieved by electron transfer that produces a redox reaction in the electroactive material according to Faraday's law.

With an electric double-layer capacitor (EDLC), the charge storage process is nonfaradaic; that is, ideally, no electron transfer takes place across the electrode interface and the storage of electric charge and energy is electrostatic. Actual electron charges are accumulated on the electrode surface with lateral repulsion and the involvement of redox chemical changes. Table 13.1 summarizes the perceived advantages and disadvantages of such ES energy storage. Because the charging and discharging of such EDLCs involve no

TABLE 13.1 Advantages and Disadvantages of Supercapacitor Energy Storage Devices

Advantages	Disadvantages
• High power density (1–10 kW/kg compared with lithium ion batteries (150 W/kg) • Can be charged and discharged rapidly in seconds whereas batteries require hours • Long life expectancy (>100,000 cycles) up to 30 years whereas batteries have only 5–10 years • Long shelf life • High efficiency • Simple principle and mode of construction • Combine state of charge indication • Can be combined with secondary battery for hybrid applications (electric vehicles) • Wide range of operating temperatures • Environmental friendliness • Safe	• Limited energy density • Poor volume energy density • Low working voltage • Requires stacking for high potential operation (electric vehicles) • High cost • High discharge rates • Industrial standards for commercialization

chemical phase and composition changes, such capacitors have a high degree of cyclability, on the order of 10^6 times, and a high specific power density, although the specific energy density is small. However in some cases of a supercapacitor based on pseudocapacitance (a redox type of supercapacitor), the essential process is faradaic; that is, the charge storage is achieved by an electron transfer that produces a redox reaction (faradaic reaction) in the electroactive materials according to Faraday's law. Supercapacitors based on pseudocapacitance have higher specific capacitance do EDLCs, owing to the redox reaction as in a battery, although the redox reaction gives rise to high internal resistance in supercapacitors, resulting in a decrease in specific power density. The typical electrodes of supercapacitors based on pseudocapacitance are metal oxides (ie, RuO_2, IrO_2, Co_3O_4) and CP (ie, PPy, polyaniline (PANI), polythiophene). A supercapacitor requires two equivalent electrodes, one of which is charged negatively with respect to the other, with the charge storage and separation being electrostatic. At each electrode, the charge storage and separation are established across the electrode interface. Usually, the electrodes of supercapacitors have a high surface area and porous matrices. However, batteries have a bipolar electrode configuration for high voltage series combinations. For a battery, the maximum Gibbs energy is the product of charge Q and the difference of potential, ΔE, between the Nernstian reversible potentials of the two electrodes: that is, $G = Q \cdot \Delta E$. In the capacitor case, for a given charge Q, G is $1/2 \cdot QV$. For a given electrode potential difference, $\Delta E = V$, it is evident that the energy stored by a two-electrode cell accommodating a given faradaic charge Q at voltage $\Delta E = V$ is twice that stored in a capacitor charged with the same Q at the same voltage. In the process of charging a pure EDLC, every additional element of charge has to do electrical work (Gibbs energy) against the charge density already accumulated on the electrodes, progressively increasing the interelectrode potential difference.

In a battery cell being charged, a thermodynamic potential (ideally) exists independently of the extent of charge Q added, as long as two components (reduced and oxidized forms) of the electroactive material remain existing together. Thus the potential difference (electromotive force) of the battery cell is ideally constant throughout the discharge or recharge half-cycles, so that $G = Q$, ΔE rather than Q, $1/2 \cdot \Delta E$ (or 1/2 V). This difference can be illustrated by the discharge curves shown schematically in Fig. 13.3, in which the voltage in the capacitor declines linearly with the extent of charge, for an ideal battery it remains constant as long as two phases remain in equilibrium. The decline in the supercapacitor voltage arises formally because $C = Q/V$ or $V = Q/C$; therefore $dV/dQ = 1/C$. The ideal battery cell voltages on discharge and recharge, as a function of state of charge, are shown as parallel lines of zero slope in Fig. 13.3. In the slope of the discharge and recharge lines for the supercapacitor in Fig. 13.3, there is a significant internal resistance drop, depending on the discharging and recharging rates.

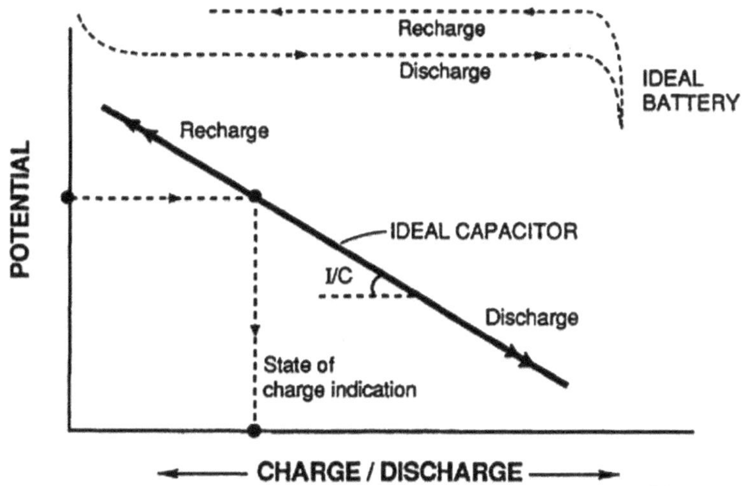

FIGURE 13.3 Difference in discharge and recharge relationships for a supercapacitor and a battery.

ELECTRODE MATERIALS FOR SUPERCAPACITORS

The capacitance of ES depends heavily on the specific surface area of the electrode materials. The electrode materials for ES can be classified into three types: carbon materials with a high specific surface area; CP; and metal oxides such as RuO_2, IrO_2, MnO_2, NiO, Co_2O_3, SnO_2, V_2O_5, and MoO.

Carbon

Carbon materials for electrochemical energy devices such as secondary batteries, fuel cells, and supercapacitors have been studied extensively. However, each type of electrochemical energy device requires different physical properties and morphology. For supercapacitor applications, the carbon material for the EDLC type must have (1) a high specific surface area, (2) good intra- and interparticle conductivity in porous matrices, (3) good electrolyte accessibility to intrapore surface area, and (4) available electrode production technologies. Carbons for supercapacitors are available with a specific surface area of up to 2500 m^2/g as powders, woven cloths, felts, or fibers. Surface conditioning of these carbon materials for supercapacitor fabrication is of substantial importance.

Activated Carbon

Carbons with a high specific surface area have many oxygen functional groups, such as ketone, phenolic, carbonyl, carboxylic, hydroquinoid, and lactone, introduced during the activation procedure for enlarging the surface area. These oxygen functional groups on activated carbons or activated carbon fibers result in one kind of electrochemical reactivity: oxidation or reduction. Oxidation or reduction of the functional groups shows pseudocapacitance, which amounts to about 5–10% of the total realizable capacitance. However, the various surface functionalities in activated carbon are one factor that increases the internal resistance (equivalent series resistance) because of the redox reaction. Activated carbon is cheaper than metal oxide and conducting polymer and it has a larger specific surface than the others. Activated carbon–based supercapacitors have been commercialized for small memory backup devices. However, activated carbon has lower conductivity than does metal oxide and conducting polymer, resulting in a large surface area that gives smaller power density.

In addition, the observed specific capacitances of carbon-based supercapacitors are about one-fourth the theoretical capacitance despite their high specific surface area, which is attributed to the existence of micropores. This is a weak point of active carbons as electrode materials in supercapacitors with a high energy density and power density. Activated carbons are famous for their surface areas of 1000–3000 m^2/g.

CARBON NANOTUBE–BASED SUPERCAPACITORS

Over the past decade, the application of activated carbons as electrode materials in supercapacitors has been investigated intensively because of their high specific surface area and relatively low cost. Because the specific capacitance of a supercapacitor is proportional to the specific surface area, the effective surface area of the electrode materials is important. Theoretically, the higher the specific surface area is of an activated carbon, the higher the specific capacitance should be. Unfortunately, the theoretical capacitance of the activated carbons is not in good agreement with the observed value because a significant part of the surface area remains in the micropores (<20 Å), which are not accessible to the electrolyte ions. Therefore, the pore size distribution, together with the surface area, is important for the determination of the EDLC. From this point of view, carbon nanotubes (CNTs) have several advantages as electrode materials for supercapacitors. CNTs have a narrow distribution of mesopore size, a highly accessible surface area, low resistivity, and high stability, attracting great interest worldwide for building supercapacitors.

CARBON NANOTUBE ELECTRODES

High-power supercapacitors based on CNT electrodes using aqueous electrolytes have been reported. Supercapacitor electrodes prepared from catalytically grown multiwalled CNTs whose surface area is 430 m^2/g have a maximum specific capacitance of 113 F/g and a power density of 8 kW/kg at an energy density of 0.56 Wh/kg in a solution of 38 wt% H_2SO_4 used as the electrolyte. CNT-based electrodes showed specific capacitances of 15 ± 25 F/g in a solution of 38 wt% H_2SO_4. The electrochemical characteristics of supercapacitors built from multiwalled nanotube (MWNT) electrodes with a specific surface area of 430 m^2/g, in a 1-M KOH aqueous solution, as well as the correlation of microtexture with the elemental composition of the materials, have also been investigated. The presence of mesopores caused by the central canal and or the entanglement of CNTs is the reason for the easy access of ions to the electrode–electrolyte interface for charging the electric double layer. The values of specific capacitance varied from 4 to 135 F/g, depending on the type of nanotube and posttreatment (acidic oxidation).

Supercapacitors using MWNT electrodes in organic electrolyte systems have also been investigated. The MWNT electrodes exhibited a specific surface area of 100 m^2/g and a measured specific capacitance of up to 18.2 F/g (16.6 F/cm^3) with 1 M $LiClO_4$ in a mixture of ethylene carbonate and propylene carbonate (1:1 volume ratio) as the organic electrolyte solution. The specific capacitance was lower than that reported by other groups using aqueous electrolytes, owing to the low specific surface area of the MWNT electrode and the organic electrolyte solution used. However, the energy density of the supercapacitor can reach 20 Wh/Kg at 10 mA discharge current density, depending on the organic electrolyte solution system. The relative volume of mesopores and macropores of the electrode used exceeds 92%, and that of micropores is nearly negligible. In the case of the organic electrolytes, because of their large molecular structures, only mesopores and macropores are accessible and are larger than those of activated carbons for supercapacitors.

Because the sizes of hydrated ions are in the range of 6 ± 7.6 Å the minimum effective pore size should be greater than 15 Å. It is known that in general, pore sizes in the range of 30 ± 50 Å are required to maximize capacitance in the EDLC. In macropores (>50 Å), the hydrated ions are usually loosely bound to the surface layer and do not contribute much to capacitance.

Fig. 13.4A shows the specific capacitances of the heat-treated electrodes at various temperatures as a function of the charging time. Capacitance increases abruptly and reaches about 80% of maximum during the initial 10 min, regardless of the heat treatment temperatures. The capacitance gradually increases further and saturates to maximum values at long charging times. A persistent increase in capacitance over a long time is generally observed from the porous electrodes and is attributed to the existence of various forms of pores and pore diameters in the electrode. The saturated capacitance increases with increasing heat treatment temperatures and saturates to 180 F/g at 1000°C. High-temperature annealing of CNT electrodes improves the quality of the sample not only by increasing the specific surface area but also by redistributing the CNT pore sizes to the smaller values near 30 ± 50 Å.

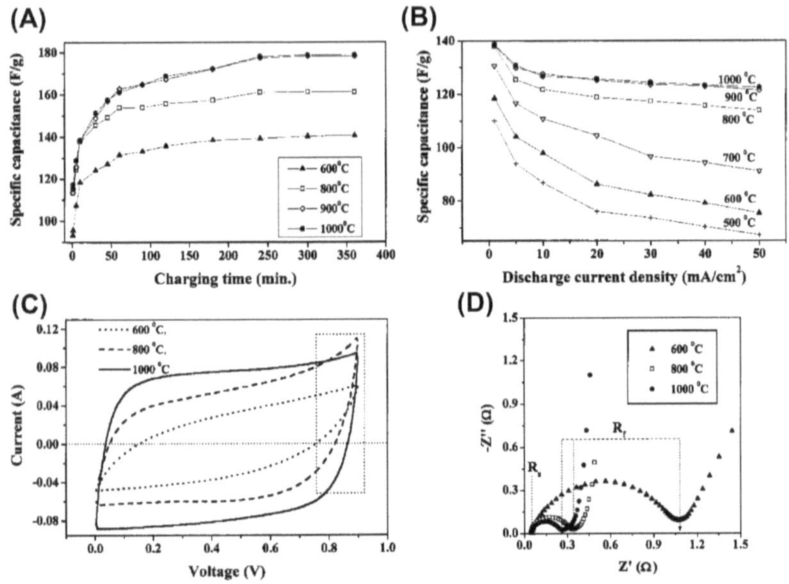

FIGURE 13.4 Electrochemical properties of supercapacitor using CNT electrodes: (A) Specific capacitances of heat-treated electrodes at various temperatures as a function of the charging time at a charging voltage of 0.9 V, in which the capacitance was measured at a discharging current of 1 mA/cm^2. (B) Specific capacitances of heat-treated electrodes at various temperatures as a function of the discharging current density at a charging voltage of 0.9 V for 10 min. (C) Cyclic voltammetric (CV) behaviors (sweep rate, 100 mV/s) for CNT electrodes at various heat treatment temperatures. (D) Complex-plane impedance plots for CNT electrodes for various heat treatment temperatures at ac-voltage amplitude of 5 mV. Z^2, imaginary impedance; $Z¢$, real impedance.

Fig. 13.4B shows the specific capacitance as a function of discharge current density at various heat -treatment temperatures, where in which the data were obtained from the samples charged at 0.9 V for 10 min. At low temperatures below 700 °C, the specific capacitance at a discharging current density of 50 mA/m^2 may drop by about 30 % of the capacitance at 1 mA/cm^2. However, at a high annealing temperature (1000°C), the capacitance drops only by only about 10 % even for a large discharging current density. The existence of the long flat region in the discharging current density is of practical importance for applications of supercapacitors to various realistic devices. Large capacitance drops at low annealing temperatures are caused by the internal resistance of the CNT electrode. Fig. 13.4C shows the cyclic voltammetric (CV) behavior with a sweep rate of 100 mV/s at various temperatures. The inner integrated area (current × voltage) is the power density, which increases with increasing heat -treatment temperatures. This power density will be larger if the equivalent series resistance (ESR), the slope of V/I (indicated by the dotted box in Fig. 13.4C), is smaller. The CV curve at 1000°C is close to the ideally rect-angular shape, indicating the smallest ESR in the CNT electrode. The magnitude of the ESR can be more clearly shown in the complex-plane impedance plots, as shown in Fig. 13.4D. The electrolyte resistance, Rs, is constant and varies with the electrolyte. The sum of the resistance of the electrode itself and the contact resistance between the electrode and the current collector is represented by R_f. The electrolyte resistance and the contact resistance are identical in all samples. Therefore, a decrease ofin the R_f indicates a decrease ofin the CNT -electrode resistance. The CNT electrode resistance decreases rapidly at high temperatures of 800 and 1000°C. The R_f is closely related to the power density, as evidenced by comparing two curves in Fig. 13.4C and D. The ideally polarizable capacitance will result ingive to a straight line along the imaginary axis (Z2). In real capacitors with a series resistance, this line has a finite slope, representing the diffusive resistivity of the electrolyte within the pore of the electrode. With increasing heat -treatment temperature, the diffusive line comes closer to an ideally straight line, as shown in Fig. 13.4D. The formation of abundant pore diameters of 30 ± 50 Å with increasing temperature may also enhance the diffusivity of the hydrated ions in the pore, which in turn reduces the CNT-electrode resistance and increases the capacitance.

CARBON NANOTUBE–COMPOSITE ELECTRODES

To increase the capacitance of nanotubes, it is possible to increase the electrode surface area or increase the pseudocapacitance effects obtained by adding special oxides or electrically conducting polymers (ECP) such as polypyrrole (PPy). ECPs have the advantage of lower costs compared with oxides. Another advantage is that the pseudocapacitance effects of ECPs are stable. Modification of carbon material by a specific additive providing quick pseudocapacitance redox reactions is another way to enhance capacitance. This is possible with metal oxides, but in this case ECP is added. ECP itself has a capacitance of about 90 F/g. The pseudocapacitance effects of ECP are relatively stable. If one can coat a nanotube with, for instance, PPy, it creates good electronic conducting properties and keeps the advantage of ionic conductivity in the opened mesoporous network of the nanotube. These are perfect conditions for a supercapacitor.

The results of capacitance measurements on the different combinations of ECPs composites working in their optimal potential range were also tested and are given in Table 13.2. Nanotubes with electrochemically deposited PPy gave higher values of capacitance than did untreated samples. The electrochemical behavior of PANI dictates its choice as a positive electrode because of the rapid loss of conductivity in the negative potential range. On the other hand, PPy as well as poly-(3,4-ethylenedioxythiophene) (PEDOT) could serve as both electrodes (+) and (−), taking into account a suitable voltage range. Higher performance is observed for a PANI/CNTs (+)//PPy/CNTs (−) capacitor that supplies 320 Fg^{-1}. An additional increase in the supercapacitor power and energy density through enhancement of the operating voltage can

TABLE 13.2 Combinations of Different Materials for Positive and Negative Electrodes of Supercapacitors

Positive (+)	Negative (−)	C (Fg^{-1})	U (V)
PANI	PPy	320	0.6
PANI	PEDOT	160	0.8
PANI	Carbon (PX21)	330	1.0
PPy	Carbon (PX21)	220	1.0
PEDOT	Carbon (PX21)	120	1.8

Electrolyte 1 mol/L sulfuric acid electrically conducting polymer—carbon nanotube composites 80; 20 wt%. *PANI*, polyaniline; *PEDOT*, poly-(3,4-ethylenedioxythiophene); *PPy*, polypyrrole.

easily be realized by applying activated carbon as a negative electrode. Instead of CNTs, acetylene black could be used as a carbon additive in such composites; however, nanotubes act as a more convenient backbone and allow better dispersion of the conducting polymer.

METAL OXIDE–HYDROXIDES

In general, metal oxide—based supercapacitors can provide higher energy density than can conventional carbon-based materials. They also have better stability than do polymer-based materials. They function as a storage medium like electrostatic carbon materials and also exhibit electrochemical faradaic reactions between electrode materials and ions within certain potential windows. General requirements for metal oxides to be considered for supercapacitor applications are that: (1) the material should be electronically conducting; (2) the metal should exist in variable valency with no phase changes involving irreversible modifications of a three-dimensional structure; and (3) the protons can freely intercalate into the oxide lattice on reduction (and out of the lattice), allowing facile interconversion of $O^{2-} \leftrightarrow OH^-$. Oxide systems that satisfy these conditions are ruthenium oxide, manganese oxide, cobalt oxide, nickel oxide, and vanadium oxide.

Ruthenium oxide is the most studied system owing to its wide potential window, highly reversible redox reactions, three distinct oxidation states accessible within 1.2-V voltage window, high proton conductivity, remarkably high specific conductance, good thermal stability, long cycle life, metallic type conductivity, and high rate capability. With an RuO_2 electrode, double-layer charging contributes to only about 10% of the accumulated charge; in parallel, pseudocapacitance mechanism can occur. Factors that have key roles in the electrochemical behavior of ruthenium oxides include: (1) a specific surface area, (2) combined water in RuO_2, (2) the crystallinity of $RuO_2 \times H_2O$, (4) the size of RuO_2, and (5) the electrolytes used. Despite its high theoretical capacitance (1360 Fg^{-1}) its cost is a practical impediment. To reduce the cost, mixed oxide composites or ruthenium-based composite materials are used in which particles of hydrous ruthenium oxide are deposited on other substrates such as carbon or conducting polymer. Factors that determine the capacitance of Ru-based oxide composites are: (1) the electrochemical kinetics of electron transfer for active materials, (2) within-particle electron-hopping resistance within every RuO_2 particulate, (3) between-particle electron hopping resistance between particulates, (4) electron hopping resistance at the interface between active materials and current collectors, and (5) barriers to proton diffusion within oxide composites and particulates.

CONDUCTING POLYMERS

CPs have many attributes that make them suitable candidates for supercapacitor applications, including their low cost, low environmental impact, high conductivity, high voltage window, high storage capacity, porosity/reversibility, and adjustable redox activity through chemical modification. Common electrical CPs in supercapacitor applications are PANI, PPy, polythiophene, and their corresponding derivatives. Unfortunately, swelling and shrinking of CPs may occur during the intercalating–deintercalating process, which is a disadvantage. To overcome these disadvantages, strategies such as improving CP material structures and morphologies and fabricating composite electrode materials have been attempted. In Table 13.3 data on conducting polymer-based composite materials are collected. It is seen from these data

TABLE 13.3 Typical Capacitance Values of Selected Conducting Polymer–Based Composites

Conducting Polymer–Based Composite	Specific Capacitance (Fg^{-1})	Electrolyte
PPy-20 wt% MWNT	320	1 M sulfuric acid
PANI-20 wt% MWNT	670	1 M sulfuric acid
PEDOT-PPy (5:1)	230	1 M $LiClO_4$
PPy-graphene	165	1 M NaOH
PANI-Si	409	0.5 M sulfuric acid
PPy-RuO_2	302	1 M sulfuric acid
PPy-MnO_2	602	0.5 M Na_2SO_4

MWNT, multiwalled nanotube; *PANI*, polyaniline; *PEDOT*, poly-(3,4-ethylenedioxythiophene); *PPy*, polypyrrole.
Data collected from G. Wang, L. Zhang, J. Zhang, Chem. Soc. Rev., 41 (2012) 797.

that the composite materials can give a wide distribution of capacitance values that depend on parameters such as the constituents of composites, the electrolytes used, and the scan rate, current load, and mass ratio of the components as well as the cell configuration. More research effort is required to optimize these parameters to achieve optimum capacitance values for CP-based composites for ES applications.

Chapter 14

Fuel Cells

INTRODUCTION

Over the past decade, fuel cells have received considerable attention across the world as potential electrical energy conversion devices. The major factor that has influenced the development of fuel cells in the past few years is worldwide concern regarding environmental consequences from the use of fossil fuels for generating electricity and driving vehicles. The dependence of industrialized nations on oil and the pressure on fossil fuels reserves are main causes of anxiety to find alternative energy conversion devices. More important, however, is increasing social awareness over the past few decades concerning issues of environmental pollution and carbon dioxide levels in the atmosphere. The combustion of fossil fuels releases harmful emissions into air that influence the greenhouse effect as well as direct health problems for human beings. Major emissions of NO_x, SO_x, CO_2, and particulate matter cause pollution in the environment, resulting in an increase in the ozone level in the lower atmosphere, acid rain, and warming of the atmosphere. Particulate matter or soot produced from combustion in vehicle engines, power plants, and industrial processes can penetrate the lungs and cause a variety of health problems for living beings. There are several ways in which emissions from human activities can be curbed and dependence on fossil fuels may be reduced without diminishing the standards of living. In addition to potential to save energy, the efficiency of end-use equipment such as vehicle propulsion engines as well as power-generating facilities can be improved. The contribution of renewable energy sources from wind, sun, and water may be increased, but the complete conversion process should be taken into account when comparing different systems, and these sources are not suited to cover the base load. Therefore, it appears at present that solar energy and other renewable energy sources cannot meet the energy demand substantially. The future of the world is critically energy dependent. Many consider the effect of new technologies to lead to the better use of fuels in addition to the use of solar and other renewable energies, including the proper and appropriate use of atomic energy. Their higher efficiencies and lower emissions make fuel cells a valuable contribution to power generation facilities, although their contribution to the total energy demand may be minimal. All of these issues indicate that fuel cells appear to be an

Energy Sources. http://dx.doi.org/10.1016/B978-0-444-56353-8.00014-9

alternative energy sources that can cater to the need for energy of the future world. The 19th century was considered to be the age of the steam engine, the 20th century was considered to be the age of the internal combustion engine, and 21st century may come to be recognized as the age of the fuel cells.

The invention of fuel cells as an electrical energy conversion system is attributed to Sir William Grove; however, the principle was discovered by Christian Friedrich Schönbein. Sir William Grove mixed hydrogen and oxygen in the presence of an electrolyte and produced electricity and water; he called it a gaseous voltaic battery. The invention, which later came to be known as a fuel cell, did not produce enough electricity to be useful. In 1889, the term "fuel cell" was first used by Ludwig Mond and Charles Langer, who attempted to build a working fuel cell using air and industrial coal gas. Although William Grove first demonstrated it in 1839, it was used only in 1960 for space applications.

WHAT IS A FUEL CELL?

A fuel cell is an electrochemical energy converter that converts the chemical energy of fuel into DC electricity, heat, and water. Typically, a process of electricity generation from fuels involves several energy conversion steps (Scheme 14.1):

1. Combustion of fuel converts chemical energy of fuel into heat.
2. This heat is then used to boil water and generate steam.
3. Steam is used to run a turbine in a process that converts thermal energy into mechanical energy.
4. Mechanical energy is then used to run a generator that generates electricity.

A fuel cell circumvents all of these processes and generates electricity in a single step without involving moving parts. Such a device must be simpler, less expensive, and far more efficient than the four-step process stated earlier.

Unlike a conventional internal combustion engine, higher efficiencies are achievable in fuel cells because they do not have Carnot's limitations. In some aspects a fuel cell is similar to a battery. It has an electrolyte and negative and positive electrodes, and it generates DC electricity through electrochemical

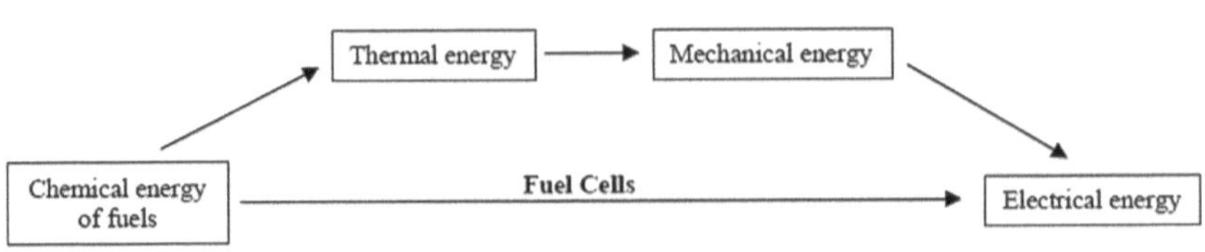

SCHEME 14.1 Direct energy conversion with fuel cells compared with conventional indirect technology.

reactions. However, unlike a battery a fuel cell requires a constant supply of fuel and oxidant. Also, unlike a battery the electrodes in a fuel cell do not undergo chemical changes. Batteries generate electricity by electrochemical reactions that involve materials that are already in batteries. Because of this, a battery may be discharged, which happens when the materials that participate in the electrochemical reactions are depleted. Some batteries are rechargeable, which means that the electrochemical reactions may proceed in reverse when external electricity is applied, a process of recharging the battery. A fuel cell cannot be recharged as long as the reactants, fuel and oxidant, are supplied.

Fuel cells have many applications that make them attractive compared with existing conventional energy conversion devices:

1. promise of high efficiency
2. promise of low or zero emissions
3. simplicity
4. no moving parts and promise of long life
5. Noiselessness
6. fuel and size flexibility

Because of their attractive properties, fuel cells have already been developed and have been projected for widespread use through three main applications: transportation, stationary power generation, and portable applications.

In Table 14.1, the history of fuel cell development is given in chronological order.

TABLE 14.1 History of Fuel Cell Development

Year	Activity
1839	Grove: hydrogen–oxygen fuel cell
1854	Proposal to use carbon and fats
1855	Bacquerel: carbon fuel/fused nitrate/Pt, Oxygen
1877	Jablackoff: carbon fuel, fused nitrate/cast iron, oxidant
1889	Mond and Lange: hydrogen–oxygen fuel cell similar to modern fuel cell using platinum electrodes–platinum black as catalyst maximum voltage developed 0.75 V
1897	Jacques: Modified fused salt cell replacing nitrate with NaOH
	Stone–Webster reported efficiency of 85% for the modified fused salt cell
	Haber–Brunner demonstrated indirect fuel cell in which hydrogen was produced by chemical reaction

Continued

TABLE 14.1 History of Fuel Cell Development—cont'd

Year	Activity
1910	Taitebaum: advocated use of porous diaphragm separator; addition of manganese dioxide to positive electrode in C/NaOH/Fe cell led to 0.53 V, current delivery 0.02 μA; addition of sawdust resulted in higher currents
1911	Beutner: used Pt tube electrodes containing hydrogen and oxygen; lead diffuser electrode was used; other electrolytes such as HF and NaCl melts were suggested
1912	Bantz and Ehrenbeg: advocated direct fuel cell with fused boride electrolyte, carbon anode, and silver air cathode; potential developed was 0.95 V but current was negligible
1913	Siegel: improved upon Mond–Lange cell; since then several molten salt systems have been tried
1911–1913	Biochemical fuel cells were strongly advocated and have continued to be supported
1954	Modern Bacon cell introduced
1960s	*Gemini* and *Apollo* launched
1980s	Space shuttle
1989	Parry energy systems with Ballard power systems demonstrated proton exchange membrane fuel cell (PEMFC)-powered submarine
1990	California Air Resources Board introduced concept of zero-emission vehicle
1993	Georgetown Bus phosphoric acid fuel cell (PAFC) onboard methanol reformer
1993	Partnership for new generation of vehicles government industry partnership
1993	Ballard: first PEMFC demonstrated
1995	Daimler–Benz NECAR1 on PEMFC
1995–2000	Several European, American, and Asian automobile industries come together
2002	Honda (FCX4), Toyota (FCHV-BUS2), Nissan (X-Trail FCV)
2002	EPRI fuel cell program; 5 kW PEMFC
2004	Honda delivered hydrogen-powered fuel cell vehicle

CHOICE OF FUEL AND OXIDANT

The choice and design of fuel cells depend on the kind of fuel and oxidant to be employed, the operating temperature, power rating/conditioning, and other use requirements. A fuel cell system that includes a fuel reformer can use hydrogen from any hydrocarbon fuel, from natural gas to methanol and even gasoline. Gaseous hydrogen has become the fuel of choice for most applications because of its high reactivity when suitable catalysts are used, its ability to be produced from hydrocarbons for terrestrial applications, and its high energy density (32 kWh/kg) when stored cryogenically for closed-environment applications, such as in space. Similarly, the most common oxidant is gaseous oxygen or air, which is readily and economically available from air for terrestrial applications and is also easily stored in a closed environment. In general, oxygen needed by a fuel cell is generally supplied in the form of air. In Table 14.2 the characteristics of various fuels that can be used in a fuel cell are listed.

HOW DOES A FUEL CELL WORK?

The basic physical structure or building block of most fuel cells consists of an electrolyte layer in contact with a porous anode and cathode electrodes on either side. All fuel cells have similar basic operating principles. The input fuel is catalytically reacted (electrons removed from the fuel) in the fuel cell to create an electric current. The input fuel passing over the anode is catalytically split into electrons and ions. Air/oxygen passing over the cathode is reduced

TABLE 14.2 Chemical and Electrochemical Data on Selected Fuels for Fuel Cell Application

Fuel	ΔG° (kcal/mol)	$E^\circ_{theoretical}$ (V)	$E^\circ_{maximum}$ (V)	Energy Density (kWh/kg)
Hydrogen	−56.69	1.23	1.15	32.67
Methanol	−166.80	1.21	0.98	6.13
Ammonia	−80.80	1.17	0.62	5.62
Hydrazine	−143.90	1.56	1.28	5.22
Formaldehyde	124.70	1.35	1.16	4.82
Carbon monoxide	−61.60	1.33	1.22	2.04
Formic acid	−68.20	1.48	1.14	1.72
Methane	−195.50	1.06	0.58	–
Propane	−503.20	1.08	0.65	–

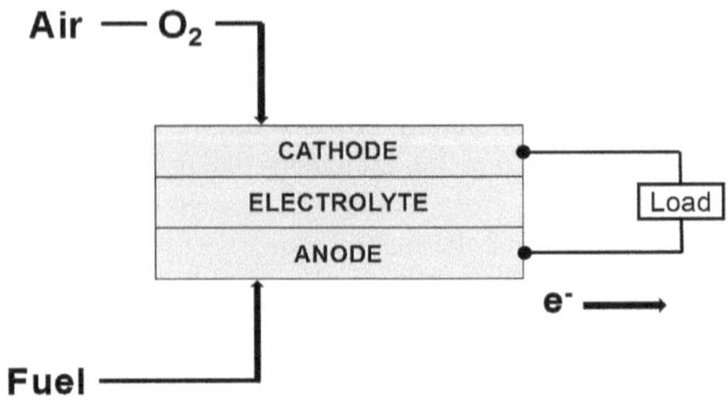

FIGURE 14.1 Simple schematic of a fuel cell.

by the electrons that are generated at anode and passed on to the cathode by an external circuit. At the cathode, the ions are formed at anode and transported to the cathode through the electrolyte; they combine with the oxide ions and generate the oxidized product. If the fuel happens to be hydrogen, water is formed (Fig. 14.1).

Thermodynamic and kinetic aspects of electrochemical energy conversion:

The energy storage and power characteristics of electrochemical energy conversion systems follow directly from the thermodynamic and kinetic formulations for chemical reactions as adapted to electrochemical reactions.

THERMODYNAMICS

Basic thermodynamic equations for a reversible electrochemical transformation are given in this section. The basic equation is:

$$\Delta G = \Delta H - T\Delta S$$

where ΔG is the Gibbs free energy, or the energy of a reaction available for useful work; ΔH is the enthalpy, or the energy released by the reaction; ΔS is the entropy; and T is the absolute temperature, where $T\Delta S$ is the heat associated with the organization or disorganization of materials. ΔG, ΔH, and ΔS are state functions and depend only on the identity of the materials and the initial and final states of the reaction.

The effect of temperature on free energy change (Gibbs–Helmholtz equation) is given by:

$$\Delta G = \Delta H + T\,(\partial(\Delta G)/\partial T)_{\mathrm{p}}$$

or

$$(\partial(\Delta G/T)/\partial T)_{\mathrm{p}} = -\Delta H/T^2$$

The effect of pressure on free energy change is given by:

$$(\partial(\Delta G)/\partial P)_{\mathrm{T}} = \Delta nRT/P$$

The maximum electrical work (W_{el}) obtainable in a fuel cell operating at constant temperature and pressure is given by the change in Gibbs free energy (ΔG) of the electrochemical reaction [Eq. (14.1)]:

$$\Delta G = -nFE \tag{14.1}$$

and [Eq. (14.2)]:

$$\Delta G^{\circ} = -nFE^{\circ} \tag{14.2}$$

where n is the number of electrons transferred per mole of reactants; F is the Faraday constant, which is equal to the charge of 1 equivalent of electrons; and E is the voltage of the cell with the specific chemical reaction; in other words, E is the electromotive force (emf) of the cell reaction. The voltage of the cell is unique for each reaction couple. The amount of electricity produced, nF, is determined by the total amount of materials available for reaction and can be thought of as a capacity factor; the cell voltage can be considered an intensity factor. The usual thermodynamic calculations for the effect of temperature and pressure apply directly to electrochemical reactions. Spontaneous processes have a negative free energy and a positive emf with the reaction written in a reversible fashion, which goes in the forward direction. The van't Hoff isotherm identifies the free energy relationship for bulk chemical reactions as:

$$\Delta G = \Delta G^{o} + RT \ln (Q); \; Q = a_{P}/a_{R}$$

where R is the gas constant, T the absolute temperature, and Q is the term dependent on the activity of reactants (a_R) and products (a_P).

Combining Eqs. (14.1) and (14.2) with the van't Hoff isotherm, one obtains the Nernst equation for electrochemical reactions:

$$E = E^{o} - RT/nF \ln (Q)$$

According to the Nernst equation for hydrogen–oxygen reaction, the ideal cell potential depends on the cell temperature and pressure of reactants. The impact of temperature on the ideal voltage, E, for the oxidation of hydrogen is shown in Fig. 14.2.

At a given temperature, the ideal cell potential can be increased by operating at higher reactant pressures, according to the equation

$$E = E^{\circ} + (RT/2F) \ln [P\, H_2/P\, H_2O] + (RT/2F) \ln [P^{1/2}\, O_2]$$

and improvements in fuel cell performance have in fact been observed at higher pressures.

KINETICS

Thermodynamics describes reactions at equilibrium and the maximum energy release for a given reaction. Useful work (electrical energy) is obtained from a fuel cell only when a reasonable current is drawn, but the actual cell potential is decreased from its equilibrium potential because of irreversible losses. Fig. 14.3 shows a typical voltage–current ($E - I$) discharge curve for a fuel

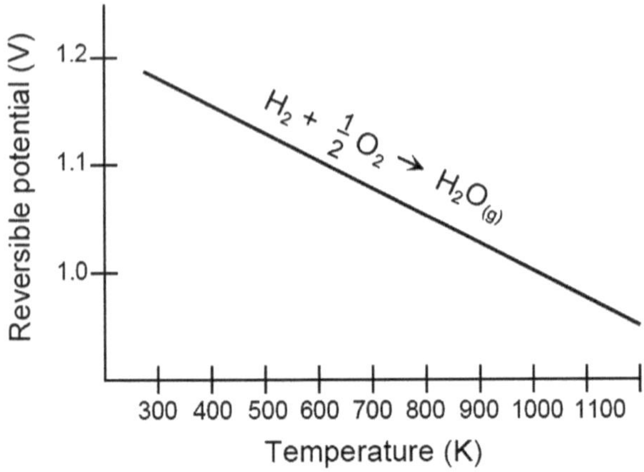

FIGURE 14.2 H_2–O_2 fuel cell ideal potential as a function of temperature.

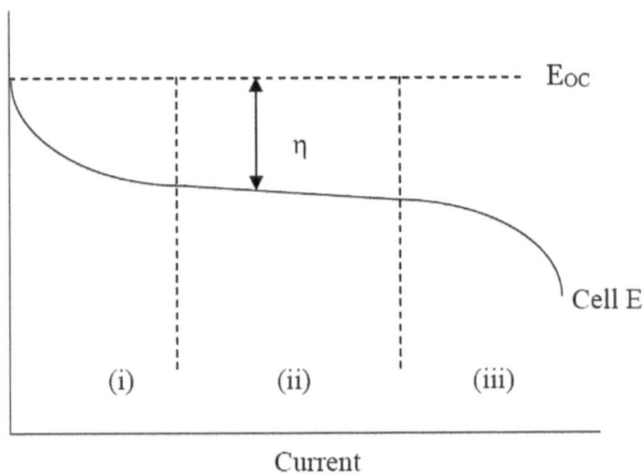

FIGURE 14.3 Typical polarization curve for a fuel cell: voltage drops caused by (i) surface reaction kinetics, (ii) electrolyte resistance, and (iii) reactant/product diffusion rates.

cell with an open-circuit voltage E_{oc}. The overpotential $\eta = (E_{oc} - E)$ reflects the resistive IR losses due to the surface reaction kinetics, the resistance to transport of the working ion, H^+ or O^{2-} between the reductant and the oxidant reactive sites, and the resistance to diffusion of the oxidant and/or reductant to the catalytic sites and their products away from these sites. At low currents, the performance of a fuel cell is dominated by kinetic losses. These losses mainly stem from the high overpotential of the reactions occurring at the anode and cathode. At intermediate currents, ohmic losses arise from ionic losses in the electrodes and separator, although contact and electronic resistances can be important under certain operating conditions. At high currents, mass transport limitations become increasingly important. These losses result from reactants not being able to reach the electro-catalytic sites. Typically, oxygen is the problem that is caused by flooding of the cathode by liquid water, but protons and electrons can also cause mass transfer limitations.

At low current densities ($i_o < 1$ mA/cm^2), electrodes give a larger R_{tr} and therefore overpotential, η, should be greater than 400 mV (at room temperature). An extremely active electro-catalyst is needed to overcome this initial voltage drop in the E versus current discharge curve. The goal of fuel cell developers is to minimize the polarization so that E_{cell} approaches E_{oc}. This goal is approached by modifications to fuel cell design (improvement in electrode structures, better electro-catalysts, more conductive electrolyte, and thinner cell components). For a given cell design, it is possible to improve cell performance by modifying the operating conditions (eg, higher gas pressure, higher temperature, change in gas composition to lower the gas impurity concentration). However, for any fuel cell compromises exist between achieving higher performance by operating at a higher temperature or pressure and problems associated with the stability or durability of cell components encountered at more severe conditions.

FUEL CELL EFFICIENCY

The ideal or maximum efficiency of an electrochemical energy converter depends on electrochemical thermodynamics whereas real efficiency depends on electrode kinetics. The thermal efficiency of an energy conversion device is defined as the amount of useful energy produced relative to the change in stored chemical energy (commonly referred to as thermal energy) that is released when a fuel is reacted with an oxidant:

$$\varepsilon = \text{Useful energy}/\Delta H$$

In the ideal case of an electrochemical converter such as a fuel cell, the change in Gibbs free energy, ΔG, of the reaction is available as useful electric energy at the temperature of conversion. The ideal efficiency of a fuel cell operating irreversibly is then:

$$\varepsilon = \Delta G/\Delta H$$

The most widely used efficiency of a fuel cell is based on the change in the standard free energy for the cell reaction; for example:

$$H_2 + \frac{1}{2}O_2 \rightarrow H_2O \tag{14.3}$$

given by:

$$\Delta G^{\circ} = G^{\circ}_{H_2O} - G^{\circ}_{H_2} - \frac{1}{2}\left(G^{\circ}_{O_2}\right)$$

where the product water is in a liquid form. At standard conditions of 25°C (298K) and 1 atm, the chemical energy (ΔH) in the hydrogen–oxygen reaction is 285.8 kJ/mol and the free energy available for useful work is 237.1 kJ/mol. Thus, the thermal efficiency of an ideal fuel cell operating reversibly on pure hydrogen and oxygen at standard conditions would be:

$$\varepsilon_{\text{ideal}} = 237.1/285.8 = 0.83$$

The efficiency of an actual fuel cell can be expressed in terms of the ratio of the operating cell voltage to the ideal cell voltage. The actual cell voltage is less than the ideal cell voltage because of the losses associated with cell polarization and IR loss. The thermal efficiency of the fuel cell can then be written in terms of the actual cell voltage:

$$\begin{aligned}\varepsilon &= \text{useful energy}/\Delta H = \text{useful power}/(\Delta G/0.83)\\ &= (\text{volts}_{\text{actual}} \times \text{current})/(\text{volts}_{\text{ideal}} \times \text{current}/0.83)\\ &= 0.83\ (\text{volts}_{\text{actual}})/(\text{volts}_{\text{ideal}})\end{aligned}$$

As mentioned, the ideal voltage of a cell operating reversibly on pure hydrogen and oxygen at 1 atm pressure and 25°C is 1.229 V. Thus, the thermal efficiency of an actual fuel cell operating at a voltage of E_{cell}, based on the higher heating value of hydrogen, is given by:

$$\varepsilon_{\text{ideal}} = 0.83 \times E_{\text{cell}}/E_{\text{ideal}} = 0.83 \times E_{\text{cell}}/1.229 = 0.675 \times E_{\text{cell}}$$

A fuel cell can be operated at different current densities, expressed as mA/cm^2 or A/ft^2. The corresponding cell voltage then determines fuel cell efficiency. Decreasing the current density increases the cell voltage, thereby increasing fuel cell efficiency. The trade-off is that as the current density is decreased, the active cell area must be increased to obtain the requisite amount of power. Thus, designing the fuel cell for higher efficiency increases the capital cost but decreases the operating cost.

WHAT ARE THE VARIOUS TYPES OF FUEL CELLS?

A variety of fuel cells are in different stages of development. They can be classified into diverse categories, depending on the type of fuel and oxidant, whether the fuel is processed outside (external reforming) or inside (internal reforming) the fuel cell, the type of electrolyte, the temperature of operation, and whether the reactants are fed to the cell by internal or external manifolds. The most common classification of fuel cells is by the type of electrolyte used in the cells and includes: (1) alkaline fuel cells (AFCs), (2) PAFCs, (3) PEMFCs, (4) direct methanol fuel cells (DMFCs), (5) molten carbonate fuel cells (MCFCs), and (6) solid oxide fuel cells (SOFCs). These fuel cells are listed in the order of approximate operating temperature, ranging from ~353K for PEMFC, 333 to 353K for DMFC, ~373K for AFC, ~273K for PAFC, ~923K for MCFC, and 1273K for SOFC. The operating temperature and useful life of a fuel cell dictate the physicochemical and thermomechanical properties of materials used in the cell components (ie, electrodes, electrolyte, interconnect, current collector). Aqueous electrolytes are limited to temperatures of about 200°C or lower because of their high water vapor pressure. The operating temperature also has an important role in dictating the type of fuel that can be used in a fuel cell. In most practical applications, low-temperature fuel cells with aqueous electrolytes are restricted to hydrogen as a fuel. In high-temperature fuel cells, CO and even CH_4 can be used because of the inherently rapid electrode kinetics and the lesser need for high catalytic activity at high temperatures (Fig. 14.4).

The characteristics of each of these fuel cells and the essential features of each are assembled in Table 14.3.

The heart of the fuel cell is the membrane electrode assembly. The important components and their functions are given in Table 14.4. A significant problem is the control of the interface at the junction of the reactant phase, the electrolyte medium, and the catalyzed conducting electrode, the so-called "three-phase boundary," where the electrolyte, electrode, and reactant all come together. A stable three-phase boundary is critical to good performance and long operation. Therefore, the porosity and the wetting behavior with the electrolyte and the electrode surface must be precisely adjusted.

The electrodes have to be gas (or liquid) permeable and therefore should possess a porous structure. The structure and content of gas diffusion electrodes

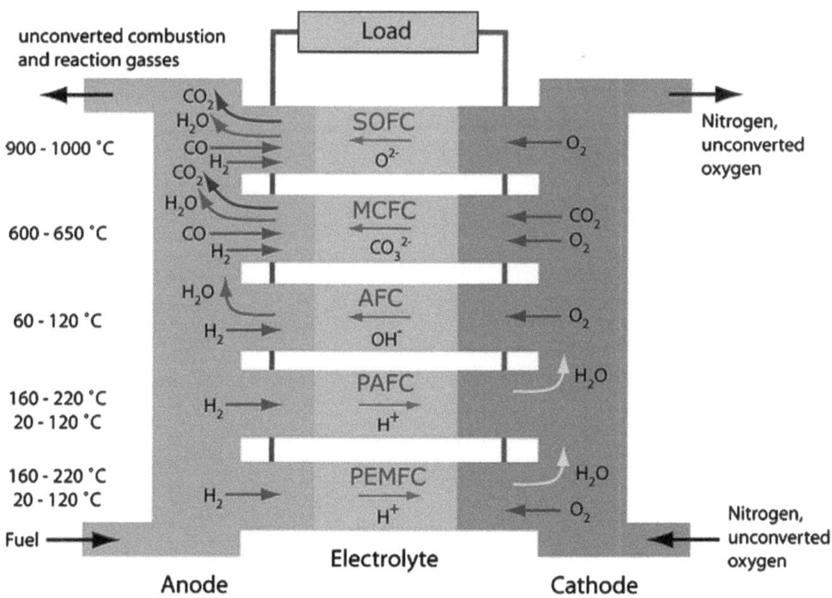

FIGURE 14.4 Classification of fuel cell according to the temperature of operation or on the basis of the electrolyte.

TABLE 14.3 Essential Features of Each Fuel Cell

Type of Fuel Cell	Operating Temperature Range (K)	Electrolyte	Charge Carrier	Electrolyte State	Fuel for Cell	Oxidant for Cell
Alkaline fuel cell	333–423	45% KOH	OH^-	Immobilized liquid	Hydrogen	Oxygen/air
Phosphoric acid fuel cell	453–493	H_3PO_4	H^+	Immobilized liquid	Hydrogen	Oxygen/air
Proton exchange membrane fuel cell	333–353	Ion exchange membrane (Nafion)	H^+	Solid	Hydrogen	Oxygen/air
Direct methanol fuel cell	333–353	Ion exchange membrane	H^+	Solid	Methanol	Oxygen/air
Molten carbonate fuel cell	923–973	Alkali carbonate mixture	CO_3^{2-}	Immobilized liquid	Hydrogen	Oxygen/air
Solid oxide fuel cell	1073–1273	Yttria-stabilized zirconia	O^{2-}	Solid	Hydrogen	Oxygen/air

TABLE 14.4 Membrane Electrode Assembly Components and Their Functions

Membrane Electrode Assembly Component	Functions/Effect
Anode substrate	Fuel supply and distribution, (hydrogen/fuel gas)
	Electron conduction
	Heat removal from reaction zone
	Water supply (vapor) into electro-catalyst
Anode catalyst layer	Catalysis of anode reaction
	Ion conduction into membrane
	Electron conduction into substrate
	Water transport
	Heat transport
Proton exchange membrane	Ion conduction
	Water transport
	Electronic insulation
Cathode catalyst layer	Catalysis of cathode reaction
	Oxygen transport to reaction sites
	Ion conduction from membrane to reaction sites
	Electron conduction from membrane to reaction sites
	Water removal from reactive zone into substrate
	Heat generation/removal
Cathode substrate	Oxidant supply and distribution (air/oxygen)
	Electron conduction toward reaction zone
	Heat removal
	Water transport (liquid/vapor)

are complex and require considerable optimization for practical application. Functions of porous electrodes in fuel cells are:

1. to provide a surface site where gas/liquid ionization or deionization reactions can take place,

2. to conduct ions away from or into the three-phase interface once they are formed (so an electrode must be made of materials that have good electrical conductance), and
3. to provide a physical barrier that separates the bulk gas phase and the electrolyte.

A corollary of first one is that, to increase the rates of reactions, the electrode material should be catalytic as well as conductive, and porous rather than solid. The catalytic function of electrodes is more important in lower-temperature fuel cells and less so in high-temperature fuel cells because ionization reaction rates increase with temperature. It is also a corollary that the porous electrodes must be permeable to both electrolyte and gases, but not such that the media can easily be "flooded" by the electrolyte or "dried" by the gases in a one-sided manner.

The reason for this is that current densities obtained from smooth electrodes are usually in the range of a single-digit mA/cm^2 or less because of rate-limiting issues such as the available area of the reaction sites. Porous electrodes used in fuel cells deliver higher current densities. These high-current densities are possible because the electrode has a high surface area, relative to the geometric area that significantly increases the number of reaction sites, and the optimized electrode structure has favorable mass transport properties. In an idealized porous gas fuel cell electrode, high current densities at reasonable polarization are obtained when the electrolyte layer on the electrode surface is sufficiently thin so that it does not significantly impede the transport of reactants to the electroactive sites, and a stable three-phase (gas–electrolyte–electrode surface) interface is established. When an excessive amount of electrolyte is present in the porous electrode structure, the electrode is considered "flooded" and the concentration polarization increases to a large value.

Porous electrodes used in low-temperature fuel cells (AFC, PAFC, PEMFC, and DMFC) consist of a composite structure that contains a platinum (Pt) electro-catalyst on a high–surface area carbon black and a polytetrafluoroethylene (PTFE) binder. In these porous electrodes, PTFE is hydrophobic (acts as a wet proofing agent) and serves as the gas-permeable phase, and carbon black is an electron conductor that provides a high surface area to support the electro-catalyst. Platinum serves as the electro-catalyst, which promotes the rate of electrochemical reactions (oxidation–reduction) for a given surface area. Carbon black is also hydrophobic, depending on the surface properties of the material. The composite structure of PTFE and carbon establishes an extensive three-phase interface in the porous electrode, which is the benchmark of PTFE-bonded electrodes.

In MCFCs, which operate at relatively high temperatures, no materials are known that are wet-proof and also retain a porous structure against ingress by molten carbonates. Consequently, the technology used to obtain a stable three-phase interface in MCFC porous electrodes is different from that used in

PAFCs. In the MCFC, the stable interface is achieved in the electrodes by carefully tailoring the pore structures of the electrodes and the electrolyte matrix ($LiAlO_2$) so that the capillary forces establish a dynamic equilibrium in the different porous structures. In an SOFC, there is no liquid electrolyte present that is susceptible to movement in the porous electrode structure, and electrode flooding is not a problem. Consequently, the three-phase interface that is necessary for an efficient electrochemical reaction involves two solid phases (solid–electrolyte–electrode) and a gas phase. A critical requirement of porous electrodes for SOFCs is that they be sufficiently thin and porous to provide an extensive electrode–electrolyte interfacial region for an electrochemical reaction.

Essential criteria for a better electrode material are:

- high electronic conductivity,
- high adsorption capacity of reactants and oxidants,
- chemical and structural stability under conditions employed in devices, ie, operating temperature, wide range of partial pressures of reactant and oxidant and concentration of electrolytes,
- chemical and thermomechanical compatibility to electrolyte and interconnector materials,
- high ionic conductivity,
- ability to decompose intermediate species formed during the oxidation–reduction process,
- tolerance to contaminants, eg, halide ions, NO_X, CO_X, SO_X, and
- low cost of materials.

The oxygen reduction reaction (ORR), which is a common cathodic reaction to all fuel cell devices, has been studied over the years because of its fundamental complexity, great sensitivity to the electrode surface, and sluggish kinetics. Sluggish kinetics of ORR under the conditions employed in electrochemical devices results from the low partial pressure of oxygen in air and the slow flow rate of oxygen (ie, less residence time for oxygen molecules on active sites) under ambient conditions. The main disadvantage in this important electrode reaction is the exchange current density (i_o) value in the region of 10^{-10} A/cm^2 in an acidic medium and 10^{-8} A/cm^2 at 298K in an alkaline solution, which is lower than the i_o value of an anodic reaction (10^{-3} A/cm^2) in all electrochemical devices. Hence [from the equation $\eta = RT/nF \ln (i/i_o)$] the oxygen reduction reaction usually contributes considerably to the overpotential and therefore results in low efficiency in the functioning of electrochemical energy devices using air as an oxidant. Understanding and exploiting electro-catalysis for this reaction is needed more than any other reactions in electrochemical devices. Oxygen undergoes a two-step indirect reduction reaction. For most electro-catalysts, oxygen reduction takes place by forming a high-energy intermediate, H_2O_2, followed by further reduction to H_2O. The stable H_2O_2 intermediate is undesirable because it lowers the cell voltage and H_2O_2 attacks and corrodes the carbonaceous electrode material commonly used. Better catalysts are needed to speed the decomposition of H_2O_2 to reduce its impact on the overall reaction. Similarly, a catalyst can enhance the fuel dissociation rate at the anode. To obtain maximum efficiency and avoid the corrosion of carbon supports and other materials by peroxide, it is desirable to achieve a four-electron reduction. Finding suitable electrocatalysts that can promote the direct four-electron reduction of oxygen molecules is an important task.

The characteristic features, advantages, and limitations of various types of fuel cells are given next.

ALKALINE FUEL CELLS

The first commercial fuel cell systems were the AFCs, which became available in the 1950s. AFCs were used to power the *Apollo* space crafts and are currently used in the space shuttles. The electrolyte in AFCs is a concentrated KOH solution. For low-temperature applications (60–90°C), the KOH concentration is 35–50 wt%. To achieve optimal performance of AFCs with KOH concentrations of 85 wt%, the operating temperature was increased to 200°C. These high-temperature cells are also operated at high pressures (4–6 atm) to prevent the electrolyte solution from boiling.

Pure H_2 and O_2 are input as the fuel and oxidizer in an AFC (Fig. 14.5). The gas diffusion electrodes are constructed of porous carbon and are doped

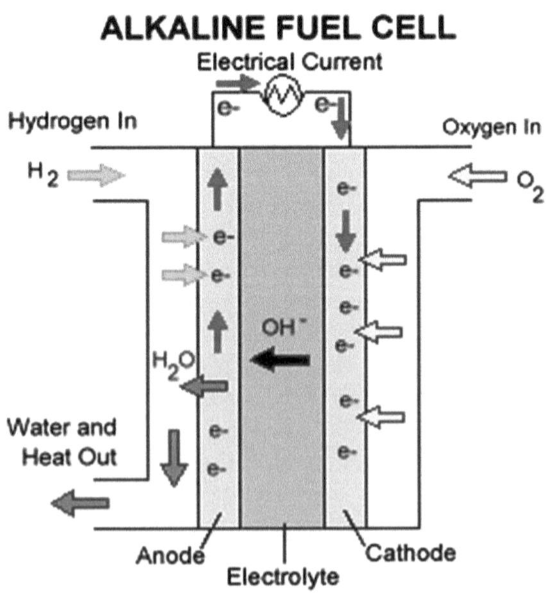

FIGURE 14.5 Typical alkaline fuel cell.

with Pt to catalyze the oxidation and reduction reactions. The anodes contain 20% Pd in addition to the Pt and the cathodes contain 10% Au and 90% Pt. For higher-temperature operations, Ni catalysts are also used. Ni is used for the interconnectors in an AFC stack. The AFC operates at up to ~1 A/cm^2 at 0.7 V.

The mobile ions in the system are the OH^- ions in the alkaline solution that are transported from the cathode, where reduction of O_2 occurs, to the anode, where oxidation of H_2 occurs. Water is produced at the anode. The reactions that define the operation of AFCs are:

$$\text{At the anode: } H_2 + 2OH^- \rightarrow 2H_2O + 2e^-$$

$$\text{At the cathode: } \frac{1}{2}O_2 + H_2O + 2e^- \rightarrow 2OH^-$$

$$\text{Cell reaction: } H_2 + \frac{1}{2}O_2 \rightarrow H_2O$$

Although AFCs have the highest electrical efficiency of all fuel cell systems (60% lower heating value), they are extremely sensitive to impurities. The presence of N_2 and impurities in the gas streams substantially reduce cell efficiency. The presence of even small amounts of CO_2 is detrimental to the long-term performance of AFCs because K_2CO_3 forms and inhibits gas diffusion through the carbon electrodes. Small amounts of CO_2 in air [~300 parts per million (ppm)] preclude the use of air as the oxidant in an AFC. This restriction limits the use of these fuel cell systems to applications such as space and military programs, where the high cost of providing H_2 and O_2 is permissible. Because pure gases are used, AFCs can generate pure, potable water for consumption during space missions.

PHOSPHORIC ACID FUEL CELLS

Other than the AFCs, PAFCs are closer to commercialization than the other fuel cell systems (Fig. 14.6). The two intended commercial uses for PAFCs are: (1) distributed power using reformed natural gas as a fuel; and (2) small-scale, onsite cogeneration. Air is used as the oxidant. In contrast to the AFC, PAFCs are tolerant of CO_2 because concentrated phosphoric acid (H_3PO_4) is used as the electrolyte. Compared with other inorganic acids, phosphoric acid has relatively low volatility at operating temperatures of 150–220°C. Protons migrate from the anode to the cathode through 100% H_3PO_4 that is immobilized in an SiC-PTFE matrix. Electrodes are made of platinized, gas-permeable graphite paper. The water produced at the cathode is removed with the excess O_2 and the N_2. PAFCs have demonstrated excellent thermal, chemical, and electrochemical stability compared with other fuel cell systems. PAFCs are defined by the following reactions:

$$\text{At the anode: } H_2 \rightarrow 2H^+ + 2e^-$$

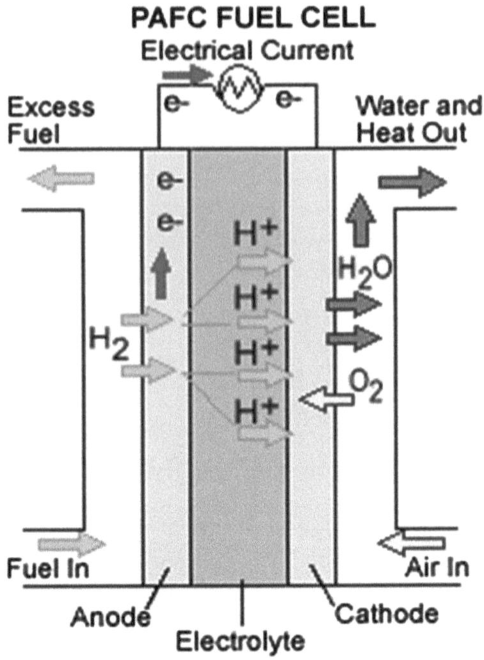

FIGURE 14.6 Typical phosphoric acid fuel cell.

$$\text{At the cathode}: \frac{1}{2}O_2 + 2H^+ + 2e^- \rightarrow H_2O$$

$$\text{Cell reaction}: H_2 + \frac{1}{2}O_2 \rightarrow H_2O$$

To optimize the ionic conductivity of the electrolyte, operating temperatures are maintained between 150 and 220°C at pressures ranging from atmospheric to ~8 atm. The reduction of oxygen is slower in an acid electrolyte than in an alkaline electrolyte, hence the need for Pt metal in the electrodes to help catalyze the reduction reactions. CO poisoning of the Pt electrodes is slower at PAFC operating temperatures than at lower temperatures, so up to 1% CO in the fuel gas produced during the reforming process can be tolerated. At lower temperatures, CO poisoning of the Pt in the anode is more severe. Currently, Pt-based materials are used as anodes and cathodes. The anode operates at a nearly reversible voltage with ~0.1 mg/cm^2 catalyst loading. The cathode requires a higher catalyst loading of ~1 mg/cm^2 of catalyst. PAFCs are semicommercially available in container packages (200 kW) for stationary electricity generation. Hundreds of units have been installed over the world.

Aside from the CO produced during hydrocarbon reforming, the concentration of other impurities must be low compared with the reactants and diluents. Sulfur gases (mainly H_2S and COS) that originate from the fuel gas can poison the anode by blocking active sites for H_2 oxidation on the Pt surface. Molecular nitrogen acts as a diluent but nitrogen compounds such as NH_3, HCN, and NO_x are potentially harmful impurities. NH_3 acts as a fuel; however, the oxidant nitrogen compounds can react with the H_3PO_4 to form a phosphate salt, $(NH_4)H_2PO_4$. Unacceptable performance losses can occur if the concentration of this phosphate salt in the electrolyte increases above 0.2 mol%.

PROTON EXCHANGE MEMBRANE FUEL CELLS

PEMFCs are a serious candidate for automotive applications but also for small-scale distributed power generation, and for portable power applications as well. PEMFCs contain a proton-conducting ion exchange membrane as the electrolyte material. The membrane material is a fluorinated sulfonic acid polymer commonly referred to by the trade name given to a material developed and marketed by DuPont: Nafion. The acid molecules are immobile in the polymer matrix; however, protons associated with these acid groups are free to migrate through the membrane from the anode to the cathode, where water is produced. Electrodes in a PEMFC are made of porous carbon cloths doped with a mixture of Pt and Nafion. The catalyst content of the anode is $\sim 0.1\ mg/cm^2$, and that of the cathode is $\sim 0.5\ mg/cm^2$. The PEMFC operates at $\sim 1\ A/cm^2$ at 0.7 V.

PEMFCs use H_2 as the fuel and O_2 as the oxidant. The PEMFC is insensitive to CO_2, so air can be used instead of pure O_2 and reforming hydrocarbon fuels can produce the H_2. Thermally integrating fuel reformers with operating temperatures of 700–800°C with PEMFCs that operate at 80°C are a considerable challenge. The PEMFC is defined by the reactions:

$$\text{At the anode: } H_2 \rightarrow 2H^+ + 2e^-$$

$$\text{At the cathode: } \frac{1}{2}O_2 + 2H^+ + 2e^- \rightarrow H_2O$$

$$\text{Cell reaction: } H_2 + \frac{1}{2}O_2 \rightarrow H_2O$$

PEMFCs have received considerable attention as the primary power source in electric vehicles, for several reasons. Because the electrolyte is a polymeric material, there is no free corrosive liquid inside the cell (water is the only liquid), hence material corrosion is kept to a minimum. PEMFCs are also simple to fabricate and have a demonstrated long life. On the other hand, the polymer electrolyte (Nafion) is expensive and Pt loadings in the electrodes are high, so the fuel cell cost is high. The power and efficiency of a PEMFC also depend on the water content of the polymer electrolyte, so water management in the membrane is critical for efficient operation. The conductivity of the membrane is a function of the number of water molecules available per acid site, and if the membrane dries out, fuel cell power and efficiency decrease. If water is not removed from the PEMFC, the cathode can become flooded, which also degrades cell performance. For high-temperature PEMFCs, polybenimidazole-based membranes will be preferred.

The required moisture content of the membrane limits the operating temperature of a PEMFC to less than 120°C. This temperature ensures that the by-product water does not evaporate faster than it is produced. Low operating temperatures equates to high Pt loadings in the electrodes to catalyze the oxidation and reduction reactions efficiently. The Pt content of the electrodes also necessitates that the CO content of the fuel gas be very low (<5 ppm) because CO blocks the active sites in the Pt catalyst. Therefore if a hydrocarbon reformer is used to produce H_2, the CO content of the fuel gas needs to be greatly reduced. This is usually accomplished by oxidizing CO to CO_2 using a water gas shift reactor, or using pressure swing adsorption to purify the hydrogen.

DIRECT METHANOL FUEL CELLS

The DMFC uses the same basic cell construction as for the PEMFC (Fig. 14.7). It has the advantage of a liquid fuel in that it is easy to store and transport. There is no need for the reformer to convert the hydrocarbon fuel into hydrogen gas. Methanol is the liquid fuel with the highest energy density (6.2 kWh/kg) among all liquid fuels next to hydrogen. The anode feedstock is a methanol and water mixture or neat methanol, depending on the cell configuration. The DMFC is under development as a power source for portable electronic devices such as notebook computers and cellular phones. The pure

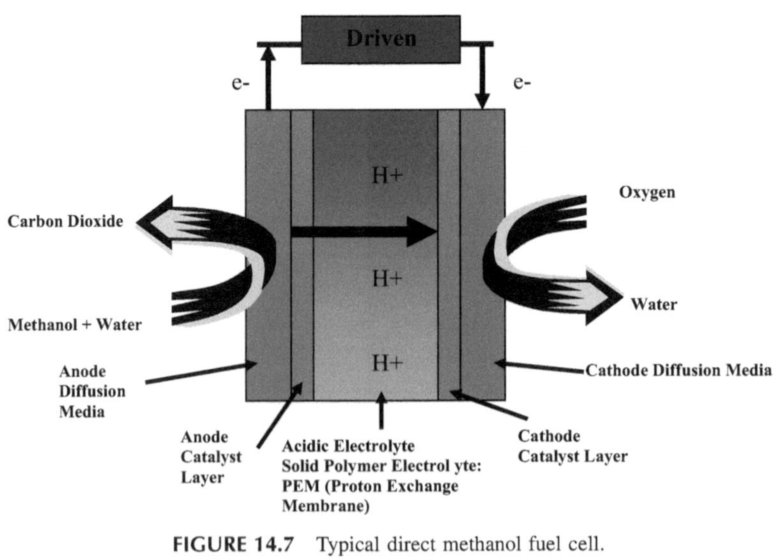

FIGURE 14.7 Typical direct methanol fuel cell.

methanol or a methanol–water mixture would be stored in a cartridge similar to that used for fountain pens. Refueling would involve the quick replacement of the cartridge. The reaction for the direct conversion of methanol has a voltage similar to that for hydrogen. DMFCs use methanol as the fuel and molecular oxygen as the oxidant. Because of the chemical similarity of water and methanol, methanol has considerable solubility in the polymer membrane, leading to significant crossover from the anode to the cathode side of the cell. Upon reaching the cathode, the methanol is oxidized. This significantly lowers the cathode voltage and the overall efficiency of the cell operation. The typical DMFC yields ~0.5 V at 400 mA/cm^2 at 60°C. The DMFC is defined by the following reactions:

$$\text{At the anode: } CH_3OH + H_2O \rightarrow CO_2 + 6H^+ + 6e^-$$

$$\text{At the cathode: } \frac{3}{2}O_2 + 6H^+ + 6e^- \rightarrow 3H_2O$$

$$\text{Cell reaction: } CH_3OH + \frac{3}{2}O_2 \rightarrow CO_2 + 2H_2O$$

The main disadvantage of the DMFC system is the relative low power density, which has to be significantly improved if the DMFC is to be a viable alternative to the PEMFC plus reformer system. The lower cell performance of a DMFC is caused by the poor kinetics of the anode reaction. The oxidation reaction proceeds through the formation of carbon monoxide as an intermediate that strongly adsorbs on the surface of a Pt catalyst. Therefore, a potential, which is more anodic than the thermodynamic value, is needed to obtain a reasonable reaction rate. In contrast to the PEMFC, in which mainly the cathode is kinetically hindered, both electrodes of a DMFC have kinetic losses. Consequently, numerous materials were studied to find an electrode material that displays enhanced catalytic activity and therefore lower overpotentials toward the methanol oxidation. To date, the most active anode catalysts are based on Pt–Ru alloys. Ruthenium reduces the poisoning effect by lowering the overpotentials at the anode and thus considerably increases the catalytic activity of pure platinum. The platinum–ruthenium catalyst loadings for the anode are higher than for the PEMFC and are in the range of 1–3 mg/cm^2. Cathode catalysts are based on Pt and Pt alloys (Pt-M, where M = Cr, Co, Fe, and Ni).

MOLTEN CARBONATE FUEL CELLS

MCFCs contain an electrolyte that is a combination of alkali (Li, Na, and K) carbonates stabilized in an $LiAlO_2$ ceramic matrix (Fig. 14.8). The electrolyte should be pure and relatively free of alkaline earth metals. Contamination by more than 5–10 mol% $CaCO_3$, $SrCO_3$, and $BaCO_3$ can lead to

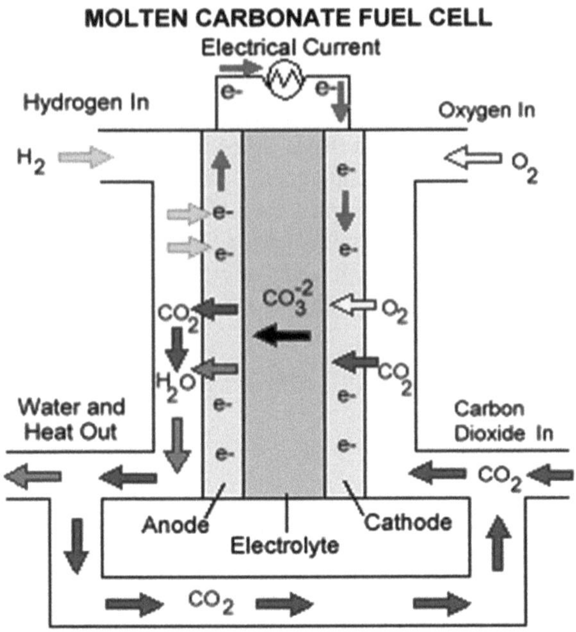

FIGURE 14.8 Typical molten carbonate fuel cell.

performance loss. Electrons are conducted from the anode through an external circuit to the cathode, and negative charge is conducted from the cathode through the electrolyte by CO_3^{2-} ions to the anode. Water is produced at the anode and removed with CO_2. The CO_2 needs to be recycled back to the fuel cell to maintain the electrolyte composition. This adds complexity to the MCFC systems. The oxidation and reduction reactions that define MCFC operation are:

$$\text{At anode: } H_2 + CO_3^{2-} \rightarrow H_2O + CO_2 + 2e^-$$

$$CO + CO_3^{2-} \rightarrow CO_2 + 2e^-$$

$$\text{The shift is: } CO + H_2O \rightarrow H_2 + CO_2$$

$$\text{At cathode: } \frac{1}{2}O_2 + CO_2 + 2e^- \rightarrow CO_3^{2-}$$

$$\text{The cell reaction is: } H_2 + \frac{1}{2}O_2 \rightarrow H_2O$$

MCFCs typically operate at temperatures between 600°C and 700°C, providing the opportunity for high overall system operating efficiencies, especially if the waste heat from the process can be used in the fuel-reforming step or for cogeneration. Operating temperatures higher than 700°C lead to diminishing gains in fuel cell performance because of electrolyte loss from evaporation and increased high temperature materials corrosion. Typical operating parameters are ~150 mA/cm^2 at 0.8 V at 600°C.

The high operating temperature of an MCFC system also provides for greater fuel flexibility; a variety of hydrocarbon fuels (eg, natural gas, alcohols, landfill gas, synthetic gas from petroleum coke, coal, and biomass) can be reformed to generate hydrogen for the fuel cell. The CO from biomass and coal gasification product gas and reformed hydrocarbons is not used directly as a fuel, but when mixed with water vapor it can produce additional hydrogen via the water–gas shift reaction. Oxygen or air is used as the oxidant. An attractive design incorporates an internal fuel reformer within the fuel cell, eliminating the need for a separate fuel processor.

A higher operating temperature also means that less expensive materials can be used for the electro-catalysts in the electrodes; Pt is not required and Ni is used as the catalyst. The Ni in the cathode becomes oxidized and lithiated (from contact with the electrolyte) during initial operation of an MCFC so that the active material is Li-doped NiO. Unfortunately, NiO is soluble in molten carbonates, which leads to the possible dissolution of the cathode and dispersion of metallic nickel in the electrolyte, and which can eventually short-circuit the electrodes. This is a materials issue that has to be addressed to improve the long-term operability of MCFC systems. $LiCoO_2$ and Li_2MnO_3 were found to be alternative materials as cathodes.

The anode contains Ni doped with 10% Cr to promote sintering. An external methane reformer is not needed in an MCFC system because the presence of Ni in the anode at MCFC operating temperatures is effective for internal methane reforming at the anode. Internal methane reforming can increase overall system efficiencies, but it can also induce unwanted temperature gradients inside the fuel cell that may cause materials issues. Catalyst poisoning is also an issue if the sulfur content of the reagent gases is greater than 10 ppm, similar to all Ni-based fuel-reforming systems. Coke formation on the anode from fuel reforming can also be an issue.

SOLID OXIDE FUEL CELLS

SOFC systems operate between 900°C and 1000°C, higher than any other type of fuel cell system (Fig. 14.9). At these operating temperatures, fuel composition is not an issue because in the presence of enough water vapor and oxygen, complete oxidation can be achieved, even in the absence of catalytic materials. High overall system efficiencies are possible with waste heat recovery. The electrolyte material in an SOFC is yttrium (8–10 mol%)-stabilized zirconia (YSZ). This material is a solid with a stable cubic structure and very high oxide ion conductivity at SOFC operating temperatures. The

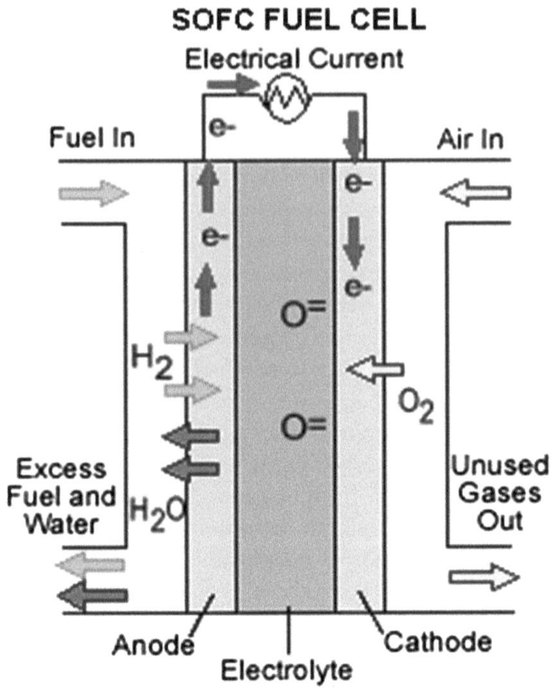

FIGURE 14.9 Typical schematic diagram of SOFC system.

mobile oxide ions migrate from the cathode to the anode where water is produced. The electrochemical reactions occurring in an SOFC system are:

$$\text{At the anode: } H_2 + O^{2-} \rightarrow H_2O + 2e^-$$

$$\text{At the cathode: } \frac{1}{2}O_2 + 2e^- \rightarrow O^{2-}$$

$$\text{Cell reaction: } H_2 + \frac{1}{2}O_2 \rightarrow H_2O$$

Similar to MCFC systems, the high operating temperatures of SOFCs provide fuel flexibility without the need for expensive catalysts in the electrodes. The cathode in an SOFC consists of mixed oxides with a perskovite crystalline structure, typically Sr-doped lanthanum manganate ($LaMnO_3$). The anode material is an Ni cermet (ceramic and metal composite). It contains metallic Ni for catalytic activity in a YSZ support. The YSZ adds mechanical, thermal, and chemical stability; chemical and thermal compatibility between the anode and the electrolyte is not an issue. Like MCFC systems, internal methane steam reforming at the Ni-based anode in the presence of water vapor is possible in SOFC systems. The cells operate at $\sim 1\ A/cm^2$ at 0.7 V.

Overall, SOFC systems can tolerate impurities because of their high operating temperatures. Sulfur tolerances can be up to two orders of magnitude higher in SOFCs than in other fuel cell systems because of the high operating temperatures. Energy-efficient, high-temperature sulfur removal methods are used to lower the sulfur content of the gas to less than 10 ppm. At the same time, the high operating temperatures of SOFCs can cause considerable materials issues such as material incompatibilities (thermal and chemical) and corrosion.

Significant research and development efforts have gone into technically and cost-effectively addressing materials issues in SOFC systems for commercial applications. This is reflected in the variety of designs for SOFC systems. There are three general types of designs for SOFC systems: tubular, bipolar monolithic, and bipolar planar. The bipolar designs have a bipolar plate that prevents reactant gases in adjacent cells from mixing, and provide serial electrical interconnectivity between cells. The single cells are stacked with interconnectors, gas channels, and sealing elements between. There are two types of tubular designs: seal-less and segmented cell in-series. A single cell in a tubular SOFC consists of a long porous YSZ ceramic tube that acts as a substrate. The cathode, followed by the electrolyte, and finally the anode are deposited on the outer surface of the tube. A portion of the tube is left with a strip of the cathode covered by the interconnector material to make the electrical connection. Individual tubes are arranged in a case and air flows inside the tubes while fuel flows around the outside of the tubes.

Developments in SOFC systems that operate at intermediate temperatures (550–800°C) are receiving considerable attention. A reduction in the operating temperature of SOFC systems is being pursued in an attempt to reduce the cost of these systems. Some benefits of a reduced operating temperature include better thermal integration with fuel reformers and sulfur removal systems, reduced material issues such as less thermal stress and more material flexibility, lower heat loss, shorter time to achieve operating temperature, and less corrosion. Capitalizing on the benefits of lower SOFC operating temperatures is an area of continued and future research and development.

ELECTRODES AND ELECTRO-CATALYSIS

One of the principal aims in fuel cell development is to find a cost-effective electrode materials in addition to formulating a sustainable membrane. Electrodes in a fuel cell perform the following functions:

1. They provide an interface for fuel and electrolytes where the charge transfer reactions take place.
2. They promote electro-catalysis.
3. They provide enhanced mechanical strength.

Redox processes involving charge transfer takes place at the interface. If the fuel is soluble, it undergoes mass transportation from the bulk to the electrode—electrolyte interface, a two-phase boundary, and experience charge transfer. Maximized current can be realized by using high concentrations of the dissolved fuel, electrodes of a large surface area, maximum agitation, and high temperature.

Gaseous fuels have limited solubility and charge transfer should be made to occur at a three-phase boundary of meta, solution, and gas for the appreciable flow of current. Only in the case of hydrogen can a diffuser electrode be made. The three-phase boundary is a thinly wetted interphase of about 1—3 μm dimension. The next development is the introduction of a biporous electrode by Justi. The electrode is made of two porous structures, one with small pores and the other with large ones. The fine pore layer is exposed to the electrolyte and a high gas pressure is applied to the back of the coarse pore layer; this pressure is sufficient to drive the liquid out of the coarse pores but not into the fine ones. Although it works well, efficient use of the electro-catalytic material is not used. Partially oxidized Raney nickel with 6% NiO catalysts has been reported to be superior in activity. Next, Haldeman et al. introduced Teflon-bonded electrode. This electrode is made by spreading a mixture of Teflon and electro-catalyst onto a metal screen, which is the current collector. Teflon is hydrophobic, so its pores remain dry; the electro-catalyst is hydrophilic and also porous. It is filled with an electrolyte, facilitating charge transfer. An optimized Teflon-bonded electrode can provide five to six times the current output of a bipolar electrode system for the same amount of platinum loading.

ELECTRO-CATALYSIS

The introduction of a electro-catalyst is essential because (1) the reduction of oxygen is an irreversible process in an aqueous medium at all electrodes at normal temperatures, and (2) the area of the electrode cannot be indefinitely increased owing to limitations in pore size. The basic requirements of an electro-catalyst are:

1. rapid adsorption of all reacting gases,
2. sufficient corrosion resistance to electrolytes,
3. chemical stability at the operating potential, and
4. high electrical conductivity.

The problem of choosing an appropriate electro-catalyst is approached in a way similar to heterogeneous chemical catalysis. Primarily the electronic structure and geometry of the atomic structure or lattice spacing of the catalyst and the interatomic distances of the reacting gas molecules should be reasonably close to each other. Many developments have taken place in dispersing the catalyst on suitable supports and partial reduction of metals such as molybdenum or tungsten bronze catalysts, which exploit a phenomenon called spillover.

The hydrogen evolution reaction is the most widely investigated electrochemical reaction. The overall process is $2H_3O^+ + 2e^- \rightarrow H_2 + 2H_2O$. Three well-known mechanisms for this process are:

1. Volmer mechanism (Hg, Cd, Pb)

$$H_3O^+ + e^- \rightarrow H_{ad} + H_2O \text{ (rate limiting)}$$
$$H_{ad} + H_3O^+ + e^- \rightarrow H_2 + H_2O$$

2. Tafel mechanism (Pt, Rh)

$$H_3O^+ + e^- \rightarrow H_{ad} + H_2O$$
$$H_{ad} + H_{ad} \rightarrow H_2 \text{ (rate limiting)}$$

3. Heyrovsky mechanism (Ni, W, Au)

$$H_3O^+ + e^- \rightarrow H_{ad} + H_2O$$
$$H_{ad} + H_3O^+ + e^- \rightarrow H_2 \text{ (rate determining)}$$

Data on reported values of exchange current densities for a hydrogen evolution reaction are listed in Table 14.5. The exchange current density is an index of the reversibility of an electrochemical process.

TABLE 14.5 Values of Exchange Current Densities (I_o) for Hydrogen Evolution Reaction at 298K

Metal	$-\log(I_o)$ in 1 M Sulfuric Acid With I_o in ACm^{-2}
Palladium	3.0
Platinum	3.1
Rhodium	3.6
Iridium	3.7
Nickel	5.2
Gold	5.4
Tungsten	4.9
Niobium	6.8
Cadmium	10.8
Manganese	10.9
Thallium	11.0
Lead	12.0
Mercury	12.3

The oxygen reduction reaction is invariably a slow or irreversible process with all electrode materials at ordinary temperatures. The overall reaction is $O_2 + 2H_2O + 4e^- = 4OH^-$ and the mechanism is complicated. As many as five different sequences have been reported that lead to the same observed cathode Tafel slopes of 0.040 V ($\alpha = 3/2$) and anode Tafel slopes 0.040 V ($\alpha = 3/2$) at the platinum electrode.

The electro-catalysis of the oxygen reduction reaction is a challenging problem in electrochemistry. The drastic reduction in the overall efficiency of a fuel cell and the lower open circuit potentials have been attributed to the formation and accumulation of a peroxide intermediate in the oxygen reduction reaction. Two or four electron reduction processes generate peroxide or oxide ions, and finding a suitable catalytic material for this step in fuel cell reactions is still a task to be completed.

Chapter 15

Biochemical Routes for Energy Conversion

The demand for energy continues to increase in every part of the world owing to speedy development in population and industrialization. Most of the energy demand is met with conventional energy resources such as coal, oil, and natural gas. The problems faced by conventional fuel sources are that the world's reserves are becoming depleted over time and that the use of these conventional fuels causes pollution. There is an urgent need for alternative energy resources that are inexpensive and renewable and that do not produce pollution; from these points of view, answers are renewable sources such as solar, wind, thermal, hydroelectric, and biomass. Biomass contributes nearly 14% of the world's primary energy consumption, such as generating electricity, fueling vehicles, and providing heat for industries. Biomass can be converted into three main types of products: electrical or heat energy, fuel for the transport sector, and feedstock for chemicals. Biomass use as a source of energy is of interest because:

1. Biomass is a renewable, potentially sustainable, and relatively environmentally friendly source of energy
2. There is a variety of new structural features to exploit.
3. Increasing the use of biomass will extend the lifetime of limited crude oil resources.
4. The sulfur content in biomass is low, so sulfur dioxide emissions will be reduced and acid rain will be contained.
5. Combustion of biomass produces less ash, and even that ash can be used as a soil additive on farms.
6. The use of biomass for energy conversion will be an effective employment of waste products, thus facilitating waste disposal problems.
7. Biomass is a domestic resource and thus not subject to price fluctuations or supply uncertainties.
8. Use of biomass as an energy source may improve our environment, economy, and energy securities.

Possible routes for biomass conversion are given schematically in Fig. 15.1.

Energy Sources. http://dx.doi.org/10.1016/B978-0-444-56353-8.00015-0

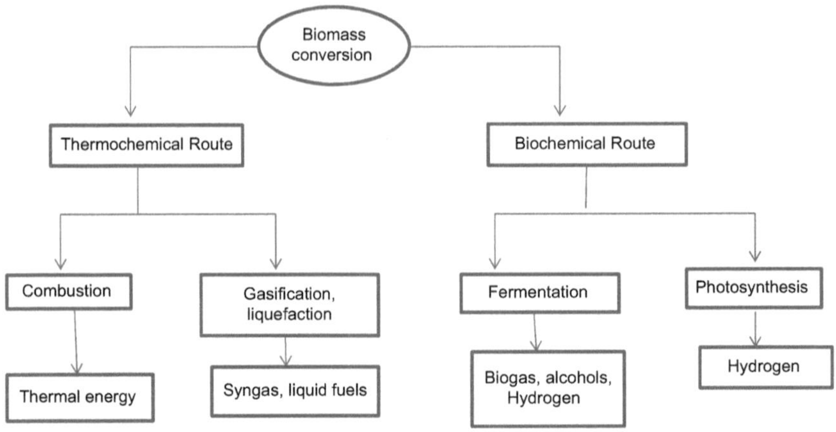

FIGURE 15.1 Routes for biomass to energy conversion.

BIOMASS COMPONENTS

The components of biomass depend on the chemical and physical properties of the large molecules that constitute the biomass. It contains varying amounts of cellulose, hemicellulose, and lignin. Cellulose is a glucose polymer containing linear chains of (1,4)-D-glucopyranose units in which the units are linked 1 to 4 in the alpha configuration with an average molecular weight of around 100,000. Alpha cellulose is a polysaccharide with the general formula ($C_6H_{10}O_5$). Hemicellulose is a complex polysaccharide that exists in association with cellulose in the cell wall. It is a mixture of polysaccharides composed of sugars such as glucose, mannose, xylose, and arabinose and methylglucoronic and galacturonic acids with average molecular weight of less than 30,000. Lignins are branched substituted mononuclear aromatic polymers in the cell walls of the certain biomass, especially woody species, and are often adjacent to cellulose fibers to form a lignocellulosic complex. On average, biomass consists of 40–50% cellulose by weight and about 20–40% hemicellulose.

A detailed diagrammatic representation of possible conversion routes is given in Fig. 15.2.

PHOTOSYNTHESIS

Photosynthesis is the process by which chlorophyll-containing organisms, including green plants, algae, and certain bacteria, capture energy in the form of solar light and convert it to chemical energy. Almost all energy available to life on earth is made through photosynthesis.

A general reaction sequence for photosynthesis can be written as:

$$CO_2 + H_2A + \text{light energy} = (CH_2O) + H_2O + A_2$$

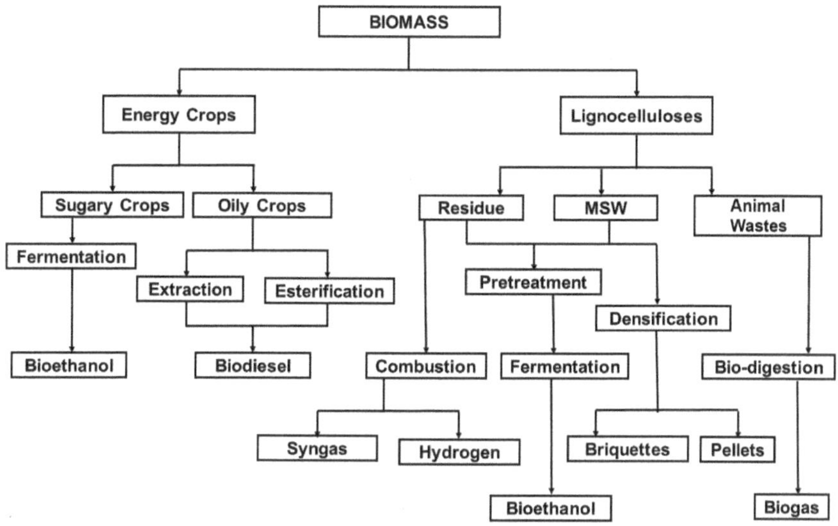

FIGURE 15.2 Diagrammatic detailed representation of the conversion system for biomass. *Reproduced from G.T. Tucho, Int. J. Eng. Res. Technol. 2 (2720) (2013).*

In this representation H_2A is mostly water, but in the case photosynthetic bacterial it is H_2S.

Photosynthesis consists of two stages: a series of light-dependent reactions that are temperature independent, and temperature-dependent reactions that are light independent. The rate of light-dependent reaction depends on the light intensity available and the second dark reaction depends on temperature (to a certain extent). The plants follow the C_3 or C_4 routes for harnessing solar energy into chemicals (carbohydrates). Details of these processes will be dealt with in a subsequent section.

BIOMASS CONVERSION PROCESS

Biomass can be converted into useful forms of energy by a number of processes. Biomass can be converted into three main products: power or heat generation, transportation fuels, and chemical feedstock.

BIOLOGICAL HYDROGEN PRODUCTION

Many photosynthetic microorganisms have been recognized for years to produce hydrogen gas. The process is connected with the light-dependent photosynthetic process mentioned previously. There are a few advantages to this route: (1) the system can be operated at low temperatures; (2) the only inputs are solar energy and a hydrogen donor, possibly water or saltwater; (3) this process does not result in pollutants; (4) the system would have potentially

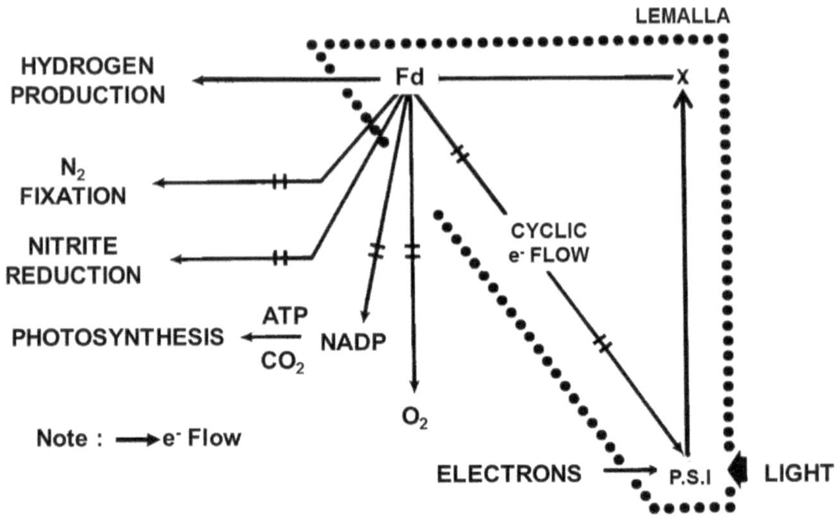

FIGURE 15.3 Pathways for electron flow in living cells.

beneficial and profitable multiple uses such as directly producing chemical by-products.

Basically, solar radiation is absorbed by some pigment (chlorophyll protein). This energy is transmitted to an electron by an electron transfer site. The source of this electron is some electron donor compound, usually water. The energized electron is the source for hydrogen gas evolution. The electron for hydrogen production follows the same path as those leading to carbon fixation and other associated functions, as shown in Fig. 15.3. The actual shunting of electrons from the photosynthetic carbon fixation pathway can occur just after ferredoxin (or any other electron acceptor) to $2H^+$, which can take place in three different ways: (1) an adenosine triphosphate (ATP)-independent hydrogenase catalyst, (2) an ATP-dependent nitrogenase system, or (3) a nicotinamide adenine dinucleotide phosphate (NADP) and hydrogenase-mediated or analogous system. Not all photosynthetic bacteria or algae contain all three hydrogen-evolving systems. Photosynthetic bacteria require an electron donor such as an organic compound and algae may use water as the electron donor.

Essentially, there seem to be some differences between photosynthetic bacteria and algae. Generally, algal biophotolysis systems usually need two light reactions and bacterial hydrogen production systems may require only one photon.

Another important aspect in hydrogen production is oxygen inhibition. In many species of algae oxygen inhibits hydrogen production by inactivating the catalysts (hydrogenase or nitrogenase) of the reaction. Because the production of hydrogen gas is an offshoot of the photosynthetic pathway, the solar energy conversion efficiency is related to the metabolic demands of the plant cells used in the system. The bioconversion efficiency depends on:

1. Metabolic potential: Different species exhibit different metabolic potentials and depend on environmental conditions such as sunlight, temperature, salinity, and nutrient availability;
2. Competing pathways for electrons, as shown in Fig. 15.1. These include oxygen fixation, oxygen reduction, nitrogen fixation, nitrite reduction, or hydrogen production;
3. Oxygen inactivation of hydrogenase or nitrogenase;
4. Rate-limiting reactions; in particular, electron flow between photosystems II and I and hydrogenase and nitrogenase catalysis is rate limiting.

These limitations are addressed to overcome them by studying manipulation of living cells and the development of cell-free production systems. The barriers encountered in the widespread application of hydrogen photoproduction are the lack of advanced production methods, storage, and use of the developed technology. These problems will be resolved by a strong research and development effort in this area.

FERMENTATIVE HYDROGEN PRODUCTION

In dark fermentation, anaerobic bacterial and some microalgae on carbohydrate-rich substrates can produce hydrogen and organic end-product carbon dioxide. Cultures that could produce hydrogen from carbohydrate substrates are *Enterobacter, Bacillus*, and *Clostridium*. Substrates used include glucose, starch, and cellulose. If glucose is the starting material and acetic acid is the organic end product, each mole gives rise to 4 mol hydrogen as per the equations:

$$C_6H_{12}O_6 + 2H_2O_2 \rightarrow 2CH_3COOH + 2CO_2 + 4H_2$$

$$C_6H_{12}O_6 \rightarrow CH_3CH_2CH_2COOH + 2CO_2 + 2H_2$$

Anaerobic fermentation in dark by heterotrophic bacteria can be depicted by the following process:

$$\text{Biomass} \rightarrow \text{carbohydrate} \rightarrow \text{(dark fermentation) organic acids} \rightarrow \text{(hydrogenase) } H_2 + CO_2$$

PHOTOSYNTHESIS PROCESS

Hydrogen production in the presence of light by algae, purple sulfur, and nonsulfur is:

$$H_2O + CO_2 \rightarrow \text{(sunlight hydrogenase and nitrogenase) carbohydrates} + H_2 + O_2$$

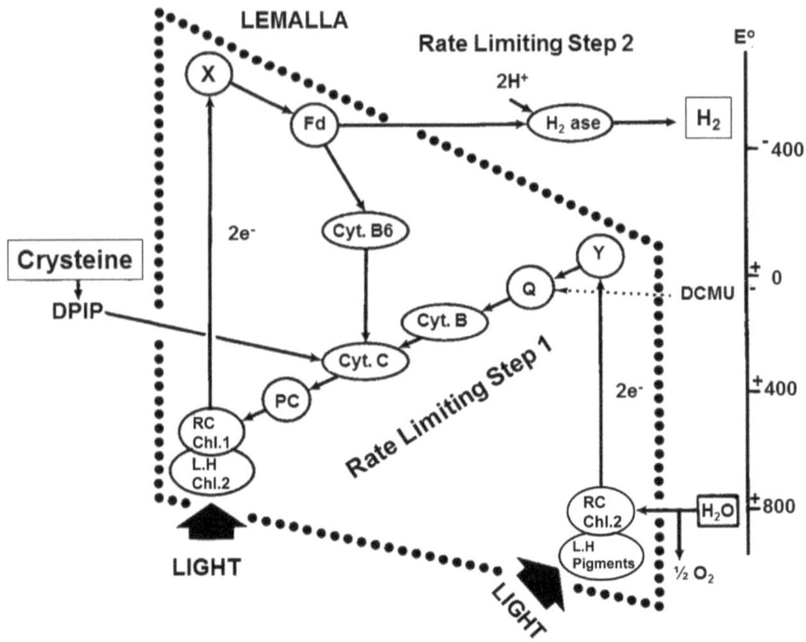

FIGURE 15.4 Electron transport system and rate-limiting steps in hydrogen photoproduction. *L.H*, light harvesting pigment; *R.C.Chl1 and Chl2*, reaction center chlorophyll protein of photosystems I and II; *Q*, plastoquinone; *cyt.B*, cytochrome B; *Cyt C*, cytochrome C; *PC*, plastocyanine, *Fd*, ferredoxine; *H₂ase*, hydrogenase. E_0' and mV redox potential in millivolts at pH 7. Components contained within the perimeter of dashed lines represent the core of the photosynthetic electron transport system of the lamella.

Many phototropic organisms such as purple bacteria, green bacteria, cyanobacteria, and several algae can product hydrogen in the presence of sunlight. Light absorption generates electrons that are transferred to ferredoxin using solar energy absorbed by the photosystem. Although the detailed mechanism may vary with respect to the type of organism, the general scheme is shown in Fig. 15.4.

Data on hydrogen production by some bacteria are assembled in Table 15.1.

There is retardation in biophotolysis because it is sensitive to oxygen; this can be overcome by producing hydrogen and oxygen at different stages. Currently the hydrogen production rate is too low to power a 1-kW proton exchange membrane fuel cell on a continuous basis.

ETHANOL AS BIOFUEL FROM BIOMASS

Ethanol has been known as a fuel from the early days of the automobile era. However in the 1970s, Brazil introduced a "Proalcool" program and ethanol

TABLE 15.1 Data on Typical Hydrogen Production by Some Bacteria

	Hydrogen Evolution Rate Per Hour Per Gram Dry Biomass	
Phototropic Microorganisms	mL	mmol
Purple bacteria		
Rhodospiritham rubrum	146	6.5
Rhodopseudomonas palustris A	54	2.8
Rhodopseudomonas capsulata SL	130–150	5.8–6.7
Rhodopseudomonas capsulata B 10	300–500	13.4
Rhodopseudomonas capsulata L B 2	178	7.9
Thiocapsa raseopersicina	13	0.6
Cyanobacteria		
Anabaena cylindrica B 629	5–40	0.2–2.0
An variabilis	32	1.3
Spirulina platensis	9	0.4
Green algae		
Chlamydomonas reinhardii 137 C	45	2.0
Chlamydomonas moevussii ICC97	8	0.4

Reproduced from R.C. Saxena, et al., Renewable Sustainable Energy Rev. 13 (167) (2009).

returned to the fuel arena. Ethyl alcohol is a comparatively cleaner-burning fuel with high octane and fuel-extending properties. Blending ethanol with petrol reduces CO emissions. There are many options for the production of ethanol from biomass.

ETHANOL FROM MOLASSES

This process involves the three steps of hydrolysis, fermentation, and ethanol recovery. Microbial degradation of lignocellulosic waste and the downstream products resulting from it are achieved by the concerted action of a number of enzymes. The most prominent one is cellulases (Fig. 15.5).

Pretreatment is a critical step that enhances the enzymatic hydrolysis of biomass. Basically it alters the physical and chemical properties of biomass and improves enzyme access and effectiveness. This step can lead to a change in crystallinity and the degree of polymerization of cellulose. Pretreatment can be broadly considered to occur in four groups: physical (mechanical

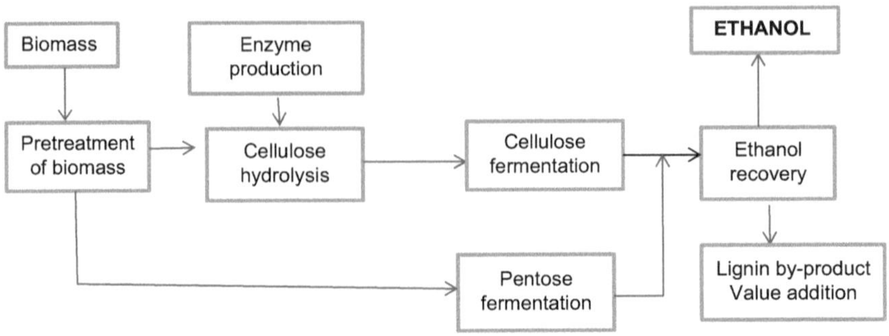

FIGURE 15.5 Schematic of bioethanol production process especially from lignocellulosic biomass.

comminution or irradiation), chemical (treatment with acids or alkalis), physiochemical (treatment with steam or steam with gases such as SO_2 and CO_2), and biological methods. The pretreated biomass is subjected to enzymatic hydrolysis using cellulose enzymes to convert the cellulose to fermentable sugars. Cellulase refers to a class of enzymes produced by fungi and bacteria that catalyzes the hydrolysis of cellulose by attacking the glycosidic linkages. Cellulase contains mainly three different functional protein groups: exo-glucanase (Exo-G) endo-glucanase (endo-G), and β-glucosidase (β-G). These functional proteins work synergistically to hydrolyze the cellulose into glucose. These sugars are further fermented using microorganisms and are converted to ethanol.

BIODIESEL

Oleaginous microalgae have increased as a nonedible biodiesel feedstock. This method provides the opportunity to use abandoned or unproductive land using brackish, salt, and wastewater. Unlike terrestrial oilseeds, microalgae are cultivated in dilute aqueous suspensions that make lipid recovery difficult. Drying and extraction steps are an energy- and cost-intensive processes. A process involving intracellular lipid hydrolysis and supercritical in situ transesterification has been developed. In the hydrolysis step, an easily filterable solid is formed that retains the lipid; in the second step, a wet fatty acid solid is subjected to supercritical in situ transesterification with ethanol to produce biodiesel in the form of fatty acid ethyl esters. Nutrients (N and P) and glycerol from processed biomass can be reused in a sterile aqueous phase free of catalyst.

A two-step hydrolysis-solvolysis process for the production of biodiesel from lipid-rich wet algal biomass has been proposed. Solvolysis of biomass and its components in liquid water results in micellar-like substructures of the biomass. This step obviates biomass drying, solvent extraction, and use of catalyst. At the same time this method provides a mechanism for nutrient recycling.

BIOGAS

Algal biomass is rich in nutrients, especially nitrogen and phosphorus. Therefore a process to recycle algal waste after lipid extraction is desirable. Anaerobic digestion may be suitable for this because the biotechnological process can mineralize algal waste containing organic nitrogen and phosphorus. The anaerobic digestion of algal waste not only recycles the nutrients but also provides biomethane. The anaerobic process occurs in three sequential stages of hydrolysis, fermentation, and methanogenesis. During the hydrolysis step, the complex compounds are broken down into soluble sugars, and then fermentative bacteria convert these into alcohols and volatile fatty acids and a gas containing hydrogen and carbon dioxide, which is metabolized into methane and CO_2, which is metabolized into methane by methanogens. The composition of algal biomass is species dependent: proteins (6–52%), lipids (7–23%), and carbohydrates (5–23%). The theoretical methane content depends on the content of lipids, but hydrolysis of lipids is slow compared with proteins and carbohydrates. The extent of biogas production during anaerobic digestion depends on the biomass composition, pH, temperature, hydraulic, and solid retention time. Data on the typical composition of different algal species and methane potential and ammonia release during anaerobic digestion are given in Table 15.2.

Energy from biomass is a renewable source and the conversion can be achieved in multiple processes and different forms of energy: solid, liquid, or gas. Most of these technologies are limited to laboratory scale and have yet to be commercialized. However some are easily adaptable to commercialization. Anaerobic digestions, densification, fermentation, and extraction are some of that are applicable to convert biomass into bioethanol and biodiesel. These technologies are mature enough and easily adaptable to any areas in which biomass resources are available. It is expected that tremendous progress will be made to make use of this source of energy in the near future if it has not already been exploited.

PHOTOSYNTHESIS

Photosynthesis in its complete form is a complex phenomenon involving various steps. Elementary details will be considered in this section. In simplest language, photosynthesis can be considered to be involve the reduction of carbon dioxide to carbohydrate (a molecular form consisting of carbon, hydrogen, and oxygen) with the simultaneous use of water and sunlight. A simple pictorial representation of this process is shown in Fig. 15.6. Photosynthesis consists of two parts: in light reactions, light energy is converted into high-energy chemicals whereas in dark reactions, carbon dioxide is converted into carbohydrate. This process requires hydrogen generated from the decomposition of water together with the evolution of

TABLE 15.2 Composition of Different Algal Species and Theoretical Methane Potentials and Ammonia Release During Anaerobic Digestion of Total Biomass

Algal Species	Protein (%)	Lipid (%)	Carbohydrates (%)	Methane (L/g)	NH_3 (mg/g)
Euglena gracilis	39–61	14–20	14–18	0.53–0.8	54.3–84.9
Chamydomanas reinhardtii	48	21	17	0.69	44.7
Chlorella pyrenoidosa	57	2	26	0.8	53.1
Chlorella vulgaris	51–58	14–32	12–17	0.63–0.79	47.5–54.0
Dunaliella salina	57	6	32	0.68	53.1
Spirulina maxima	60–71	6–7	13–16	0.63–0.74	55.9–66.1
Spirulina platensis	46–63	4–9	8–14	0.47–0.69	42.8–58.7
Scenedesmus obliquus	50–56	12–14	10–17	0.59–0.69	46.6–42.3

Data reproduced from A. Singh, S.I. Olsen, Appl. Energy 88 (3548) (2011).

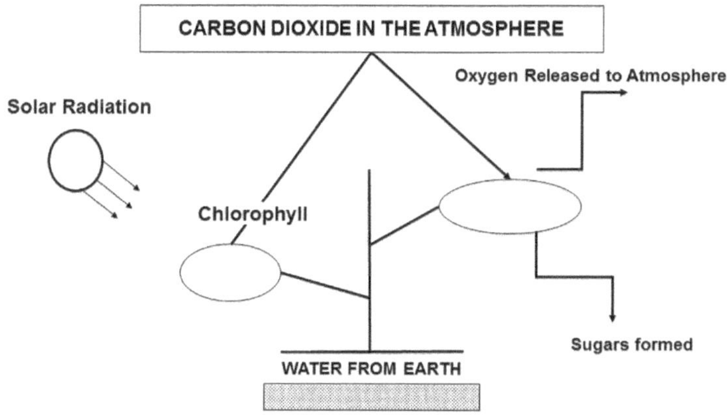

FIGURE 15.6 Pictorial representation of the photosynthesis process.

molecular oxygen. There are some governing principles for the light reactions:

1. The pigment protein must be capable of absorbing available photons and generating an excited state.
2. Excess energy available in the protein from the photon must be capable of being transformed to the adjacent molecule, and this energy transfer path necessarily facilitates the ultimate energy transfer and reduction normally difficult to reduce carbon dioxide.
3. Molecules that receive excitation energy are contained in a thylakoid membrane and thus probably create the necessary reaction centers in a highly structured arrangement.

Two kinds of energy-containing molecules are formed in the light reactions of photosynthesis: ATP and reduced NADP (NADPH), which are the sources for the reducing power. The photosystems and other reaction centers are connected by a series of compounds that act as electron carriers. An excited electron in a photosystem II reaction center enters this chain of carriers and moves from one to the next until it reaches a photosystem I reaction center. The next step in this process is when an excited electron in photosystem I is transferred to a molecule of NADP along with a proton, thereby reducing it to NADPH. The proton source is achieved by the decomposition of water, which sustains the process. What has been achieved is the conversion of photon energy into chemical energy in the form of ATP and NADPH with the concomitant formation of molecular oxygen. The conversion of carbon dioxide into carbohydrate and other biological molecules is referred to as dark reactions, although only in dark these reactions cannot take place. In essence, ATP and NADPH coming from light reactions are used to convert atmospheric carbon dioxide into carbohydrates, a form of energy storage molecule.

Only the essential minimum on photosynthesis is covered in this presentation. Interested readers should refer to the authoritative treatment on this subject. This understanding is essential for the further development of energy conversion devices mimicking nature's process.

CONCEPT OF BIOREFINERY

In coming years oil refineries have to undergo changes because the inventory of raw material will have to change to biomass-based material. A conceptual design of a biorefinery is shown in Fig. 15.7. This concept has to be evolved and the process design has to undergo a refined change.

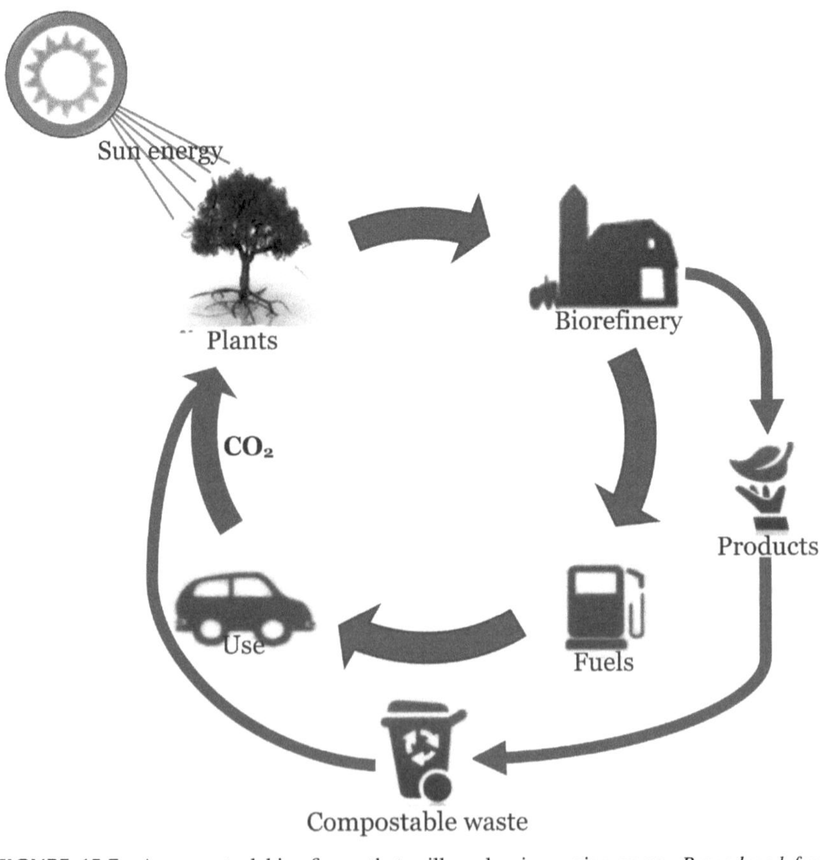

FIGURE 15.7 A conceptual biorefinery that will evolve in coming years. *Reproduced from V.P. Soudham, Biochemical Conversion of Biomass to Biofuels (Ph.D. thesis), UMA University, Sweden, 2015.*

Chapter 16

Energy Conversion Routes: An Evaluation

INTRODUCTION

This book was not created to serve as a comprehensive or authoritative textbook on chemical and electrochemical energy sources. Two decades ago, the author of this book compiled such a book on that subject [1] and also developed an elective course for Indian postgraduate students at the Indian Institute of Technology, Madras, under the same title. Although the course was popular, the author was not fully satisfied with the content of the book or course and even more so after having completed this compilation. This dissatisfaction arises for these reasons:

1. The contents of the book are a selection of topics fully governed by the preference of the author and are not based on the desirability of topics under this heading.
2. Coverage of each topic could have been more comprehensive, because the literature on each of these topics has grown enormously over the past 4 decades, but the author could not do full justice to discussing each topic thoroughly.
3. Each topic could have been viewed with an alternative viewpoint but the author chose one particular viewpoint so that the presentation is clear and more precisely covered.
4. Subtopics taken up for consideration are purely arbitrary; there are a variety of other possible routes. Also, other devices may emerge and even existing knowledge is not fully justifiably covered in this presentation.

These and other limitations that one may find in this coverage reflect the knowledge domain of the author.

Energy conversion is a topic that cannot be covered in a single volume, nor is it desirable. Although many energy conversion routes have been researched and adopted, electrochemical and photo-electrochemical routes have received considerable attention over the past 4 or 5 decades for reasons such as non- (or zero) pollution systems, but success in making these processes commercially viable has not yet been fully achieved. Although the

Energy Sources. http://dx.doi.org/10.1016/B978-0-444-56353-8.00016-2

efficiencies of photo-electrochemical routes in the laboratory have reached double-digit numbers, on a large scale they have not yet crossed the single-digit figures. This appears to be a material problem and designing and fabricating the successful material in the form in which they will yield the desired levels of efficiency. Electrochemical energy conversion is green compared with thermal-based energy conversion processes. In addition, electrochemical energy conversion processes are not limited to the Carnot efficiency whereas thermal-based techniques are subject to this limitation. Despite of these distinct advantages, electrochemical energy conversions have not reached the stage of extensive exploitation because of various issues in their development. The main stumbling blocks appear to be identifying designing and fabricating appropriate materials, whether in fuel cells, supercapacitors, or batteries.

Solar energy or photovoltaic technology is one of the new energy conversion devices that has reached successful deployment. Despite this adaptation, this technology still faces material and technological issues in deployment. Although this technology is proven and feasible, the cost comparison with respect to electricity generation through other means such as hydro or wind is unfavorable. Also, this technology is site specific, and between the southern and northern hemispheres there is a considerable natural difference. This situation is similar to that of fossil fuels, especially crude oil and natural gas, which are found in certain regions of the earth and thus affect the economy of other regions in the earth. However, despite this difference, this appears to be a leveling ground for energy availability by nature even though it is not equitable.

Nuclear energy based on fission technology is being practiced in most countries, although periodically (remember the Chernobyl, and, more recently Fukushima disasters) there is a warning against it. In addition, the cost of constructing nuclear power stations and maintaining them appears to a heavy burden on many countries. Nuclear power is feasible, but the desirability has not been unequivocally established. Fusion technology is still far in the future and a technology adaptable for common humanity has yet to be established.

Fossil fuel—based energy conversion processes are fairly well established and can be adopted for use by common citizens. However, the site-specific nature of the availability and control over these resources by the whims of certain people (remember the oil shock of the 1970s) are some of the points to be considered. Energy availability must be equally distributed among all; otherwise conflicts may arise that may cause disparity among humans. In addition, these forms of energy conversion will cause environmental degradation and pollution.

Current Status of Electrochemical Energy Systems

As stated, electrochemical energy conversions have certain advantages over other forms of energy conversion technologies. Despite higher efficiency and

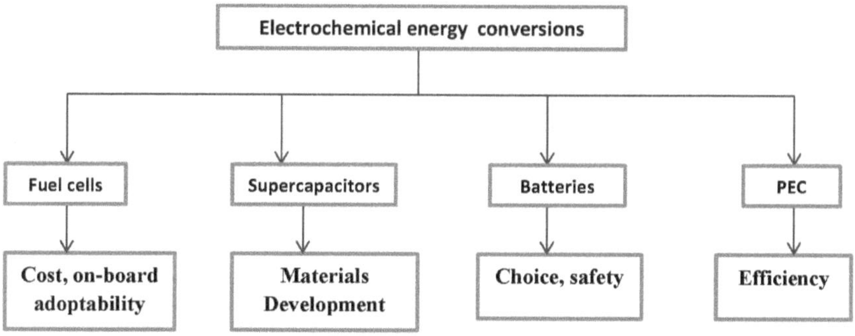

FIGURE 16.1 Electrochemical energy conversion devices and their stumbling blocks.

least polluting systems, extensive deployment of these technologies still has hurdles (Fig. 16.1).

Although electrochemical energy conversion devices are more efficient than energy conversions involving a thermal step, development and exploitation of these energy sources have taken more time (for example, fuel cells) than were expected and deserved. This may be because energy sources based on fossil fuels are economical than other sources, but this trend cannot continue for long.

These electrochemical devices are critical and enabling technologies for renewable energy, conservation and storage, pollution control, and greenhouse gas reduction. The growing demand for electrochemical energy systems along with the increasing maturity of these technologies will have a significant impact on lifestyle and also on research and development. The important role that renewable energy will have in the future energy mix is more obvious. As increased levels of renewable energy are integrated into the national grid, one has to pay attention to the intermittent nature of these resources. From the points of view of response time, power capability, and energy density, hybrid solutions will be expected to have a role in the future; in this regard electrochemical energy technologies will be critical.

Hydrogen can be considered to be an important energy carrier and storage medium for the future society based on hydrogen economy. A number of different electrochemical technologies are in the development stage because this route gives storage media for load leveling and peak load shaving. Both low-temperature and high-temperature electrolyzers are under examination with an efficiency in the range of 50–55%. A conceptual renewable energy system based on hydrogen generation by coupling an electrolyzer to solar-based photovoltaics and/or a wind generator is shown in Fig. 16.2. Even if electricity and water are the only inputs, the cost of electricity will be in the range of 0.06 c/kWh. Although this is feasible, the additional costs associated with compression, transportation, and distribution make the conversion of high-grade electrical power from the grid to hydrogen uneconomical.

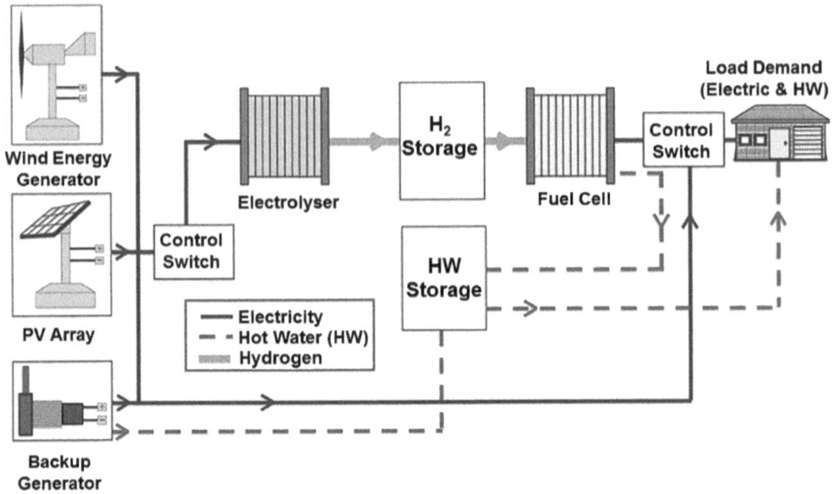

FIGURE 16.2 Conceptual diagram of a hydrogen renewable energy system for distributed power system. *Reproduced from S.P.S. Badwal, et al., Front. Chem. 2 (79) (2014).*

Hydrogen generation from natural gas (NG) or coal involves three steps requiring separate reactors and temperatures in excess of 773 K: (1) NG reforming or coal gasification, (2) a water gas shift reaction, and (3) H_2/CO_2 separation and gas cleaning. The participation of carbon in the anodic reaction of electrolysis results in thermoneutral voltage from 1.48 to 1.45 V required for the electrolysis of water around 298K. This situation can be more favorable if carbon-assisted electrolysis is carried out at higher temperatures, which can result in the further reduction of input electric energy. In Figs. 16.3 and 16.4 the schematic diagrams of electrochemical reactions involved in carbon-assisted electrolysis are shown. However, research in this area is at an early stage and a full understanding of the science is required before designing a prototyped device.

Fuel Cells: Next-Generation Energy Conversion Devices

Details of the various types of fuel cells are summarized in Table 16.1.

Although fuel cell systems are becoming available, there are still technical challenges that need to be overcome before they can be widely adopted. These challenges related to lifetime, cost, and a suitable fuel supply. There is considerable progress on all of these fronts. However, new developments are required to make the revolutionary advancements necessary to allow fuel cells to displace a significant fraction of conventional power generation capacity. Fuel cells based on solid fuels are attractive, and the two fuel cells, microbial fuel cells and direct carbon fuel cells, may stand out according to these considerations.

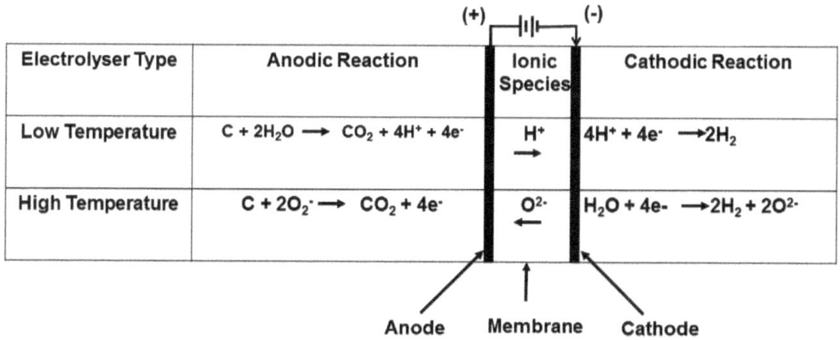

FIGURE 16.3 Schematic representation of reactions involved in low- and higher-temperature carbon-assisted electrolysis process for hydrogen production. *Reproduced from S.P.S. Badwal, et al., Front. Chem. 2 (79) (2014).*

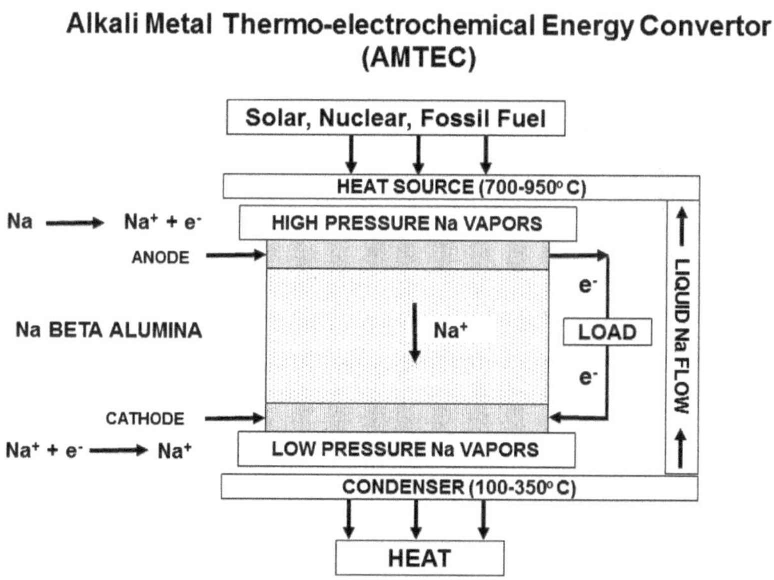

FIGURE 16.4 Operating principle of an alkali metal thermoelectrochemical energy converter. *Reproduced from S.P.S. Badwal, et al., Front. Chem. 2 (79) (2014).*

TABLE 16.1 Theoretical Electrical Efficiencies of Fuel Cells

					Actual Efficiency (%)	
Fuel Cell	Fuel	Overall Reaction	Operating Temperature (°C)	Theoretical Efficiency (% Electric)	Electric	Combined Heat and Power
PEMFC	H_2	$H_{2(g)} + 1/2O_{2(g)} = H_2O_{(l)}$	60–80	83	45–50	80–90
PEMFC	NG	$CH_{4(g)} + 2O_{2(g)} = CO_{2(g)} + 2H_2O_{(l)}$	60–80	–	35–40	80–90
Direct methanol fuel cells	CH_3OH	$CH_3OH_{(l)} + 3/2O_{2(g)} = CO_{2(g)} + 2H_2O_{(l)}$	20–60	97	20–25	–
Alkaline fuel cells	H_2	$H_{2(g)} + 1/2O_{2(g)} = H_2O_{(l)}$	70	83	45–60	–
Phosphoric acid fuel cells	NG	$CH_{4(g)} + 2O_{2(g)} = CO_{2(g)} + 2H_2O_{(g)}$	200	–	40	90
Solid oxide fuel cells	NG	$CH_{4(g)} + 2O_{2(g)} = CO_{2(g)} + 2H_2O_{(g)}$	600–1000	92	45–60	90
Molten carbonate fuel cells	NG	$CH_{4(g)} + 2O_{2(g)} = CO_{2(g)} + 2H_2O_{(g)}$	650	92	45–55	90
Direct carbon fuel cells	Carbon	$C_{(s)} + O_{2(g)} = CO_{2(g)}$	500–1000	100	70–80	90

NG, Natural gas; *PEMFC*, proton exchange membrane fuel cells.

Alkali Metal Thermoelectrochemical Energy Converters

Alkali metal thermoelectrochemical energy converters use heat sources from the sun, nuclear power, or the combustion of fossil fuels to generate electricity; in this sense they are similar to the Carnot cycle and have an efficiency in the range of 20–40%. These devices can be employed when small-scale power generation is required, when power supplies are required in remote places, or when vehicle (aerospace) propulsion is demanded. The technology is complex. Several issues still contributing to the cost, system efficiency, and lifetime, including the stability of the electrodes, electrolytes, and other materials of construction, and sodium fluid flow management including heat removal from the cathode side and heat input on the anode side.

Energy Storage

Main issues regarding energy storage are a reduction in the cost of renewable energy generation and issues with grid stability, load leveling, and the cost of supplying load. Essentially, two versions of rechargeable lithium–air cells, both aqueous and nonaqueous, are under development. The main focus in this development is to understand the oxygen reduction step at the cathode, which is a porous conductive carbon. Simultaneously, the development of sodium–air cells, lithium sulfur batteries, and flow batteries is being pursued with renewed vigor. These developments can mitigate some of the issues associated with energy storage problems.

Supercapacitors

The issues of supercapacitors as energy conversion and storage devices are the threshold for the specific energy (beyond 5 Wh/kg) and the lifecycle values (up to a million cycles). However, considerable progress has been made regarding these two issues using manganese oxide or conducting polymers. Hybrid capacitors have been successful with lithium titanium oxide and carbon. Carbon nanotubes and graphene have shown to be good templates for the mesoporous electrode morphologies that are essential to balance the dual requirements of conductivity and ion transport. There is strong indications to believe that further developments in high-power devices with enhanced energy storage capability will take place in the near future. In Fig. 16.5, the type of requirements, power, and response time and the types of applications of electrochemical energy conversion devices are shown.

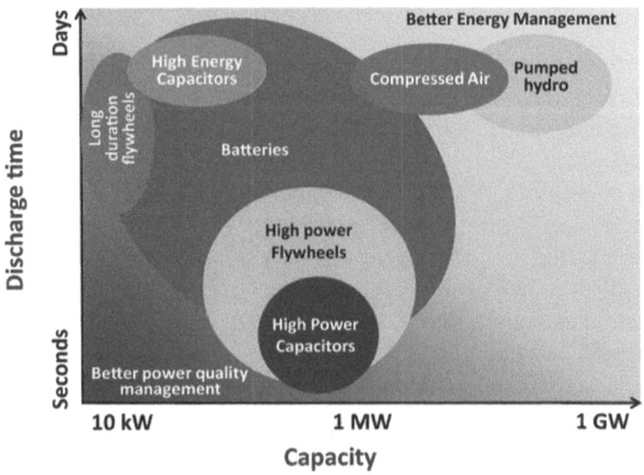

FIGURE 16.5 Power and energy performance of different storage technologies (Some types, especially batteries, encompass or overlap many technologies within a general shape). *Reproduced from S.P.S. Badwal, et al., Front. Chem. 2 (79) (2014).*

CONCLUSION

Electrochemical energy conversion devices have already contributed to a reduction in pollution and account for increased energy conversion efficiency. There is a growing demand for stabilized power generation and delivery. This has led to an increasing number of studies to develop conventional devices with improved performance and cost reduction. However as with any new technologies, many technical challenges need to be faced to develop future electrochemical power systems. At this stage it is not correct to predict which technology will emerge as the most attractive for the power market. One possibility is the development of integrated hybrid systems that will exploit multiple technology features. Similar comparisons and evaluations are not desirable because these systems have adverse effects on the environment and on living beings. These two factors outweigh all of the advantages that can be derived from chemical energy sources.

Index

'*Note*: Page numbers followed by "f" indicate figures and "t" indicate tables.'

C

D

E

F

G

I

K

L

M

N

O

P

Q

R

S

T